ENCYCLOPÆDIA
Britannica

Encyclopædia Britannica, Inc., is a leader in reference and education publishing whose products can be found in many media, from the Internet to mobile phones to books. A pioneer in electronic publishing since the early 1980s, Britannica launched the first encyclopedia on the Internet in 1994. It also continues to publish and revise its famed print set, first released in Edinburgh, Scotland, in 1768. Encyclopædia Britannica's contributors include many of the greatest writers and scholars in the world, and more than 110 Nobel Prize winners have written for Britannica. A professional editorial staff ensures that Britannica's content is clear, current, and correct. This book is principally based on content from the encyclopedia and its contributors.

Introducer

Robert McCredie May, Lord May of Oxford, OM AC Kt FRS, holds a Professorship jointly at Oxford University and Imperial College, London and is a Fellow of Merton College, Oxford. He was until recently President of The Royal Society (2000–05), and before that Chief Scientific Adviser to the UK Government and Head of the UK Office of Science and Technology (1995–2000). He was awarded a Knighthood in 1996, and appointed a Companion of the Order of Australia in 1998, both for "Services to Science". In 2001 he was one of the first 15 Life Peers created by the House of Lords Appointments Commission. In 2002, the Queen appointed him to the Order of Merit (the fifth Australian in its 100-year history). In 2007 he received The Royal Society's Copley Medal, its oldest (1731) and most prestigious award, given annually for "outstanding
branch of science".

Also available

The Britannica Guide to the 100 Most Influential Americans
The Britannica Guide to the 100 Most Influential Scientists
The Britannica Guide to the Brain
The Britannica Guide to the Ideas
that Made the Modern World
The Britannica Guide to India
The Britannica Guide to Modern China

ENCYCLOPÆDIA
THE Britannica® GUIDE TO

CLIMATE CHANGE

An unbiased guide to
the key issue of our age

Introduction by Robert M. May

ROBINSON   RUNNING PRESS
PHILADELPHIA · LONDON

Constable & Robinson Ltd
3 The Lanchesters
162 Fulham Palace Road
London W6 9ER
www.constablerobinson.com

Encyclopædia Britannica, Inc.
www.britannica.com

First published in the UK by Robinson,
an imprint of Constable & Robinson Ltd, 2008

UK ISBN 978-1-84529-867-8
1 3 5 7 9 10 8 6 4 2

First published in the United States in 2008 by Running Press Book Publishers

US Library of Congress Control Number 2007938549
US ISBN 978-0-7624-3392-6

Running Press Book Publishers
2300 Chestnut Street
Philadelphia, PA 19103-4371

www.runningpress.com

Printed and bound in the EU

CONTENTS

Introduction   vii

Part 1   The big picture

1 Climate change: a short introduction   3
2 What is climate?   40
3 A history of weather forecasting   159

Part 2   The changing world

4 The changing planet: land   187
5 The changing planet: hydrosphere   212
6 The changing planet: atmosphere   285
7 The decline in biodiversity   335

Part 3   Ideas, arguments, and progress

8 Environmentalism: past, present, and future   387
9 Key environmental thinkers   409

10 Responses to change   420

   Index   449

INTRODUCTION

Robert M. May

During the billions of years of our planet's existence, its climate has varied a lot. At times the entire planet may have been wholly or mostly enveloped by snow and ice ("Snowball Earth" or at least "Slushball Earth"), whilst at other times tropical animals inhabited the polar regions. Even in the roughly hundred-thousand years of *Homo sapiens'* tenancy, ice ages have come and gone. The most recent 8,000 years or so, since the beginnings of agriculture and the first cities, however, have been unusually steady.

Over this time, ice-core records show clearly that levels of carbon dioxide (CO_2) in the atmosphere have been around 280 parts per million (ppm), give or take 10 ppm. CO_2 is, of course, the principal "greenhouse gas" in the atmosphere, and the density of this "blanket" plays a crucial, if complex, role in determining the Earth's climate. Some indeed have argued that the beginnings of agriculture, and the subsequent development of cities and civilizations, is not a coincidence, but is a consequence of this unusual steadiness over many millennia.

Be this is it may, things began to change with the advent of the Industrial Revolution, which may be said to have begun with James Watts' development of the steam engine around 1780. As industrialization began to drive up the burning of fossil fuels in the developed world, CO_2 levels rose. At first the rise was slow. It took about a century and a half to reach 315 ppm, moving outside the multi-millennial envelope. Accelerating during the twentieth century, levels reached 330 ppm by the mid-1970s, 360 ppm by the 1990s, 380 ppm today. This change of magnitude by 20 ppm over only a decade has not been seen since the most recent ice age ended, ushering in the Holocene epoch, around 10,000 years ago. And if current trends continue, by 2050 atmospheric CO_2 levels will have reached more than 500 ppm, roughly double pre-industrial levels.

There are long time lags involved here, which are not easily appreciated by those unfamiliar with physical systems. Once in the atmosphere, the characteristic "residence" time of a CO_2 molecule is a century. And the time taken for the oceans' expansion to come to equilibrium with a given level of greenhouse warming is several centuries; it takes a very long time for water-expanding heat to reach abyssal depths. It is worth noting that the last time our planet settled to greenhouse gas levels as high as 500 ppm was some 20–40 million years ago, when sea levels were around 300 ft higher than today. The Dutch Nobelist, Paul Crutzen, has suggested that we should recognize that we are now entering a new geological epoch, the Anthropocene, which began around 1780, when industrialization began to change the geochemical history of our planet.

As discussed in great detail, and with great clarity, in this book, such increases in the concentrations of the greenhouse gases which blanket our planet will cause global warming, albeit with the time lags just noted. In their most recent report in 2007, the Intergovernmental Panel on Climate Change (IPCC), which brings together the world's top climate scientists from some 170 countries, estimates that this warming will be in the range of 1.1 to 6.4°C by 2100 (with the most likely range 2.0 to 2.8°C). This assumes that we will manage to stabilize greenhouse gas concentrations at around 450–550 ppm by that date (which could be optimistic); things get much worse at higher concentrations. This would be the warmest period on Earth for at least the last 100,000 years.

It should be emphasised that these IPCC estimates are probably conservative. This is evidenced by looking at the major projections made in the IPCC reports of 1990, 1995, and 2001 with respect to increases in global average temperatures, sea level rise, and other similar indices, and then comparing them with what has actually happened. Such a detailed comparison was published in late 2007. It showed that, without exception, in every such measure of the consequences of the thickening greenhouse gas blanket they had increased somewhat faster than the "best guess" (median range) projection made by the IPCC. I add my personal opinion that the IPCC in 1990, 1995, 2001 – and I would bet also in 2007 – has been significantly inclined toward conservatism. This fact deserves emphasis.

Many people find it hard to grasp the significance of such seemingly small temperature changes, given that temperatures can differ from one day to the next by 10°C. But there is a huge difference between daily fluctuations, and global averages sustained year-on-year. The difference in average global temperature between today and the depth of the last ice age is only around 5°C.

The impacts of a rise of around 2–3°C in global average temperatures are many and serious. And they fall disproportionately on the inhabitants, human and non-human, of developing countries. Sea-level rise derives both from warmer water expanding, and also from ice melting at the poles. This threatens not only low-lying islands and countries (such as Bangladesh), but also – at the higher levels of estimated temperature increase – major cities such as London, Shanghai, New York, and Tokyo.

There will also be significant changes in the availability of fresh water, in a world where human numbers already press hard on available supplies in many countries. Conversely, some countries will be winners here, although often offset by floods, as we have just noted. More generally, we are already seeing increased incidence of "extreme events" – droughts, floods, hurricanes, heat waves – the serious consequences of which are rising to levels which invite comparison with "weapons of mass destruction". In particular, recent studies, made before Hurricane Katrina, suggest that increasing ocean surface temperature (the primary source of a hurricane's energy) has little effect on the frequency of hurricanes, but strong effects on their severity; indeed the damage inflicted by Katrina in 2005 has been estimated as equivalent to 1.7 per cent of US GDP that year. In essentials, a thicker greenhouse gas blanket means a warmer Earth, and such warmth simply means more energy in the climate system. The consequence is that, based on simple projections of trends, estimates of the increasing annual costs of damage from extreme weather amount to 0.5–1 per cent of global GDP by 2050, and will keep rising as the world continues to warm.

As carefully reviewed in this book, the timescales and magnitudes of other important and nonlinear processes associated with climate change are less certain (nonlinear means, roughly, that doubling the cause does not simply double the effect; huge, and often irreversible, "tipping points" can occur). As seen for example in Chapter 5, as the polar ice caps melt, the surface reflectivity is altered – dazzling white ice or snow giving way to dark oceans – causing more warming and fast melting; the timescale for the ice cap to disappear entirely (a few decades? a century? longer?) is unclear. Such melting or collapse of ice sheets would eventually threaten land which today is home to 1 in every 20 people. As northern permafrost thaws, large amounts of methane gas are released, further increasing global warming; methane is a more efficient greenhouse gas than carbon dioxide. Nearer home for the British, increased precipitation in the North Atlantic region, and increased fresh water run-off, will reduce the salinity of surface water. Water will therefore be less dense and will not sink so readily. Such changes in marine salt balance have, in the past, modified the fluid dynamical processes which ultimately drive the Gulf Stream, turning it off on decadal timescales. Although current thinking sees this as unlikely within the next century or so, it is worth reflecting that the Gulf Stream, in effect, transports "free" heat towards the British Isles amounting to roughly 30,000 times the total power generation capacity of the UK. These and other nonlinear and potentially catastrophic events are less well understood than is the direct warming caused by increased greenhouse gases. But their potential impacts are great, and should be included in risk assessments.

Chapter 4 addresses the interplay between climate change and crop production. In particular, a recent report by the Royal Society – the British Commonwealth's premier scientific academy, founded in 1660, and among other things the originator of peer-reviewed journals – unhappily emphases that "Africa is consistently predicted to be among the worst hit areas across a range of future climate change scenarios". This echoes the disconnect between two central themes in global politics. On the one hand, solemn promises have been made to increase aid and support development in Africa. On the other hand, the lack of agreement on measures to curb greenhouse gas emissions means that increasing amounts of aid will be spent on tackling the consequences of climate change.

Moving beyond ourselves, Chapter 7 deals with concern for the other living things – plants and non-human animals – that share the planet with us. Seen through a wider-angle lens, the impending diminution of the planet's diversity of plant and animal species – which derives from human impacts, and pre-dates the effects of climate change – could be an even greater threat than climate change itself.

Currently around 1.5 to 1.6 million distinct species of plants and multicellular animals have been named and recorded. Even this number is uncertain to within around 10 per cent, because the majority of species are invertebrate animals of one kind or another, for most of which the records are still on file cards in separate museums and other institutions. We know even less about how many species exist on Earth today. Plausible estimates range around 5–10 million, and some experts would argue for higher or lower numbers; either way, the number remaining to be discovered exceeds the total number identified by us. Given this lamentable ignorance, we clearly cannot say much about the number of species that are likely to become extinct this century. We can note that the IUCN Red Data Books in 2004, using specific and sensible criteria, estimate 20 per cent of recorded mammal species are threatened with extinction, and likewise 12 per cent of birds, 4 per cent of reptiles, and 3 per cent of fish. However, when these figures are re-expressed in terms of the number of species whose status has been evaluated (as distinct from dividing the number known to be threatened by the total

number known – however slightly – to science), the corresponding numbers are not much different for mammals and birds at 23
per cent and 12 per cent respectively, but are 61 per cent of
reptiles and 46 per cent of fish. This says a lot about how much
attention reptiles and fish have received. The corresponding
numbers for the majority of plants species, dicots and monocots,
are respectively 4 per cent and 1 per cent of known species, which
contrasts with the corresponding figures of 74 per cent and 68 per
cent of those evaluated. Most dramatic are the two numbers for
the most numerous group of species, insects: 0.06 per cent of all
known species are threatened, versus 73 per cent of those actually
evaluated. The same pattern holds true for other invertebrate
groups. For these small things, which arguably run the world, we
know too little to make any rough estimate of the proportions
that have either become extinct or are threatened with it.

As explained in detail in Chapter 7, it is perhaps surprising
that we can nevertheless say some relatively precise things about
current and likely future rates of extinction in relation to the
average rate seen over the roughly 550 million year sweep of the
fossil record. Humans have much greater emotional resonance
with furry and feathery creatures – mammals and birds – than
with other species, and they consequently have been relatively
well studied. Over the previous century, documented extinctions
within these two groups have been at a rate roughly 1,000 times
greater than that seen in the fossil record. And four different
lines of argument suggest a further rough upswing of extinction
rates, by a factor of around 10, over the coming centuries. So, if
mammals and birds are typical (and there is no reason to
suppose they are not), we are looking at an acceleration in
extinction rates which is of the magnitude which characterised
the "Big Five" mass extinction events in the fossil record, such as
the one which did in the dinosaurs. There is, however, a crucial
difference between the Sixth Wave of mass extinction on whose
breaking tip we stand and the previous Big Five: the earlier
extinctions stem from external environmental events; the Sixth,
set to unfold over the next several centuries (seemingly long to
us, but a blink of the eye in geological terms), derives directly
from human impact.

The main causes of extinction have been habitat loss, over-exploitation, and introduction of alien species. Often two, or all three, combine. But an increasing number of recent studies show clearly that the effects of climate change are compounding these more direct effects of human activities. As set out in Chapter 7, if global average surface temperatures rise by 1.5 to 2.5°C by 2100, an estimated 20 per cent to 33 per cent of all plant and animal species are likely to be threatened with extinction. This temperature interval is in the lower range of those projected by the IPCC, and species-loss estimates rise to 40 per cent at the IPCC's higher levels of projected temperature increase. Furthermore, ocean acidification, a direct result of rising carbon dioxide levels, will have major effects on marine ecosystems, with possible adverse consequences on fish stocks.

The UN-sponsored Millennium Ecosystem Assessment (MEA), published in 2005, integrated ecological studies with economic and social considerations, and concluded that approximately 60 per cent of the ecosystem services that support life on Earth – such as fresh water, fisheries, air and water regulation, pollinators for crops, along with the regulation of regional climate, pests, and certain kinds of natural hazards – are being degraded and/or used unsustainably. These ecosystem services are not counted in conventional economic measures of global GDP, but necessarily rough estimates suggest their monetary value (at around £20–30 trillion in 1996) is of the same order as the conventional economists' GDP.

Despite the growing weight of evidence of climate change and loss of biological diversity, along with growing awareness of the manifold adverse consequences, there remains an active and well-funded "denial lobby". It shares many features with the lobby that for so long denied smoking is the major cause of lung cancer. For climate change, the plain fact is, as shown in a recent study of the 1,000 or so papers on the subject published in peer-reviewed scientific journals in recent years, that not one denies that climate change is real, and primarily caused by us. The loss of biodiversity is equally clearly evident, although its long-term consequences are less well understood.

In order to emphasise the scientific consensus on climate

change, in the summer of 2005 the Royal Society took the unprecedented step of producing a brief statement on the science of climate change, signed by the Science Academies of all the G8 countries – USA, Japan, Germany, France, UK, Italy, Canada, Russia – along with China, India and Brazil. Making it clear that climate change is real, caused by human activities, and with serious consequences, this statement called on the G8 nations to, "Identify cost-effective steps that can be taken now to contribute to substantial and long-term reductions in net global greenhouse gas emission [and to] recognise that delayed action will increase the risk of adverse environmental effects and will likely incur a greater cost."

The Stern Review[1] – the most detailed and authoritative review of its kind to date, in my opinion – underlines these many and varied concerns, and elaborates the likely economic impacts, and the cost of actions to ameliorate them.

Under "Business As Usual" scenarios for greenhouse gas emissions, and using the most recent scientific evidence on global warming and its consequences, the Stern Review estimates the economic impact over the next two centuries to lie in the range of a 5–7 per cent reduction of global GDP or "personal consumption". If direct impacts on the environment and human health (sometimes called "non-market" impacts, and not usually included in such calculations, despite their obvious relevance) are included these figures increase to 11–14 per cent or more.

Stabilization of greenhouse gas concentrations, whatever the level, requires that annual emissions be reduced to a level where they can be balanced by the Earth's natural capacity to remove them from the atmosphere (see Chapter 1). The longer emissions remain above this level, the higher the final stabilization level will be. The Stern Review therefore focuses on the "feasibility and costs of stabilization of greenhouse gas concentrations in the atmosphere in the range of 450–550 ppm CO_2".

[1] The Stern Review: The Economics of Climate Change is published by HMSO (2006) and is available at: www.hm-treasury.gov.uk/stern_review_climate_change.htm

Exploring this in detail, the Stern Review concludes that stabilization at or below 550 ppm requires global emissions to peak in the next 10–20 years, then fall at least 1–3 per cent each year. By 2050, this would put global emissions around 25 per cent below current levels, even though the world economy may be 3–4 times larger than today (thus requiring emissions per unit of GDP to be one quarter of today's by 2050). To achieve a much more desirable stabilization at 450 ppm would require that emissions peak in the next decade, then fall 5 per cent each year, attaining 70 per cent of current levels by 2050. None of this will be easy. But the future looks much worse if we "overshoot", and allow atmospheric greenhouse gas concentrations to rise above an eventual stabilization level in the above range.

What is the estimated economic cost of stabilizing at around 450–550 ppm? The Stern Review surveys various such estimates, concluding they range from a cost of 5 per cent of global GDP, to an uplift of 2 per cent, by 2050. A best guess is that the cost of "stabilization at around of 550 ppm CO_2 is likely to be around 1 per cent of GDP by 2050."

More specifically, what actions should we be taking? One thing is very clear, the magnitude of the problem we face with climate change is such that there is no single answer – no silver bullet – but rather a wide range of actions must be pursued.

I think these can be usefully divided into four categories, all of which are discussed in this book (some in more detail than others; see Chapters 4, 6, 7, 8 and particularly Chapter 10). First, we can *adapt* to change: stop building on flood plains; start thinking more deliberately about coastal defences and flood protection, recognizing that some areas should, in effect, be given up. Second, we can *reduce wasteful consumption*, in the home, marketplace, and workplace: there are studies, for example, which demonstrate we can design housing that consumes roughly half of current energy levels without significantly reducing living standards. Third, and necessary in the medium term while we continue to burn fossil fuels, we could *capture some of the carbon dioxide thus emitted* at the source, and sequester it (burying it on land or under seabed). Fourth, we could move more rapidly toward *renewable sources of energy*, which simply do not put greenhouse gases into the atmosphere: these include geothermal, wind, wave, and water energy; solar energy (from physics-based or biology-based devices); fission (currently generating 7 per cent of all the world's energy, and – despite its problems – surely playing a necessary role in the medium-term); fusion (a realistic long-term possibility); biomass (assuming that the carbon dioxide you put into the atmosphere was carbon dioxide you took out when you grew the fuel). Some of these renewables are already being used, others are more futuristic.

In looking toward such actions, we do well to recognize that all countries need to act, but that fairness demands they act in equitable proportions. The world's nation states can be very roughly divided into three categories: rich, transitional, poor. By one such accounting, the rich countries contain one-seventh of the global population, but account for a bit more than half global GDP, energy use, and emissions of greenhouse gases. The transitional countries contain roughly one-quarter of the population, and account for one-quarter of global GDP, energy use, and greenhouse gas output. The remaining quarter of each of these three quantities comes from the poor countries, which contain two-thirds of the global population. And, looking backwards in time, we should explicitly recognize that today's rich countries are responsible for roughly 80 per cent of the carbon dioxide currently adding to pre-industrial levels in the atmosphere. These and other important ethical issues are dealt with in Chapter 8 (see also Chapters 1, 9 and 10).

Such actions to reduce wasteful consumption and more generally address climate change also will help humanity address another, related major problem, which is reducing the rate at which biological diversity is being lost. Here we need to be learning how to reconcile human needs with the rights of other living creatures; how to co-exist better. We also need more actively to be curbing the growth and "industrialization" of the bushmeat trade, and in other ways establishing and properly administering protected areas.

Will we do all this? The answer, ultimately, is up to each and every one of us – thinking globally, acting locally.

PART I

THE BIG PICTURE

I

CLIMATE CHANGE: A SHORT INTRODUCTION

In 2007 the Intergovernmental Panel on Climate Change (IPCC) released its Fourth Assessment Report. Previous assessments, in 1990, 1995, and 2001, had provided strong indications that by various measures Earth's climate was becoming warmer – but with the latest report the picture had become clearer: "Warming of the climate system is unequivocal, as is now evident from observations of increases in global average air and ocean temperatures, widespread melting of snow and ice, and rising global average sea level."

The IPCC was established in 1988 by the United Nations Environment Programme and the World Meteorological Organization (a UN agency) in recognition of the potential importance of climate change. The IPCC is charged with reviewing comprehensive scientific climate-change studies and providing an objective understanding of climate change, its potential impacts, and options for adaptation and mitigation. Hundreds of climatologists, meteorologists, and other scientists from around the world are involved in the preparation of IPCC reports as authors, contributors, and expert reviewers. The Fourth Assessment was compiled by three IPCC working groups: this chapter is an overview of their findings on the effects of climate change on physical, biological, and human systems. It also includes a discussion of the scientific background of global warming.

Climate undergoes natural changes and cycles. To understand Earth's overall warming, therefore, scientists examine the balance of the energy that reaches Earth from the sun and the energy that is radiated away from Earth. They then identify "radiative forcings" – that is, human or natural factors that drive the energy balance up or down. The Fourth Assessment established that anthropogenic factors (those originating in human activity) are responsible for most of the current global warming, with the radiative forcing from anthropogenic sources being over ten times larger than that from all natural components combined. The primary anthropogenic source is the emission of greenhouse gases such as carbon dioxide, which is produced mainly by the burning of fossil fuels. (Greenhouse gases are trace gases with certain chemical properties, which allow sunlight to pass through but trap heat radiated from Earth as it is warmed by the sunlight.) Land-use change, such as the burning or clearing of forests, makes a lesser contribution.

Effects of climate change

Effects on the physical world

The Fourth Assessment Report documented that 11 of the past 12 years have been the warmest on record since 1850, which is when global instrumental record-keeping began. The world has not been warming uniformly as climate changes, however. In general, average land surface temperatures have been

Map of projected surface temperature changes
(relative to 1980–99 and based on a midrange emissions scenario.)

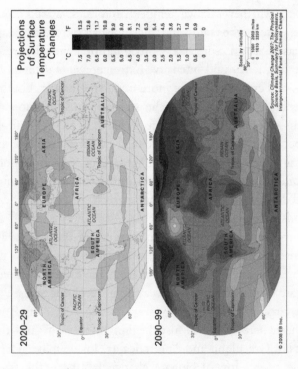

Figure 1.1 © Encyclopædia Britannica, Inc.

increasing more rapidly than ocean surface temperatures (although the oceans absorb 80 per cent of the heat that the world is gaining). The Arctic has been the region with the most rapid rate of warming – two to three times the global average. In contrast, surface temperatures of Antarctica have not risen significantly (Figure 1.1).

With warmer surface temperatures and warmer oceans, more water evaporates and the moisture in the atmosphere increases. Storms with heavy precipitation have occurred with more frequency and intensity. Extreme events such as hurricanes and cyclones are not more frequent globally, but there is evidence of an increase in the strength and duration of the storms since 1970 that is consistent with increases in ocean temperature. Increases in the extent of spring melting and in storms with heavy precipitation have resulted in more flooding in some areas. Warmer temperatures can also mean more rapid drying, however, and some areas have experienced more periods marked by drought.

With the advent of satellite imagery in the late 1970s, it became possible to monitor snow and ice coverage on a global scale. Snow pack, sea ice, and glaciers have been melting, and the rate of melt has been increasing in recent decades. Permafrost (ground that normally stays frozen year-round) in the northern hemisphere is also beginning to melt, and the ice sheets of Greenland and Antarctica are losing mass. The most visible expression of climate change has been the seasonal retreat of Arctic sea ice. The summertime sea ice minimum in the Arctic has shown a declining trend, and in 2007 the minimum was 23 per cent less than the record minimum that was set in 2005 (Figure 1.2).

The melting of land-based ice and the expansion of the oceans as they have become warmer account almost equally for the observed increases in sea level. (Melting of sea ice does

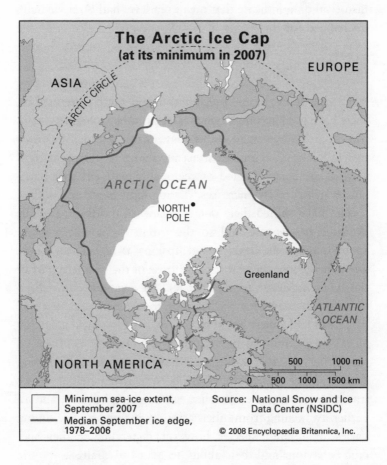

The Arctic Ice Cap
(at its minimum in 2007)

EUROPE

ASIA

ARCTIC CIRCLE

ARCTIC OCEAN

NORTH POLE

Greenland

ATLANTIC OCEAN

NORTH AMERICA

| 0 | 500 | 1000 mi |
| 0 | 500 | 1000 | 1500 km |

Minimum sea-ice extent, September 2007

Median September ice edge, 1978–2006

Source: National Snow and Ice Data Center (NSIDC)

© 2008 Encyclopædia Britannica, Inc.

Figure 1.2 © Encyclopædia Britannica, Inc.

not raise sea level, since floating ice already displaces its equivalent in melt water.) Although the rise is relatively small, historical data indicate that mean sea level had been virtually unchanged for the previous 2,000 years.

Effects on biological systems

As temperatures rise and precipitation and storm patterns shift, there have been accompanying changes in the biological world. The Fourth Assessment states that: "Observational evidence from all continents and most oceans shows that many natural systems are being affected by regional climate changes, particularly temperature increases." Some land plant and animal species have shifted their ranges poleward (northward in the northern hemisphere and southward in the southern hemisphere), and some have moved upslope to higher elevations, where it is cooler. Boreal forests (those of the northern part of the globe adjacent to the Arctic), for example, have been observed encroaching northward on the Arctic tundra at a rate of 12 km (7.5 miles) per year.

In general, mid- to high-latitude regions have experienced the earlier onset of spring and have had a longer growing season. Other changes that have been reported include earlier leaf production in trees, earlier egg hatching in birds, and an earlier awakening from hibernation by mammals. The timing varies for different species, however, depending on their specific behaviour and their ability to adapt to change.

As climate and some ecosystems have shifted, there has been some loss and fragmentation of terrestrial habitats – compounding the habitat loss already caused by humans' economic activities. Climate change is thought to play a role in the population decrease and ultimately the extinction of some species, by such mechanisms as constricting habitat, affecting reproductive patterns, and providing an advantage to competing species. Particularly at risk are species that have a restricted range and low adaptive capacity.

Some marine and freshwater biological ecosystems have also shifted poleward, apparently because of rising water temperatures, loss of ice cover, and changes in ocean circulation and water chemistry. Examples of affected organisms include algae, plankton, and fish in high-latitude regions and in high-altitude lakes. Warming of the southern oceans has been associated with a decline in the population of krill (a small crustacean of the open sea), which in turn has been linked to a decrease in seabird and seal populations in the region. Loss of habitat is also expected to affect those species that are dependent on Arctic sea ice, such as the polar bear, walrus, and several species of seals and seabirds. Overall biological abundance in the oceans is difficult to determine, but satellite imagery of chlorophyll levels, from marine plant life, indicates that primary ocean production (the production of biomass by photosynthetic organisms) has gone down 6 per cent globally since the early 1980s. Another potential impact on marine life is related to the increase of carbon dioxide in the atmosphere because some of the gas is absorbed by the oceans. The extra dissolved carbon dioxide in seawater has made it more acidic, a quality indicated by an average decrease in pH. There is evidence that the acidity may be exacerbating the coral bleaching – a loss of colour in coral caused by the expulsion of symbiotic algae – already caused by ocean warming.

Effects on human society

The effects of climate change are beginning to appear in the human sphere, although in general they are not as evident as the impacts on the natural world. Problems related to water supply are projected to increase in many regions, resulting from shrinking glaciers and snowpack, drought, evaporation, and the infiltration of salt water in low-lying areas through rising sea levels. Lack of access to usable water is a key vulnerability, especially in less-developed countries (LDCs). Climate change is expected to have a mixed impact on agriculture. With spring occurring sooner in mid- to high-latitude regions, a longer growing season would benefit crop yields. Agricultural productivity, however, is vulnerable to other potential consequences of climate change, such as heat waves, floods, and droughts. Agricultural production in low-latitude regions has already been adversely affected by global warming. The Sahel region of Africa has seen crop failures because of intense and more frequent droughts. The situation has resulted in famines and has been exacerbated by other stresses in the region. Unfortunately, crop yields are expected to continue to drop in coming decades as a result of climate change.

Like agriculture, forestry is expected to be both positively and negatively affected. Forests of the northern hemisphere would benefit from an extended growing season but might also experience adverse effects from other factors. For example, forests from British Columbia to Alaska have been subjected to severe infestations of tree-killing beetles that have proliferated with a warmer regime. Dead trees in turn increase the risk of wildfires.

Coastal cities and infrastructure, especially low-lying delta regions and small islands, are vulnerable to sea storms. A rise in sea level in combination with more intense, or extreme, weather could cause severe damage. The costs associated with such damage are not necessarily incremental, because the severity could dramatically increase when structures are subjected to forces that exceed what they have been designed to withstand.

Climate change might also have adverse effects on human health. There is evidence that the ranges of mosquitoes and other disease carriers have increased, although there is no clear indication of any corresponding increase in the incidence of the diseases they transmit. Cold-related injury and deaths are projected to decrease, but heat-related increases would out-weigh them. Heatwaves can be very serious, as shown by the 2003 heatwave in Europe, in which 35,000 excess deaths were recorded. Increased stress on water and food resources would result in a higher incidence of malnutrition. The hardest-hit areas are likely to be those with a low capacity for adaptation – in other words, regions that do not have spare economic resources and that are subject to various kinds of stress in addition to any created by climate change.

Causes of global warming

The term "global warming" is used specifically to refer to any warming of near-surface air during the past two centuries that can be traced to anthropogenic causes. Global warming is related to the more general phenomenon of climate change, which refers to changes in the totality of attributes that define climate. These include air temperature, precipitation patterns, winds, and ocean currents.

Normally, climate change can be viewed as the combination of various natural forces occurring over diverse timescales. To understand the concepts of global warming and climate change properly, it is necessary to recognize that the climate of the Earth has varied across many timescales, ranging from an individual human lifespan to billions of years (see "Biocli-

matology and palaeoclimatology", Chapter 2, page 123).
Since the advent of human civilization, however, climate
change has involved an exclusively anthropogenic element,
and this has become more important in the industrial period of
the past two centuries.

A vigorous debate is in progress over the extent and ser-
iousness of rising surface temperatures, the effects of past and
future warming on human life, and the need for action to
reduce future warming and deal with its consequences.

Giving voice to a growing conviction of most of the scientific
community, the IPCC's Fourth Assessment Report stated that
the twentieth century saw an increase in global average surface
temperature of approximately 0.6°C (1.1°F). It went on to
state that most of the warming observed over the second half
of the twentieth century could be attributed to human activities, and predicted that by the end of the twenty-first century
the average surface temperature would increase by another
1.8–4.0°C (3.2–7.2°F), depending on a range of possible
scenarios. Many climate scientists agree that significant economic and ecological damage would result if global average
temperatures rose by more than 2°C (3.6°F) in such a short
time. Such damage might include the increased extinction of
many plant and animal species, shifts in patterns of agriculture, and rising sea levels. The IPCC also reported that the
global average sea level rose by some 17 cm (6.7 inches) during
the twentieth century, that sea levels rose faster in the second
half of that century than in the first half, and that – again
depending on a wide range of scenarios – the global average
sea level could rise by another 18–59 cm (7–23 inches) by the
end of the twenty-first century. Furthermore, it reported that
average snow cover in the northern hemisphere declined by 4
per cent, or 1.5 million square km (580,000 square miles),
between 1920 and 2005.

The scenarios referred to above depend mainly on future concentrations of greenhouse gases. These gases have been injected into the lower atmosphere in increasing amounts through the burning of fossil fuels for industry, transportation, and residential uses. Modern global warming is widely believed to be the result of an increase in magnitude of the so-called greenhouse effect – a warming of Earth's surface and lower atmosphere caused by the presence of water vapour, carbon dioxide, methane, and other greenhouse gases. Of all these gases, carbon dioxide (CO_2) is the most important, both for its role in the greenhouse effect and for its role in the human economy. It has been estimated that, at the beginning of the industrial age in the mid-eighteenth century, the concentration of CO_2 in the atmosphere was roughly 280 parts per million (ppm). By the end of the twentieth century, the concentration of CO_2 reached 369 ppm (possibly the highest concentration in at least 650,000 years), and by the mid-twenty-first century, if fossil fuels continue to be burned at current rates, it is projected to reach 560 ppm – essentially a doubling of atmospheric carbon dioxide in 300 years. It has been calculated that an increase of this magnitude alone (that is, not accounting for the possible effects of other greenhouse gases) would be responsible for adding 2–5°C (3.6–9°F) to the global average surface temperatures that prevailed at the beginning of the industrial age.

The greenhouse effect

The average surface temperature of the Earth is maintained by a balance of various forms of solar and terrestrial radiation. Solar radiation is often called "short-wave" radiation, because the frequencies of the radiation are relatively high and the wavelengths relatively short – close to the visible portion of the electromagnetic spectrum. Terrestrial radiation, on the other hand, is often called "long-wave" radiation, because the frequencies are relatively low and the wavelengths relatively long – somewhere in the infrared part of the spectrum. Downward-moving solar energy is typically measured in watts per square metre. The energy of the total incoming solar radiation at the top of Earth's atmosphere (the so-called "solar constant") amounts roughly to 1,366 watts per square metre annually. Adjusting for the fact that only half of the planet's surface receives solar radiation at any given time, the average surface radiation received (known as insolation) is 342 watts per square metre annually.

The amount of solar radiation absorbed by the Earth's surface is only a small fraction of the total solar radiation entering the atmosphere. For every 100 units of incoming solar radiation, roughly 30 units are reflected back to space by either clouds, the atmosphere, or reflective regions of Earth's surface. This reflective capacity is referred to as Earth's "planetary albedo", and it need not remain fixed over time, since the spatial extent and distribution of reflective formations, such as clouds and ice cover, can change. The 70 units of solar radiation that are not reflected may be absorbed by the atmosphere, clouds, or the surface. In the absence of further complications, in order to maintain thermodynamic equilibrium, Earth's surface and atmosphere must radiate these same 70 units back to space. The magnitude of this emission is proportional to Earth's surface temperature (and to that of the lower layer of the atmosphere essentially in contact with the surface).

The Earth's energy budget is further complicated by the greenhouse effect. Greenhouse gases – mainly carbon dioxide, methane (CH_4), and nitrous oxide (N_2O) – absorb some of the infrared radiation produced by Earth's surface. Because of this absorption, some fraction of the original 70 units does not escape directly to space. Because greenhouse gases emit the same amount of radiation they absorb and because this radiation is emitted equally in all directions (that is, as much downward as upward), the net effect of absorption by greenhouse gases is to increase the total amount of radiation emitted downward toward Earth's surface and lower atmosphere. To maintain equilibrium, Earth's surface and lower atmosphere must emit more radiation than the original 70 units. Consequently, the surface temperature must be higher. This process is not quite the same as that which occurs in an actual greenhouse, but the end effect is similar. The presence of greenhouse gases in the atmosphere leads to a warming of the surface and lower part of the atmosphere (and a cooling higher up in the atmosphere) relative to what would be expected in the absence of greenhouse gases.

It is essential to distinguish the "natural", or background, greenhouse effect from the "enhanced" greenhouse effect associated with human activity. The natural greenhouse effect is associated with surface warming properties of natural constituents of Earth's atmosphere, especially water vapour, carbon dioxide, and methane. The existence of this effect is accepted by all scientists. Indeed, in its absence, Earth's average temperature would be approximately 33°C (59°F) colder than it is today, and Earth would be a frozen and likely uninhabitable planet. What has been the subject of controversy is the enhanced greenhouse effect, which is associated with increased concentrations of greenhouse gases from anthropogenic sources. In particular, the burning of fossil fuels raises the concentrations of the major greenhouse gases in the atmosphere, and these higher concentrations have the potential to warm the atmosphere by several degrees.

Radiative forcing

The temperature of Earth's surface and lower atmosphere may be modified in three ways: (1) through a net increase in the solar radiation entering at the top of Earth's atmosphere, (2) through a change in the proportion of the radiation reaching the surface, and (3) through a change in the concentration of greenhouse gases in the atmosphere. In each case the changes can be thought of in terms of "radiative forcing". As defined by the IPCC, radiative forcing is a measure of the influence a given climatic factor has on the amount of downward-directed radiant energy impinging upon Earth's surface. Climatic factors are divided into those caused primarily by human activity (such as greenhouse gas emissions and aerosol emissions) and those caused by natural forces (such as solar irradiance); then, for each factor, so-called forcing values are calculated for the time period between 1750 and the present day. "Positive forcing" is exerted by climatic factors that contribute to the warming of Earth's surface, whereas "negative forcing" is exerted by factors that cool Earth's surface.

As we have seen, an average of 342 watts of solar radiation strike each square metre of Earth's surface per year. This quantity can in turn be related to a rise or fall in Earth's surface temperature. Temperatures at the surface may also rise or fall through a change in the distribution of terrestrial radiation (radiation emitted by Earth) within the atmosphere. In some cases radiative forcing has a natural origin, such as during explosive eruptions from volcanoes where vented gases and ash block some portion of solar radiation from the surface. In other cases its origin is exclusively human. For example, anthropogenic increases in carbon dioxide, methane, and nitrous oxide are estimated to account for 2.3 watts per square metre of positive radiative forcing.

When all values of positive and negative radiative forcing are taken together and all interactions between climatic factors are accounted for, the total net increase in surface radiation due to human activities since the beginning of the Industrial Revolution is 1.6 watts per square metre.

Greenhouse gases

As discussed above, greenhouse gases warm the Earth's surface by increasing the net downward long-wave radiation reaching the surface. The relationship between atmospheric concentration of greenhouse gases and the associated positive radiative forcing of the surface is different for each gas. A complicated relationship exists between the chemical properties of each greenhouse gas and the relative amount of long-wave radiation that each can absorb. What follows is a discussion of the radiative behaviour of each major greenhouse gas.

Water vapour

Water vapour is the most potent of the greenhouse gases in Earth's atmosphere, but its behaviour is fundamentally different from that of the other greenhouse gases. The primary role of water vapour is not as a direct agent of radiative forcing but rather as a climate feedback – that is, as a response within the climate system that influences the system's continued activity (see "Water vapour feedback", page 34). This distinction arises from the fact that the amount of water vapour in the atmosphere cannot, in general, be directly modified by human behaviour but is instead set by air temperatures. The warmer the surface, the greater the evaporation rate of water from the surface. Increased evaporation leads to a greater concentration of water vapour in the lower atmosphere, which is capable of absorbing long-wave radiation and emitting it downward.

Carbon dioxide

Of the greenhouse gases, carbon dioxide (CO_2) is most significant. Natural sources of atmospheric CO_2 include outgassing (the slow release of absorbed or trapped gasses) from volcanoes, the combustion and natural decay of organic matter, and respiration by aerobic (oxygen-using) organisms. These sources are balanced, on average, by a set of physical, chemical, or biological processes, called "sinks", that tend to remove CO_2 from the atmosphere. Significant natural sinks include terrestrial vegetation, which takes up CO_2 during the process of photosynthesis (see "The carbon cycle", Chapter 4).

A number of oceanic processes also act as carbon sinks. One such process, called the "solubility pump", involves the descent of surface sea water containing dissolved CO_2. Another process, the "biological pump", involves the uptake of dissolved CO_2 by marine vegetation and phytoplankton (small, free-floating, photosynthetic organisms) living in the upper ocean, or by other marine organisms that use CO_2. As these organisms expire and fall to the ocean floor, the carbon they contain is transported downward and eventually buried at depth. A long-term balance between these natural sources and sinks leads to the background, or natural, level of CO_2 in the atmosphere.

In contrast, human activities increase atmospheric CO_2 levels primarily through the burning of fossil fuels (principally oil and coal, and also natural gas, for use in transportation, heating, and the generation of electrical power) and through the production of cement (notably in China). Other anthropogenic sources include the burning of forests and the clearing of land. Anthropogenic emissions currently account for the annual release of about 7 gigatons (7 billion tons) of carbon into the atmosphere. Anthropogenic emissions are equal to approximately 3 per cent of the total emissions of CO_2 by natural sources, and this amplified carbon load from human activities far exceeds the offsetting capacity of natural sinks (by perhaps as much as 2–3 gigatons per year). Consequently, CO_2 has accumulated in the atmosphere at an average rate of 1.4 parts per million (ppm) by volume per year between 1959 and 2006, and this rate of accumulation has been linear (that is, uniform over time). However, certain current sinks, such as the oceans, could become sources in the future (see "Carbon cycle feedbacks", page 37). This may lead to a situation in which the concentration of atmospheric CO_2 builds at an exponential rate.

The natural background level of carbon dioxide varies on a timescale of millions of years due to slow changes in outgassing through volcanic activity. For example, roughly 100 million years ago, during the Cretaceous Period, CO_2 concentrations appear to have been several times higher than today (perhaps close to 2,000 ppm). Over the past 700,000 years, CO_2 concentrations have varied over a far smaller range (between roughly 180 and 300 ppm), in association with the same orbital effects of Earth that are linked to the coming and going of the Pleistocene ice ages, 1,800,000 to 11,800 years ago (see "Glacial and interglacial changes of the Pleistocene", Chapter 2, page 144). By the early twenty-first century, CO_2 levels reached 384 ppm, which is approximately 37 per cent above the natural background level of roughly 280 ppm that existed at the beginning of the Industrial Revolution. According to ice core measurements, this level of 384 ppm is believed to be the highest in at least 650,000 years.

Radiative forcing caused by carbon dioxide varies in an approximately logarithmic fashion with the concentration of that gas in the atmosphere. The logarithmic relationship occurs as the result of a saturation effect, wherein it becomes increasingly difficult, as CO_2 concentrations increase, for additional CO_2 molecules to further influence the "infrared window" (a certain narrow band of wavelengths in the infra-red region that is not absorbed by atmospheric gases). The logarithmic relationship predicts that the surface warming potential will rise by roughly the same amount for each doubling of CO_2 concentration. At current rates of fossil-fuel use, a doubling of CO_2 concentrations over pre-industrial levels is expected to take place by the middle of the twenty-first century – when CO_2 concentrations are projected to reach 560 ppm. A doubling of CO_2 concentrations would represent an increase of roughly 4 watts per square metre of radiative forcing. Given typical estimates of "climate sensitivity" in the absence of any offsetting factors (see "Feedback mechanisms and climate sensitivity", page 33), this energy increase would lead to a warming of 2–5°C (3.6–9°F) over pre-industrial times. The total radiative forcing by anthropogenic CO_2 emissions since the beginning of the industrial age is approxi-mately 1.66 watts per square metre.

Methane

Methane (CH_4) is the second most important greenhouse gas. CH_4 is more potent than CO_2 because the radiative forcing produced per molecule is greater. In addition, the infrared window is less saturated in the range of wave-lengths of radiation absorbed by CH_4, so more molecules may fill in the region. However, CH_4 exists in a far lower concentration than CO_2 in the atmosphere, and its con-centration by volume in the atmosphere is generally measured in parts per billion (ppb) rather than ppm. CH_4 also has a considerably shorter residence time in the atmosphere than CO_2 – roughly ten years, compared with hundreds of years for CO_2.

Natural sources of methane include tropical and northern wetlands, methane-oxidizing bacteria that feed on organic material consumed by termites, volcanoes, seepage vents of the sea floor in regions rich with organic sediment, and methane hydrates trapped along the continental shelves of the oceans and in polar permafrost. The primary natural sink for methane is the atmosphere itself, as methane reacts readily with the hydroxyl radical (OH^-) within the troposphere (the lowest region of the Earth's atmosphere) to form CO_2 and water vapour (H_2O). When CH_4 reaches the stratosphere (the second layer of the atmosphere), it is destroyed. Another natural sink is soil, where methane is oxidized by bacteria.

As with CO_2, human activity is increasing the CH_4 concentration in the atmosphere faster than it can be offset by natural sinks. Anthropogenic sources currently account for approximately 70 per cent of total annual methane emissions, leading to substantial increases in its concentration over time. The major anthropogenic sources of atmospheric CH_4 are rice cultivation, livestock farming, the burning of coal and natural gas, the combustion of biomass, and the decomposition of organic matter in landfills. Future trends are particularly difficult to anticipate, partly due to an incomplete understanding of the climate feedbacks associated with CH_4 emissions. In addition, as the human population grows, it is difficult to predict how possible changes in livestock raising, rice cultivation, and energy utilization will influence CH_4 emissions.

It is believed that a sudden increase in the concentration of methane in the atmosphere was responsible for a warming event roughly 55 million years ago that raised average global temperatures by 4–8°C (7.2–14.4°F) over a few thousand years. The rise in CH_4 appears to have been related to a massive volcanic eruption that interacted with methane-containing flood deposits. As a result, large amounts of gaseous CH_4 were injected into the atmosphere. It is difficult to know precisely how high these concentrations were or how long they persisted. At very high concentrations, residence times of CH_4 in the atmosphere can be much greater than the nominal ten-year residence time that applies today. Nevertheless, it is likely that these concentrations reached several parts per million during this warming episode.

As with carbon dioxide, methane concentrations have varied over a smaller range (roughly 350–800 ppb) in association with the ice-age cycles in the Pleistocene Epoch (see "Glacial and interglacial cycles of the Pleistocene", Chapter 2, page 144). Pre-industrial levels of CH_4 in the atmosphere were approximately 700 ppb, whereas early twenty-first-century levels exceeded 1,770 ppb. This concentration is well above the natural level observed for at least the past 650,000 years. The net radiative forcing by anthropogenic CH_4 emissions is approximately 0.5 watt per square metre – roughly one third the radiative forcing of anthropogenic CO_2 emissions.

Surface-level ozone and other compounds

The next most significant greenhouse gas is surface, or low-level, ozone (O_3). Surface O_3 is a result of air pollution: it must be distinguished from naturally occurring stratospheric O_3, which has a very different role in the planetary radiation balance. The primary natural source of surface O_3 is the subsidence of stratospheric O_3 from the upper atmosphere (see "Stratospheric ozone depletion", page 29). In contrast, the primary anthropogenic source of surface O_3 is photochemical reactions involving the atmospheric pollutant carbon monoxide (CO). The best estimate of the concentration of surface O_3 is 50 ppb. The net radiative forcing due to anthropogenic emissions of surface O_3 is approximately 0.35 watt per square metre.

Nitrous oxides and fluorinated gases

Additional trace gases produced by industrial activity that have greenhouse properties include nitrous oxide (N_2O) and fluorinated gases (halocarbons), the latter including sulphur hexafluoride, hydrofluorocarbons (HFCs), and perfluorocarbons (PFCs). Nitrous oxide is responsible for a radiative forcing of 0.16 watt per square metre, while fluorinated gases are collectively responsible for 0.34 watt per square metre. Nitrous oxides have small background concentrations caused by natural biological reactions in soil and water, whereas the fluorinated gases owe their existence almost entirely to industrial sources.

Aerosols

An aerosol is the suspension of fine particles, solid or liquid, in a gas. The production of aerosols represents an important anthropogenic radiative forcing of climate. Collectively, aerosols block (reflect and absorb) a portion of incoming solar radiation, and this creates a negative radiative forcing – unlike greenhouse gases, most aerosols have a cooling rather than a warming influence on Earth's surface. Aerosols are second only to greenhouse gases in relative importance regarding their impact on near-surface air temperatures. Unlike the decade-long residence times of the "well-mixed" (compounds that react with the air or become evenly distributed) greenhouse gases, such as CO_2 and CH_4, aerosols are readily flushed out of the atmosphere within days, either by rain or snow (wet deposition) or by settling out of the air (dry deposition). They must therefore be continually generated to produce a steady effect on radiative forcing.

Aerosols have the ability to influence climate directly by absorbing or reflecting incoming solar radiation, but they can also produce indirect effects on climate by modifying cloud formation or cloud properties. Most aerosols serve as condensation nuclei (surfaces upon which water vapour can condense to form clouds); however, darker-coloured aerosols may hinder cloud formation by absorbing sunlight and heating up the surrounding air. Aerosols can be transported thousands of kilometres from their sources of origin by winds and upper-level circulation in the atmosphere.

Perhaps the most important type of anthropogenic aerosol in radiative forcing is sulphate aerosol. It is produced from sulphur dioxide (SO_2) emissions associated with the burning of coal and oil. Since the late 1980s, global emissions of SO_2 have decreased from about 73 million tons to about 54 million tons of sulphur per year.

Nitrate aerosol is not as important as sulphate aerosol, but it has the potential to become a significant source of negative forcing. One major source of nitrate aerosol is smog (the combination of ozone with oxides of nitrogen in the lower atmosphere) released from the incomplete burning of fuel in internal-combustion engines. Another source is ammonia (NH_3), which is often used in fertilizers or released by the burning of plants and other organic materials. If greater amounts of atmospheric nitrogen are converted to ammonia, and agricultural ammonia emissions continue to increase as projected, the influence of nitrate aerosols on radiative forcing is expected to grow.

Both sulphate and nitrate aerosols act primarily by reflecting incoming solar radiation, thereby reducing the amount of sunlight reaching the Earth's surface. Although most aerosols have a cooling effect on the Earth, prominent exceptions are carbonaceous aerosols such as carbon black or soot, which are produced by the burning of fossil fuels and biomass. Carbon black tends to absorb rather than reflect incident solar radiation, and so it has a warming impact on the lower atmosphere, where it resides. Because of its absorptive properties, carbon black is also capable of having an additional indirect effect on climate. Through its deposition in snowfall, it can decrease the albedo of snow cover. This reduction in the amount of solar radiation reflected back to space by snow surfaces creates a minor positive radiative forcing.

Natural forms of aerosol include windblown mineral dust generated in arid and semi-arid regions and sea salt produced by the action of waves breaking in the ocean. Changes to wind patterns as a result of climate modification could alter the emissions of these aerosols. The influence of climate change on regional patterns of aridity could shift both the sources and the destinations of dust clouds. In addition, since the concentration of sea-salt aerosol, or sea aerosol, increases with the strength of the winds near the ocean surface, changes in wind speed caused by global warming and climate change could influence this concentration. For example, some studies suggest that climate change might lead to stronger winds over parts of the North Atlantic Ocean. Areas with stronger winds may experience an increase in the concentration of sea aerosol.

Other natural sources of aerosols include volcanic eruptions, which produce sulphate aerosol, and biogenic sources (living things, e.g. phytoplankton), which produce dimethyl sulphide (DMS). Other important biogenic aerosols, such as terpenes, are produced naturally by certain kinds of trees or other plants. For example, the dense forests of the Blue Ridge Mountains of Virginia in the United States emit terpenes during the summer months, which in turn interact with the high humidity and warm temperatures to produce a natural photochemical smog. Anthropogenic pollutants such as nitrate and ozone, both of which serve as precursor molecules for the generation of biogenic aerosol, appear to have increased the rate of production of these aerosols by several times. This process seems to be responsible for some of the increased aerosol pollution in regions undergoing rapid urbanization.

Human activity has greatly increased the amount of aerosol in the atmosphere compared with the background levels of pre-industrial times. In contrast to the global effects of greenhouse gases, the impact of anthropogenic aerosols is confined primarily to the northern hemisphere, where most of the world's industrial activity occurs. The pattern of increases in anthropogenic aerosol over time is also somewhat different from that of greenhouse gases. During the middle of the twentieth century, there was a substantial increase in aerosol emissions. This appears to have been at least partly responsible for a cessation of surface warming that took place in the northern hemisphere from the 1940s to the 1970s. Since that time, aerosol emissions have levelled off due to antipollution measures undertaken in the industrialized countries since the 1960s. Aerosol emissions may rise in the future, however, as a result of the rapid emergence of coal-fired electric power generation in China and India.

The total radiative forcing of all anthropogenic aerosols is approximately -1.2 watts per square metre. Of this total, -0.5 watt per square metre comes from direct effects (such as the reflection of solar energy back into space), and -0.7 watt per square metre from indirect effects (such as the influence of aerosols on cloud formation). This negative radiative forcing represents an offset of roughly 40 per cent from the positive radiative forcing caused by human activity. However, the relative uncertainty in aerosol radiative forcing (approximately 90 per cent) is much greater than that of greenhouse gases. In addition, future emissions of aerosols from human activities, and the influence of these emissions on future climate change, are not known with any certainty.

Nevertheless, it can be said that, if concentrations of anthropogenic aerosols continue to decrease as they have since the 1970s, a significant offset to the effects of greenhouse gases will be reduced, exposing our climate to further warming in future.

Land-use change

There are a number of ways in which changes in land use can influence climate. The most direct effect is through the alteration of Earth's albedo, or surface reflectance. For example, the replacement of forest by cropland and pasture in the middle latitudes over the past several centuries has led to an increase in albedo, and thereby a greater reflection of incoming solar radiation, in those regions. This replacement of forest by agriculture has been associated with a change in global average radiative forcing of approximately -0.2 watt per square metre since 1750. In Europe and other major agricultural regions, such land-use conversion began more than 1,000 years ago and has proceeded nearly to completion. For Europe, the negative radiative forcing due to land-use change has probably been substantial, perhaps approaching -5 watts per square metre. The influence of early land use on radiative forcing may help to explain a long period of cooling in Europe that followed a period of relatively mild conditions roughly 1,000 years ago. It is generally believed that the mild temperatures of this "medieval warm period", which was followed by a long period of cooling, rivalled those of twentieth-century Europe.

Land-use changes can also influence climate through their effect on the exchange of heat between Earth's surface and the atmosphere. For example, vegetation helps to facilitate the evaporation of water into the atmosphere through transpiration. This is the process by which plants take up liquid water from the soil through their root systems and eventually release it as water vapour, through the stomata in their leaves, into the atmosphere. Evapotranspiration is the process by which water is transferred from the land to the atmosphere through transpiration and evaporation from the soil. Evapotranspiration has a cooling effect, due to the uptake of latent heat – the heat needed to evaporate the liquid. While deforestation generally leads to surface cooling due to the albedo factor discussed above, the land surface may also be warmed due to the cessation of the transpiration process. The relative importance of these two factors, one exerting a cooling effect and the other a warming effect, varies by both season and region. While the albedo effect is likely to dominate in middle latitudes, especially during the period from autumn to spring, the effect of the lack of transpiration may dominate during the summer in the mid-latitudes and year-round in the tropics. The latter case is particularly important in assessing the potential impacts of continued tropical deforestation.

The rate at which tropical regions are deforested is also relevant to the process of carbon sequestration – the long-term storage of carbon in underground cavities and biomass rather than in the atmosphere (see "The carbon cycle", Chapter 4, and "Carbon cycle feedbacks", page 37). By removing carbon from the atmosphere, carbon sequestration acts to mitigate global warming. Deforestation contributes to global warming since fewer plants are available to take up carbon dioxide from the atmosphere. In addition, as fallen trees, shrubs, and other plants are burned or allowed slowly to decompose, they release as carbon dioxide the carbon they stored during their lifetimes. Furthermore, any land-use change that influences the amount, distribution, or type of vegetation in a region can affect the atmospheric concentration of biogenic aerosols, although the impact of such changes on climate is indirect and relatively minor.

Stratospheric ozone depletion

Since the 1970s the loss of ozone from the stratosphere has led to a small amount of negative radiative forcing of the surface. This negative forcing represents a competition between two distinct effects caused by the fact that ozone absorbs solar radiation. In the first case, as ozone levels in the stratosphere are depleted, more solar radiation reaches Earth's surface. In the absence of any other influence, this rise in insolation would represent a positive radiative forcing of the surface. However, as the amount of ozone in the stratosphere is decreased, less long-wave radiation emitted by Earth's surface is absorbed, leading to a corresponding decrease in the downward re-emission of radiation. This second effect overwhelms the first and results in a modest negative radiative forcing; it has led to a modest cooling of the lower stratosphere by approximately 0.5°C (0.9°F) per decade since the 1970s.

Natural influences on climate

There are a number of natural factors that influence Earth's climate. These include external influences such as explosive volcanic eruptions, natural variations in the output of the sun, and slow changes in the configuration of Earth's orbit relative to the sun. In addition, there are natural oscillations in Earth's climate that alter global patterns of wind circulation, precipitation, and surface temperatures. One such phenomenon is the El Niño–Southern Oscillation (ENSO), a coupled atmospheric and oceanic event that occurs in the Pacific Ocean every three to seven years (see Chapter 5, page 246). The Atlantic

Multidecadal Oscillation (AMO) is a similar phenomenon that occurs over decades in the North Atlantic Ocean. Other types of oscillatory behaviour that produce dramatic shifts in climate may occur across timescales of centuries and millennia. For more details see "Bioclimatology and palaeoclimatology", Chapter 2, page 123.

Volcanic aerosols

Explosive volcanic eruptions have the potential to inject substantial amounts of sulphate aerosols into the lower stratosphere. In contrast to aerosol emissions in the lower troposphere (see "Aerosols", page 23), aerosols that enter the stratosphere may remain for several years before settling out, because of the relative absence of turbulent motions there. Consequently, aerosols from explosive volcanic eruptions have the potential to affect Earth's climate. Less explosive eruptions, or eruptions that are less vertical in orientation, have a lower potential for substantial climatic impact. Furthermore, because of large-scale circulation patterns within the stratosphere, aerosols injected within tropical regions tend to spread out over the globe, whereas aerosols injected within mid-latitude and polar regions generally remain confined to the middle and high latitudes of that hemisphere. Tropical eruptions, therefore, tend to have a greater climatic impact than eruptions occurring toward the poles. In 1991 the moderate eruption of Mount Pinatubo in the Philippines provided a peak radiative forcing of approximately -4 watts per square metre and cooled the climate by about 0.5°C (0.9°F) over the following few years. Similarly, the 1815 Mount Tambora eruption in present-day Indonesia, typically implicated for the 1816 "year without a summer" in Europe and North America, is believed to have been associated with a radiative forcing of approximately -6 watts per square metre.

While in the stratosphere, volcanic sulphate aerosol actually absorbs long-wave radiation emitted by the Earth's surface, and absorption in the stratosphere tends to result in a cooling of the troposphere below. This vertical pattern of temperature change in the atmosphere influences the behaviour of winds in the lower atmosphere, primarily in winter. Thus, while there is essentially a global cooling effect for the first few years following an explosive volcanic eruption, changes in the winter patterns of surface winds may actually lead to warmer winters in some areas, such as Europe. Some modern examples of vertically oriented eruptions include Krakatoa (Indonesia) in 1883, El Chichón (Mexico) in 1982, and Mount Pinatubo in 1991. There is also evidence that volcanic eruptions may influence other climate phenomena such as ENSO.

Variations in solar output

Direct measurements of solar irradiance, or solar output, have been available from satellites since only the late 1970s. These measurements show a very small peak-to-peak variation in solar irradiance (roughly 0.1 per cent of the 1,366 watts per square metre received at the top of the atmosphere, or approximately 0.14 watt per square metre). However, indirect measures of solar activity are available from historical sunspot measurements dating from the early seventeenth century. Attempts have been made to reconstruct graphs of solar irradiance variations from historical sunspot data by calibrating them against the measurements from modern satellites; however, since the modern measurements span only a few of the most recent 11-year solar cycles, estimates of solar output variability on 100-year and longer timescales are poorly correlated. Different assumptions regarding the relationship between the amplitudes of 11-year solar cycles and long-period solar output changes can lead to considerable differences in the resulting solar reconstructions.

These differences in turn lead to considerable uncertainty in estimating positive forcing by changes in solar irradiance since 1750. (Estimates range from 0.06 to 0.3 watt per square metre.)

Even more challenging, given the lack of any modern analogue, is the estimation of solar irradiance during the so-called Maunder minimum, a period lasting from the mid-seventeenth century to the early eighteenth century when very few sunspots were observed. While it is likely that solar irradiance was reduced at this time, it is difficult to calculate by how much. However, additional proxies of solar output exist that match reasonably well with the sunspot-derived records following the Maunder minimum, and these may be used as crude estimates of the solar irradiance variations.

In theory it is possible to estimate solar irradiance even further back in time, over at least the past millennium, by measuring levels of cosmogenic isotopes such as carbon-14 and beryllium-10. Cosmogenic isotopes are isotopes that are formed by interactions of cosmic rays with atomic nuclei in the atmosphere and that subsequently fall to Earth, where they can be measured in the annual layers found in ice cores. Since their production rate in the upper atmosphere is modulated by changes in solar activity, cosmogenic isotopes may be used as indirect indicators of solar irradiance. However, as with the sunspot data, there is still considerable uncertainty in the amplitude of past solar variability implied by these data.

Solar forcing (radiative forcing associated with energy directly produced by the sun) also affects the photochemical reactions that manufacture ozone in the stratosphere. Through this modulation of stratospheric ozone concentrations, changes in solar irradiance (particularly in the ultraviolet portion of the electromagnetic spectrum) can modify how both short-wave and long-wave radiation in the lower stratosphere are absorbed.

As a result, the vertical temperature profile of the atmosphere can change, and this change can in turn influence phenomena such as the strength of the winter jet streams.

Variations in Earth's orbit

On a timescale of tens of millennia, the dominant radiative forcing of Earth's climate is associated with slow variations in the geometry of Earth's orbit about the sun. These variations include the precession of the equinoxes (a gradual change in the orientation of Earth's rotational axis, resulting in changes in the timing of summer and winter), occurring on a roughly 26,000-year timescale; changes in the tilt angle of Earth's rotational axis relative to the plane of Earth's orbit, occurring on a roughly 41,000-year timescale; and changes in the eccentricity (the departure from a perfect circle) of Earth's orbit, occurring on a roughly 100,000-year timescale.

Changes in eccentricity slightly influence the mean annual solar radiation at the top of Earth's atmosphere, but the primary influence of all the orbital variations listed above is on the seasonal and latitudinal distribution of incoming solar radiation over Earth's surface. The major ice ages of the Pleistocene Epoch were closely related to the influence of these variations on summer insolation at high northern latitudes. Orbital variations thus exerted a primary control on the extent of continental ice sheets. However, Earth's orbital changes are generally believed to have had little impact on climate over the past few millennia, and so they are not considered to be significant factors in present-day climate variability.

Feedback mechanisms and climate sensitivity

There are a number of feedback processes important to Earth's climate system and, in particular, its response to external radiative forcing. The most fundamental of these feedback mechanisms involves the loss of long-wave radiation to space from the surface. Since this radiative loss increases with increasing surface temperatures, it represents a stabilizing factor (that is, a negative feedback) with respect to near-surface air temperature.

Climate sensitivity can be defined as the amount of surface warming resulting from each additional watt per square metre of radiative forcing. Alternatively, it is sometimes defined as the warming that would result from a doubling of CO_2 concentrations and the associated addition of 4 watts per square metre of radiative forcing. In the absence of any additional feedbacks, climate sensitivity would be approximately 0.25°C (0.45°F) for each additional watt per square metre of radiative forcing.

Put another way, if the CO_2 concentration of the atmosphere present at the start of the industrial age (280 ppm) were doubled (to 560 ppm), the resulting additional 4 watts per square metre of radiative forcing would translate into a 1°C (1.8°F) increase in air temperature. However, there are additional feedbacks that exert a destabilizing, rather than a stabilizing, influence (see below), and these feedbacks tend to increase the sensitivity of climate to somewhere between 0.5°C and 1.0°C (0.9–1.8°F) for each additional watt per square metre of radiative forcing.

Water vapour feedback

Unlike concentrations of other greenhouse gases, the concentration of water vapour in the atmosphere cannot freely vary. Instead, it is determined by the temperature of the lower atmosphere and surface through a physical relationship known as the Clausius-Clapeyron equation, named for the nineteenth-century German physicist Rudolf Clausius and nineteenth-century French engineer Émile Clapeyron. Under the assumption that there is a liquid water surface in equilibrium with the atmosphere, this relationship indicates that an increase in the capacity of air to hold water vapour is a function of increasing temperature of that volume of air. This assumption is relatively good over the oceans, where water is plentiful, but not over the continents. For this reason the relative humidity (the percentage of water vapour the air contains relative to its capacity) is approximately 100 per cent over ocean regions and much lower over continental regions (approaching 0 per cent in arid regions).

Not surprisingly, the average relative humidity of Earth's lower atmosphere is similar to the proportion of Earth's surface covered by the oceans (that is, roughly 70 per cent). This quantity is expected to remain approximately constant as Earth warms or cools. Slight changes to global relative humidity may result from human land-use modification, such as tropical deforestation and irrigation, which can affect the relative humidity over land areas up to regional scales.

The amount of water vapour in the atmosphere will rise as the temperature of the atmosphere rises. Since water vapour is a very potent greenhouse gas, even more potent than CO_2, the net greenhouse effect actually becomes stronger as the surface warms, which leads to even greater warming. This positive feedback is known as the "water vapour feedback". It is the primary reason that climate sensitivity is substantially greater than the previously stated theoretical value of 0.25°C (0.45°F) for each increase of 1 watt per square metre of radiative forcing.

Cloud feedbacks

It is generally believed that as Earth's surface warms and the atmosphere's water-vapour content increases, global cloud cover increases. However, the effects of clouds on near-surface air temperatures are complicated. In the case of low clouds, such as marine stratus clouds, the dominant radiative feature of the cloud is its albedo. Here any increase in low cloud cover acts in much the same way as an increase in surface ice cover: more incoming solar radiation is reflected and Earth's surface cools. On the other hand, high clouds, such as the towering cumulus clouds that extend up to the boundary between the troposphere and stratosphere, have a quite different impact on the surface radiation balance. The tops of cumulus clouds are considerably higher in the atmosphere and colder than their undersides. Cumulus cloud tops emit less long-wave radiation out to space than the warmer cloud bottoms emit downward toward the surface. The end result of the formation of high cumulus clouds is greater warming at the surface.

The net feedback of clouds on rising surface temperatures is therefore somewhat uncertain. It represents a competition between the impacts of high and low clouds, and the balance is difficult to determine. Nonetheless, most estimates indicate that clouds on the whole represent a positive feedback and thus additional warming.

Ice albedo feedback

Another important positive climate feedback is the so-called ice albedo feedback. This arises from the simple fact that ice is more reflective (has a higher albedo) than land or water surfaces. Therefore, as global ice cover decreases, the reflectivity of Earth's surface decreases, more incoming solar radiation is absorbed by the surface, and the surface warms. This feedback is considerably more important when there is relatively extensive global ice cover, such as during the height of the last ice age roughly 25,000 years ago. On a global scale the importance of ice albedo feedback decreases as

Earth's surface warms and there is relatively less ice available to be melted.

Carbon cycle feedbacks

The global carbon cycle comprises another important set of climate feedbacks. In particular, the two main reservoirs of carbon in the climate system are the oceans and the terrestrial biosphere (the life-supporting part of Earth's surface). These reservoirs have historically taken up large amounts of anthropogenic CO_2 emissions. Roughly 50–70 per cent is removed by the oceans, while the remainder is taken up by the terrestrial biosphere. Global warming, however, could decrease the capacity of these reservoirs to sequester atmospheric CO_2. Reductions in the rate of carbon uptake by these reservoirs would increase the pace of CO_2 build-up in the atmosphere and represent yet another possible positive feedback to increased greenhouse gas concentrations.

In the world's oceans, this feedback effect might take several paths. First, as surface waters warmed, they would hold less dissolved CO_2. Second, if more CO_2 was added to the atmosphere and taken up by the oceans, bicarbonate ions (HCO_3^-) would multiply and ocean acidity would increase. Since calcium carbonate ($CaCO_3$) is broken down by acidic solutions, rising acidity would threaten ocean-dwelling fauna that incorporate $CaCO_3$ into their skeletons or shells. As it became increasingly difficult for these organisms to absorb oceanic carbon, there would be a corresponding decrease in the efficiency of the biological pump that helps to maintain the oceans as a carbon sink (as described in the section on carbon dioxide earlier in this chapter). Third, rising surface temperatures might lead to a slowdown in the so-called thermohaline circulation (see Chapter 5, page 239), a global pattern of oceanic flow that partly drives the sinking of surface waters near the poles and is responsible for much of the burial of carbon in the deep ocean. A slowdown in this flow due to an influx of melting fresh water into what are normally saltwater conditions might also cause the solubility pump, which transfers CO_2 from shallow to deeper waters, to become less efficient. Indeed, it is predicted that if global warming continued to a certain point, the oceans would cease to be a net sink of CO_2 and would become a net source.

As large sections of tropical forest are lost because of the warming and drying of regions such as Amazonia, the overall capacity of plants to sequester atmospheric CO_2 would be reduced. As a result, the terrestrial biosphere, though currently a carbon sink, would become a carbon source. This is because ambient temperature is a significant factor affecting the pace of photosynthesis in plants, and many plant species that are well adapted to their local climatic conditions have maximized their photosynthetic rates. As temperatures increase and conditions begin to exceed the optimal temperature range for both photosynthesis and soil respiration, the rate of photosynthesis would decline. As dead plants decompose, microbial metabolic activity (a CO_2 source) would increase and would eventually outpace photosynthesis.

Under sufficient global warming conditions, methane sinks in the oceans and terrestrial biosphere also might become methane sources. Annual emissions of methane by wetlands might either increase or decrease, depending on temperatures and input of nutrients, and it is possible that wetlands could switch from source to sink. There is also the potential for increased methane release as a result of the warming of Arctic permafrost (on land) and further methane release at the continental margins of the oceans (a few hundred metres below sea level). The current average atmospheric methane concentration of 1,750 ppb is equivalent to 3.5 gigatons (3.5 billion tons) of carbon. There are at least 400 gigatons of carbon equivalent (the amount of material [methane, etc.] that could be converted into an equal amount of carbon dioxide if released) stored in Arctic permafrost and as much as 10,000 gigatons (10 trillion tons) of carbon equivalent trapped on the continental margins of the oceans in a hydrated crystalline form known as clathrate. It is believed that some fraction of this trapped methane could become unstable with additional warming, although the amount and rate of potential emission remain highly uncertain.

2

WHAT IS CLIMATE?

From its earliest usage in English, "climate" has been understood to mean the atmospheric conditions that prevail in a given region or zone. The best modern definitions of climate regard it as constituting the total experience of weather and atmospheric behaviour over a number of years in a given region. Climate is not just the "average weather" (an obsolete, and always inadequate, definition). It should include not only the average values of the climatic elements (solar radiation, temperature, humidity, precipitation, atmospheric pressure, and wind) that prevail at different times but also their extreme ranges and variability and the frequency of various occurrences. Just as one year differs from another, decades and centuries are found to differ from one another by a smaller, but sometimes significant, amount. Climate is therefore time dependent, and climatic values or indexes should not be quoted without specifying to what years they refer.

This chapter examines the factors that produce weather and climate and the complex processes that cause variations in both; global climatic types and microclimates; the impact of climate on human activities and vice versa; and climates in past geological ages.

Solar radiation and temperature

Solar radiation

Air temperatures have their origin in the absorption of radiant energy from the sun. They are subject to many influences, including those of the atmosphere, ocean, and land, and are modified by them. As variation of solar radiation is the single most important factor affecting climate, it is considered here first.

Distribution of radiant energy from the sun

Nuclear fusion deep within the sun releases a tremendous amount of energy that is slowly transferred to the solar surface, from which it is radiated into space. The planets intercept minute fractions of this energy, the amount depending on their size and distance from the sun. For example, a 1-square-metre (11-square-foot) area perpendicular (90°) to the rays of the sun at the top of Earth's atmosphere receives about 1,365 watts of solar power. (This amount is comparable to the power consumption of a typical electric heater.) Because of the slight ellipticity of Earth's orbit around the sun, the amount of solar energy intercepted by Earth steadily rises and falls by 3.4 per cent throughout the year, peaking on January 3, when Earth is closest to the sun. Although about 31 per cent of this energy is not used, as it is scattered back to space, the remaining amount is sufficient to power the movement of atmospheric winds and oceanic currents and to sustain nearly all biospheric activity (Figure 2.1).

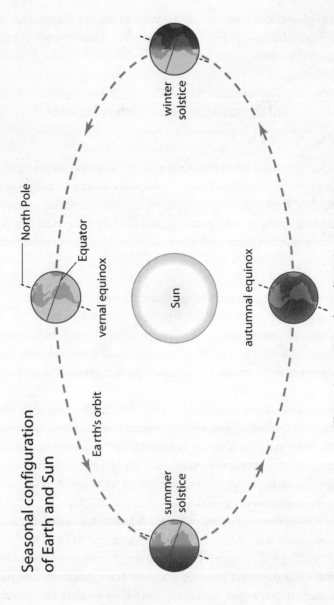

Seasonal configuration
of Earth and Sun

North Pole

Equator

vernal equinox

Sun

Earth's orbit

summer
solstice

winter
solstice

autumnal equinox

Figure 2.1 © Encyclopaedia Britannica, Inc.

Most surfaces are not perpendicular to the sun, and the energy they receive depends on their solar elevation angle. (The maximum solar elevation is 90° for the overhead sun.) This angle changes systematically with latitude, the time of year, and the time of day. The noontime elevation angle reaches a maximum at all latitudes north of the Tropic of Cancer (23.5°N) around June 22 and a minimum around December 22. South of the Tropic of Capricorn (23.5°S), the opposite holds true, and between the two tropics the maximum elevation angle (90°) occurs twice a year. When the sun has a lower elevation angle, the solar energy is less intense because it is spread out over a larger area. Variation of solar elevation is thus one of the main factors that accounts for the dependence of climatic regime on latitude. The other main factor is the length of daylight. For latitudes poleward of 66.5°N and S, the length of day ranges from zero (winter solstice) to 24 hours (summer solstice), whereas the equator has a constant 12-hour day throughout the year. The seasonal range of temperature consequently decreases from high latitudes to the tropics, where it becomes less than the diurnal range of temperature.

Effects of the atmosphere

Of the radiant energy reaching the top of the atmosphere, on average 46 per cent is absorbed by the Earth's surface, but this value varies significantly from place to place – depending on cloudiness, surface type, and elevation. If there is persistent cloud cover, as in some equatorial regions, much of the incident solar radiation is scattered back to space and very little is absorbed by the Earth's surface. Water surfaces have low reflectivity (4–10 per cent), except in low solar elevations, and are the most efficient absorbers. Snow surfaces, on the other hand, have high reflectivity (40–80 per cent) and so are the poorest absorbers. High-altitude desert regions consistently absorb higher-than-average amounts of solar radiation because of the reduced effect of the atmosphere above them.

On average, an additional 23 per cent or so of the incident solar radiation is absorbed in the atmosphere, especially by water vapour and clouds at lower altitudes and by ozone (O_3) in the stratosphere. Absorption of solar radiation by ozone shields the terrestrial surface from harmful ultraviolet light and warms the stratosphere, producing maximum temperatures of -15–10°C (5–50°F) at an altitude of 50 km (30 miles). Most atmospheric absorption takes place at ultraviolet and infrared wavelengths, so more than 90 per cent of the visible portion of the solar spectrum, with wavelengths of 0.4–0.7 μm (0.00002–0.00003 inch), reaches the surface on a cloud-free day. Visible light, however, is scattered in varying degrees by cloud droplets, air molecules, and dust particles. Blue skies and red sunsets are attributable to the preferential scattering of short (blue) wavelengths by air molecules and small dust particles. Cloud droplets scatter visible wavelengths impartially (hence, clouds usually appear white) but very efficiently, so the reflectivity of clouds to solar radiation is typically about 50 per cent and may be as high as 80 per cent for thick clouds.

The constant gain of solar energy by the Earth's surface is systematically returned to space in the form of thermally emitted radiation in the infrared portion of the spectrum. The emitted wavelengths are mainly 5–100 μm (0.0002–0.004 inch), and they interact differently with the atmosphere compared with the shorter wavelengths of solar radiation. Very little of the radiation emitted by Earth's surface passes directly through the atmosphere. Most of it is absorbed by clouds, carbon dioxide, and water vapour and is then re-emitted in all directions. The atmosphere thus acts as a radiative blanket over Earth's surface, hindering the loss of heat to space. The blanketing effect is greatest in the presence of low clouds and weakest for clear, cold skies that contain little water vapour. Without this effect, the mean surface temperature of 15°C (59°F) would be some 30°C colder. Conversely, as atmospheric concentrations of carbon dioxide, methane, chlorofluorocarbons, and other absorbing gases continue to increase, largely owing to human activities, surface temperatures would be expected to rise because of the capacity of such gases to trap infrared radiation. The exact amount of this temperature increase, however, remains uncertain because of unpredictable changes in other atmospheric components, especially cloud cover.

An extreme example of such a greenhouse effect is that produced by the dense atmosphere of the planet Venus, which results in surface temperatures of about 475°C (887°F). This condition exists in spite of the fact that the high reflectivity of the Venusian clouds causes the planet to absorb less solar radiation than Earth.

Average radiation budgets

The difference between the solar radiation absorbed and thermal radiation emitted to space determines Earth's "radiation budget". Since there is no appreciable long-term trend in planetary temperature, it may be concluded that this budget is essentially zero on a global long-term average. Latitudinally, it has been found that much more solar radiation is absorbed at low latitudes than at high latitudes. On the other hand, thermal emission does not show nearly as strong a dependence on latitude, so the planetary radiation budget decreases systematically from the equator to the poles. It changes from positive to negative at latitudes of about 40°N and 40°S. The atmosphere and oceans, through their general circulation, act as vast heat engines, compensating for this imbalance by providing nonradiative mechanisms for the transfer of heat from the equator to the poles.

While the Earth's surface absorbs a significant amount of thermal radiation because of the blanketing effect of the atmosphere, it loses even more through its own emission and thus experiences a net loss of long-wave radiation. This loss is only about 14 per cent of the amount emitted by the surface and is less than the average gain of total absorbed solar energy. Consequently, the surface has on average a positive radiation budget.

By contrast, the atmosphere emits thermal radiation both to space and to the surface, yet it receives long-wave radiation back from only the latter. This net loss of thermal energy cannot be compensated for by the modest gain of absorbed solar energy within the atmosphere. The atmosphere thus has a negative radiation budget, equal in magnitude to the positive radiation budget of the surface but opposite in sign. Non-radiative heat transfer again compensates for the imbalance, this time largely by vertical atmospheric motions involving the evaporation and condensation of water.

Surface-energy budgets

The rate of temperature change in any region is directly proportional to the region's energy budget and inversely proportional to its heat capacity (the amount of energy required to raise its temperature by a given interval). While the radiation budget may dominate the average energy budget of many surfaces, nonradiative energy transfer and storage are also generally important when local changes are considered.

Foremost among the cooling effects is the energy required to evaporate surface moisture, which produces atmospheric water vapour. Most of the latent heat contained in water vapour is subsequently released to the atmosphere during the formation of precipitating clouds, although a minor amount may be returned directly to the surface during dew or frost deposition. Evaporation increases with rising surface temperature, decreasing relative humidity, and increasing surface wind speed. Transpiration by plants also increases evaporation rates, which explains why the temperature in an irrigated field is usually lower than that over a nearby dry road surface.

Another important nonradiative mechanism is the exchange of heat that occurs when the temperature of the air is different from that of the surface. Depending on whether the surface is warmer or cooler than the air next to it, heat is transferred to or from the atmosphere by turbulent air motion (more loosely, by convection). This effect also increases with increasing temperature difference and with increasing surface wind speed. Direct heat transfer to the air may be an important cooling mechanism that limits the maximum temperature of hot dry surfaces. Alternatively, it may be an important warming mechanism that limits the minimum temperature of cold surfaces. Such warming is sensitive to wind speed, so calm conditions promote lower minimum temperatures.

In a similar way, whenever a temperature difference occurs between the surface and the medium beneath the surface, there is a transfer of heat to or from the medium. In the case of land surfaces, heat is transferred by conduction, a process where energy is conveyed through a material from one atom or molecule to another. In the case of water surfaces, the transfer is by convection and may consequently be affected by the horizontal transport of heat within large bodies of water.

Average values of the different terms in the energy budgets of the atmosphere and surface are given in Figure 2.2. The individual terms may be adjusted to suit local conditions and may be used as an aid to understanding the various temperature characteristics discussed in the next section.

Average exchange of energy between the surface, the atmosphere, and space, as percentages of incident solar radiation (1 unit = 3 watts per square metre).

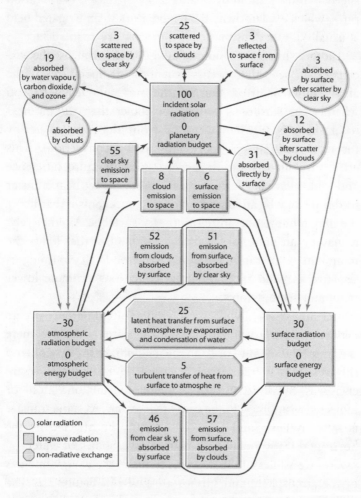

Figure 2.2 © Encyclopædia Britannica, Inc.

Temperature

Global variation of mean temperature

Global variations of average surface-air temperatures are largely due to latitude, continentality, ocean currents, and prevailing winds.

The effect of latitude is evident in the large north–south gradients in average temperature (i.e. differences in average temperature) that occur at middle and high latitudes in each hemisphere in winter. These gradients are due mainly to the rapid decrease of available solar radiation but also in part to the higher surface reflectivity at high latitudes associated with snow and ice and low solar elevations. A broad area of the tropical ocean, by contrast, shows little temperature variation.

Continentality is a measure of the difference between continental and marine climates and is mainly the result of the increased range of temperatures that occurs over land compared with water. This difference is a consequence of the much lower effective heat capacities (ratio of heat absorbed by a material to the temperature change) of land surfaces as well as of their generally reduced evaporation rates. Heating or cooling of a land surface takes place in a thin layer, the depth of which is determined by the ability of the ground to conduct heat. The greatest temperature changes occur for dry, sandy soils, because they are poor conductors with very small effective heat capacities and contain no moisture for evaporation. By far the greatest effective heat capacities are those of water surfaces, owing to both the mixing of water near the surface and the penetration of solar radiation that distributes heating to depths of several metres. In addition, about 90 per cent of the radiation budget of the ocean is used for evaporation. Ocean temperatures are thus slow to change.

The effect of continentality may be moderated by proximity to the ocean, depending on the direction and strength of the prevailing winds. Contrast with ocean temperatures at the edges of each continent may be further modified by the presence of a north- or south-flowing ocean current. For most latitudes, however, continentality explains much of the variation in average temperature at a fixed latitude as well as variations in the difference between the Earth's average lowest and highest temperatures (which occur in January and July, respectively).

Diurnal, seasonal, and extreme temperatures

The diurnal range of temperature generally increases with distance from the sea and toward those places where solar radiation is strongest – namely in dry tropical climates and on high mountain plateaus (owing to the reduced thickness of the atmosphere to be traversed by the sun's rays). The average difference between the day's highest and lowest temperatures is 3°C (5°F) in January and 5°C (9°F) in July in those parts of the British Isles nearest the Atlantic. At Tashkent, Uzbekistan, it is 9°C (16°F) in January and 15.5°C (28°F) in July. At Kandahar, Afghanistan, which lies more than 1,000 metres (about 3,300 feet) above sea level, it is 14°C (25°F) in January and 20°C (36°F) in July. There, the average difference between the day's highest and lowest temperatures exceeds 23°C (41°F) in September and October, when there is less cloudiness than in July.

Similarly, the seasonal variation of temperature and the magnitudes of the differences between the same month in different years and different epochs generally increase toward high latitudes and with distance from the ocean. Extreme temperatures observed in different parts of the world are listed in Table 2.1.

Table 2.1 World temperature extremes

Highest recorded air temperature			
Continent or region	**Place (with elevation*)**	**Degrees Celsius**	**Degrees Fahrenheit**
Africa	Al-'Aziziyah, Libya (112 m or 367 ft)	57.7	136
Antarctica	Lake Vanda 77 degrees 32 minutes S, 161 degrees 40 minutes E (99 m or 352 ft)	15	59
Asia	Tirat Zevi, Israel (-300 m or -984 ft)	53.9	129
Australia	Cloncurry, Queensland (193 m or 633 ft)	53.1	127.5
Europe	Sevilla, Spain (39 m or 128 ft)	50	122
North America	Death Valley (Greenland Ranch), California, US (-54 m or -177 ft)	56	134.5
South America	Rivadavia, Argentina (205 m or 672 ft)	48.9	120
Tropical Pacific islands	Echague, Luzon, Republic of the Philippines (78 m or 257 ft)	40.5	105

Lowest recorded air temperature			
Continent or region	**Place (with elevation*)**	**Degrees Celsius**	**Degrees Fahrenheit**
Africa	Ifrane, Morocco (1,635 m or 5, 363 ft)	-23.9	-11
Antarctica	Vostok 78 degrees 27 minutes S, 106 degrees 52 minutes E (3,420 m or 11,218 ft)	-89.2	-128.6
Asia	Oymyakon, Russia (806 m or 2,644 ft)	-67.7	-89.9
Australia	Charlotte Pass, New South Wales (1,780 m or 5,840 ft)	-22.2	-8
Europe	Ust-Shchuger, Russia (85 m or 279 ft)	-55	-67
North America	Snag, Yukon Territory, Canada (646 m or 2,119 ft)	-62.8	-81
South America	Colonia, Sarmiento, Argentina (268 m or 879 ft)	-33	-27
Tropical Pacific islands	Haleakala, Hawaii, US (2,972 m or 9,748 ft)	-7.8	18

*Above or below sea level

Temperature variation with height

There are two main levels where the atmosphere is heated: at
the Earth's surface and at the top of the ozone layer (about 50
km, or 30 miles, up) in the stratosphere. The radiation balance
(the ratio of the amount of incoming versus outgoing radiation) shows a net gain at these levels in most cases. Prevailing
temperatures tend to decrease with distance from these heating
surfaces, apart from in the ionosphere – part of the thermo-
sphere, the top layer of the atmosphere – and the outer
atmospheric layers, where other processes are at work. The
world's average lapse rate of temperature (change with altitude) in the lower atmosphere is 0.6–0.7°C per 100 metres
(about 1.1–1.3°F per 300 feet).

Lower temperatures prevail with increasing height above
sea level for two reasons: (1) because there is a less favourable
radiation balance in the free air, and (2) because rising air –
whether lifted by convection currents above a relatively warm
surface or forced up over mountains – undergoes a reduction
of temperature – adiabatic cooling – associated with its expansion as the pressure of the overlying atmosphere declines.
Adiabatic temperature changes are those that take place without the addition or subtraction of heat; they occur in the
atmosphere when bundles of air are moved vertically. The
adiabatic lapse rate of temperature is about 1°C per 100
metres (about 1°F per 150 feet) for dry air and 0.5°C per
100 metres (about 1°F per 300 feet) for saturated air, in which
condensation (with liberation of latent heat) is produced by
adiabatic cooling.

The difference between these rates of change of temperature,
and therefore density, of rising air currents and the state of the
surrounding air determines whether the upward currents are
accelerated or retarded – i.e. whether the air is unstable, so
vertical convection with its characteristically attendant tall

cumulus cloud and shower development is encouraged, or whether it is stable and convection is damped down.

For these reasons, the air temperatures observed on hills and mountains are generally lower than on low ground, except in the case of extensive plateaus, which present a raised heating surface (and on still, sunny days, when even a mountain peak is able to warm appreciably the air that remains in contact with it).

Circulation, currents, and ocean–atmosphere interaction

The circulation of the ocean is a key factor in air temperature distribution. Ocean currents that have a northward or south-ward component, such as the warm Gulf Stream in the North Atlantic or the cold Peru (Humboldt) Current off South America, effectively exchange heat between low and high latitudes. In tropical latitudes the ocean accounts for a third or more of the poleward heat transport; at latitude 50°N, the ocean's share is about one seventh. In the particular sectors where the currents are located, their importance is of course much greater than these figures, which represent hemispheric averages, indicate.

A good example of the effect of a warm current is that of the Gulf Stream in January, which causes a strong east–west gradient in temperatures across the eastern edge of the North American continent. The relative warmth of the Gulf Stream affects air temperatures all the way across the Atlantic, and prevailing westerlies extend the warming effect deep into northern Europe. As a result, January temperatures of Trom-sø, Norway (69°40' N), for example, average 24°C (43°F) above the mean for that latitude. The Gulf Stream maintains a warming influence in July, but it is not as noticeable because of the effects of continentality (see "Global variation of mean temperature", above).

The ocean, particularly in areas where the surface is warm, also supplies moisture to the atmosphere. This in turn contributes to the heat budget of those areas in which the water vapour is condensed into clouds, liberating latent heat in the process. This set of events occurs frequently in high latitudes and in locations remote from the ocean where the moisture was initially taken up.

The great ocean currents are themselves wind-driven – set in motion by the drag of the winds over vast areas of the sea surface, especially where the tops of waves increase the friction with the air above. At the limits of the warm currents, particularly where they abut directly upon a cold current – as at the left flank of the Gulf Stream in the neighbourhood of the Grand Banks off Newfoundland, and at the subtropical and Antarctic convergences in the oceans of the southern hemisphere – the strong thermal gradients in the sea surface result in marked differences in the heating of the atmosphere on either side of the boundary. These temperature gradients tend to position and guide the strongest flow of the jet stream (narrow bands of upper-level wind – see Chapter 6, page 320) in the atmosphere above, and thereby influence the development and steering of weather systems.

Interactions between the ocean and the atmosphere proceed in both directions. They also operate at different rates. Some interesting lag effects, which are of value in long-range weather forecasting, arise through the considerably slower circulation of the ocean. Thus, enhanced strength of the easterly trade winds over low latitudes of the Atlantic north and south of the equator impels more water toward the Caribbean and the Gulf of Mexico, producing a stronger flow and greater warmth in the Gulf Stream approximately six months later. Anomalies in the position of the Gulf Stream–Labrador Current boundary, which produce a greater or lesser extent of warm water near the Grand Banks, so affect the energy supply to the atmosphere and the development and steering of weather systems from that region that they are associated with rather persistent anomalies of weather pattern over the British Isles and northern Europe. Anomalies in the equatorial Pacific and in the northern limit of the Kuroshio Current (also called the Japan Current) seem to have effects on a similar scale. Indeed, through their influence on the latitude of the jet stream and the wavelength (that is, the spacing of cold trough and warm ridge regions) in the upper westerlies, these ocean anomalies exercise an influence over the atmospheric circulation that spreads to all parts of the hemisphere.

Sea-surface temperature anomalies that recur in the equatorial Pacific at variable intervals of two to seven years can sometimes produce major climatic perturbations. One such anomaly is known as El Niño (Spanish for "The Child"; it was so named by Peruvian fishermen who noticed its onset during the Christmas season). During an El Niño event, warm surface water flows eastward from the equatorial Pacific, in at least partial response to weakening of the equatorial easterly winds, and replaces the normally cold upwelling surface water off the coast of Peru and Ecuador that is associated with the northward propagation of the cold Peru Current. The change in sea-surface temperature transforms the coastal climate from arid to wet. The event also affects atmospheric circulation in both hemispheres and is associated with changes in precipitation in regions of North America, Africa, and the western Pacific. (For more about El Niño and ocean–atmosphere interactions, see Chapter 5, page 264).

Short-term temperature changes
Many interesting short-term temperature fluctuations also occur, usually in connection with local weather disturbances.

The rapid passage of a mid-latitude cold front (the advancing edge of a cooler mass of air), for example, can drop temperatures by 10°C (18°F) in a few minutes and, if followed by the sustained movement of a cold air mass, by as much as 50°C (90°F) in 24 hours, with life-threatening implications for the unwary. Temperature increases of up to 40°C (72°F) in a few hours are also possible downwind of major mountain ranges when air that has been warmed by the release of latent heat on the windward side of a range is forced to descend rapidly on the other side. Such a wind is variously called chinook, foehn, or Santa Ana. (See "Local wind systems", Chapter 6, page 299.)

Atmospheric pressure

Atmospheric pressure and wind are both significant controlling factors of Earth's weather and climate. Although these two physical variables may at first glance appear to be quite different, they are in fact closely related. Wind exists because of horizontal and vertical differences (gradients) in pressure, yielding a correspondence that often makes it possible to use the pressure distribution as an alternative representation of atmospheric motions. Pressure is the force exerted on a unit area, and atmospheric pressure is equivalent to the weight of air above a given area on Earth's surface or within its atmosphere. This pressure is usually expressed in millibars (mb; 1 mb equals 1,000 dynes, or 10,000 micronewtons, per square cm) or in kilopascals (kPa; 1 kPa equals 10,000 dynes, or 100,000 micronewtons, per square cm). Distributions of atmospheric pressure on a map are depicted by a series of curved lines called isobars, each of which connects points of equal pressure. At sea level the mean pressure is about 1,000 mb

(100 kPa), varying by less than 5 per cent from this value at any given location or time.

Mean sea-level pressure values for the midwinter months in the northern hemisphere are summarized in Figure 2.3a, and mean sea-level pressure values for the midsummer months are illustrated in the Figure 2.3b. Since charts of atmospheric pressure often represent average values over several days, pressure features that are relatively consistent day after day emerge, while more transient, short-lived features are removed. Those that remain are known as semipermanent pressure centres and indicate the source regions for major, relatively uniform bodies of air known as air masses.

Warm, moist maritime tropical air forms over tropical and subtropical ocean waters in association with the high-pressure regions prominent there. Cool, moist maritime polar air, on the other hand, forms over the colder subpolar ocean waters just south and east of the large, winter oceanic low-pressure regions. Over the continents, cold dry continental polar air and extremely cold dry continental arctic air forms in the high-pressure regions that are especially pronounced in winter, while hot dry continental tropical air forms over hot desert-like continental domains in summer in association with low-pressure areas, which are sometimes called heat lows.

A closer examination of Figure 2.3 reveals some interesting features. First, it is clear that sea-level pressure is dominated by closed high- and low-pressure centres, which are largely caused by differential surface heating between low and high latitudes and between continental and oceanic regions. High pressure tends to be amplified over the colder surface features. Second, because of seasonal changes in surface heating, the pressure centres exhibit seasonal changes in their characteristics. For example, the Siberian High, Aleutian Low, and Icelandic Low that are so prominent in the winter virtually

World distribution of mean sea-level pressure (in millibars) for a) January and b) July, and primary and secondary storm tracks; the general character of the global winds is also shown.

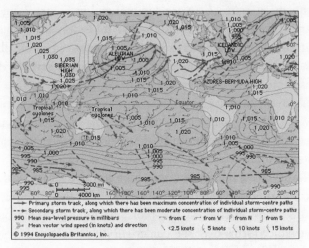

Figure 2.3 a

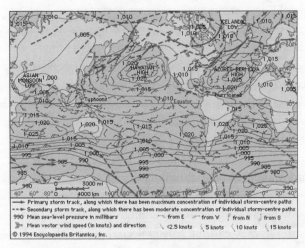

Figure 2.3 b © Encyclopædia Britannica, Inc.

disappear in summer as the continental regions warm relative to surrounding bodies of water. At the same time, the Pacific and Atlantic highs amplify and migrate northward.

At altitudes well above Earth's surface, the monthly average pressure distributions show much less tendency to form in closed centres but rather appear as quasi-concentric circles around the poles. This more symmetrical appearance reflects the dominant role of meridional (north–south) differences in radiative heating and cooling. Excess heating in tropical latitudes, in contrast to polar areas, produces higher pressure at upper levels in the tropics as thunderstorms transfer air to higher levels. In addition, the greater heating/cooling contrast in winter yields stronger pressure differences during this season.

Perfect symmetry between the tropics and the poles is interrupted by wavelike atmospheric disturbances associated with migratory and semipermanent high- and low-pressure surface weather systems. These weather systems are most pronounced over the northern hemisphere, with its more prominent land–ocean contrasts and orographic (high-elevation) features.

Further details of the Earth's wind systems, the forces that govern them, and their influence on the climate of different parts of the planet, are given in Chapter 6.

Humidity

Atmospheric humidity, which is the amount of water vapour or moisture in the air, is another major climatic element, as is precipitation. All forms of precipitation – including drizzle, rain, snow, ice crystals, and hail – are produced as a result of the condensation of atmospheric moisture that forms clouds in which some of the particles, by growth and aggregation, attain sufficient size to fall from the clouds and reach the ground.

The water vapour content may vary from one air parcel to another, but the vapour capacity is determined by temperature. Vapour in the air is a determinant of weather, because it first absorbs thermal radiation that leaves and cools the Earth's surface, and then emits thermal radiation that warms the planet.

Relative humidity

There are different ways of measuring humidity. The moisture content of a volume of air is its "absolute humidity". When a volume of air at a given temperature holds the maximum amount of water vapour, the air is said to be saturated. "Relative humidity" is the water-vapour content of the air relative to its content at saturation. "Specific humidity" is the amount of water vapour in a unit mass of air (usually per kilogram).

Of the three indexes of humidity, relative humidity is a widely used environmental indicator. It can be defined as the ratio of the vapour pressure of a sample of air to the saturation pressure at the existing temperature. Within a pool of liquid water, some molecules are continually escaping from the liquid into the space above, while more and more vapour molecules return to the liquid as the concentration of vapour rises. Finally, equal numbers are escaping and returning; the vapour is then saturated, and its pressure is known as the saturation vapour pressure.

Humidity and climate

The small amount of water in atmospheric vapour (simply, the amount of water vapour in the atmosphere), relative to the amount of water on Earth, belies its importance. Compared with one unit of water in the air, the seas contain at least 100,000 units, the great glaciers 1,500, the porous Earth nearly 200, and the rivers and lakes 4 or 5. The effect of the vapour in the air is magnified, however, by its role in transferring water from sea to land by the media of clouds and precipitation and in that of absorbing radiation.

The atmospheric vapour is the invisible conductor that carries water from sea to land, making terrestrial life possible. Fresh water is distilled from the salt seas and carried over land by the wind. Although it is also true that water evaporates from vegetation, and that rain falls on the sea, the sea is the bigger source. Water vapour becomes visible near the Earth's surface as fog when the air cools to the dew point (the temperature at which condensation begins). The usual nocturnal cooling will produce fog patches in cool valleys. Or the vapour may move as a tropical air mass over cold land or sea, causing widespread and persistent fog, such as occurs over the Grand Banks off Newfoundland. The delivery of water by means of fog or dew is slight, however.

When air is lifted, it is carried to a region of lower pressure, where it will expand and cool. It may rise up a mountain slope or over the front of a cooler, denser air mass. If condensation nuclei are absent, the dew point may be exceeded by the cooling air, and the water vapour becomes supersaturated. If nuclei are present or if the temperature is very low, however, cloud droplets or ice crystals form.

The invisible vapour has another climatic role: absorbing and emitting radiation. The Earth's temperature and its daily variation are determined by the balance between incoming and outgoing radiation. The wavelength of the incoming radiation from the sun is mostly shorter than 3 μm (0.0001 inch). It is scarcely absorbed by water vapour, and its receipt depends largely upon cloud cover. The radiation exchanged between the atmosphere and Earth's surface, and the eventual loss to space, is in the form of long waves. These long waves are strongly absorbed in the 3–8.5 mm band and in the range greater than 11 mm, where vapour is either partly or wholly opaque. As noted above, much of the radiation that is absorbed in the atmosphere is emitted back to Earth, and the surface receipt of long waves, primarily from water vapour and carbon dioxide in the atmosphere, is slightly more than twice the direct receipt of solar radiation at the surface. Thus, the atmospheric vapour combines with clouds and the advection (horizontal movement) of air from different regions to control the surface temperature.

Evaporation and humidity

Evaporation, mostly from the sea and from vegetation, replenishes the humidity of the air. Evaporation is the change of liquid water into a gaseous state, its rate being proportional to the difference between the pressure of the water vapour in the free air and the vapour that is next to, and saturated by, the evaporating liquid. It is also proportional to the conductivity of the medium between the evaporator and the free air. If the evaporator is open water, the conductivity will increase with ventilation. But if the evaporator is a leaf, the diffusing water must pass through the still air within the minute pores between the water inside and the dry air outside. In this case the porosity may modify the conductivity more than ventilation.

Vapour pressure is directly related to temperature. However, the temperature of the evaporator is rarely the same as the air temperature, because each gram of evaporation consumes about 600 calories (2,500 joules) and thus cools the evaporator. The availability of energy to heat the evaporator is therefore as important as the saturation deficit and conductivity of the air. Outdoors, some of this heat may be transferred from the surrounding air by convection, but much of it must be furnished by radiation. Evaporation is faster on sunny days than on cloudy ones not only because the air may be drier on a sunny day but also because the sun warms the evaporator and thus raises the vapour pressure at the surface. In fact, this loss of water is essentially determined by the net radiation balance during the day.

Precipitation

Precipitation is one of the three main processes (evaporation, condensation, and precipitation) that constitute the hydrologic cycle, the continual exchange of water between the atmosphere and Earth's surface. Water evaporates from ocean, land, and freshwater surfaces, is carried aloft as vapour by the air currents, condenses to form clouds, and ultimately is returned to the surface as precipitation. The average global stock of water vapour in the atmosphere is equivalent to a layer of water 2.5 cm (1 inch) deep covering the whole Earth. Because Earth's average annual rainfall is about 100 cm (39 inches), the average time that the water spends in the atmosphere, between its evaporation from the surface and its return as precipitation, is about 1/40 of a year, or about nine days. Of the water vapour that is carried at all heights across a given region by the winds, only a small percentage is converted into precipitation and reaches the ground in that area. In deep and extensive cloud systems, the conversion is more efficient, but even in thunderclouds only some 10 per cent of the moisture within is converted into rain and hail.

In the measurement of precipitation, it is necessary to distinguish between the amount – defined as the depth of precipitation (calculated as though it were all rain) that has fallen at a given point during a specified interval of time – and the rate or intensity, which specifies the depth of water that has fallen at a point during a particular interval of time. Persistent moderate rain, for example, might fall at an average rate of 5 mm per hour (0.2 inch per hour) and thus produce 120 mm (4.7 inches) of rain in 24 hours. A thunderstorm might produce this total quantity of rain in 20 minutes, but at its peak intensity the rate of rainfall might become much greater – perhaps 120 mm per hour (4.7 inches per hour), or 2mm (0.08 inch) per minute – for a minute or two.

The amount of precipitation falling during a fixed period is measured regularly at many thousands of places on Earth's surface by rather simple rain gauges. Measurement of precipitation intensity requires a recording rain gauge, in which water falling into a collector of known surface area is continuously recorded on a moving chart or a magnetic tape. Investigations are under way into the feasibility of obtaining continuous measurements of rainfall over large catchment areas by means of radar.

Apart from the trifling contributions made by dew, frost, and rime, as well as desalination plants, the sole source of fresh water for sustaining rivers, lakes, and all life on Earth is provided by precipitation from clouds. Precipitation is therefore indispensable and overwhelmingly beneficial to humankind, but extremely heavy rainfall can cause great harm: soil erosion, landslides, and flooding. Hailstorm damage to crops, buildings, and livestock can prove very costly.

The origin of precipitation in clouds

Cloud formation

Clouds are formed by the lifting of damp air, which cools by expansion as it encounters the lower pressures existing at higher levels in the atmosphere. The relative humidity increases until the air has become saturated with water vapour, and then condensation occurs on any of the aerosol particles suspended in the air. A wide variety of these exist, in concentrations ranging from only a few per cubic centimetre, in clean maritime air, to perhaps 1 million per cubic cm (16 million per cubic inch) in the highly polluted air of an industrial city. For continuous condensation leading to the formation of cloud droplets, the air must be slightly supersaturated. Among the highly efficient condensation nuclei are sea-salt particles and the particles produced by combustion (e.g. natural forest fires and man-made fires). Many of the larger condensation nuclei over land consist of ammonium sulphate. These are produced by cloud and fog droplets absorbing sulphur dioxide and ammonia from the air. Condensation onto the nuclei continues as rapidly as water vapour is made available through cooling; droplets about 10 μm (0.0004 inch) in diameter are produced in this manner. Droplets of this size constitute a nonprecipitating cloud.

Mechanisms of precipitation release

Growing clouds are sustained by upward air currents, which may vary in strength from a few centimetres per second to several metres per second. Considerable growth of the cloud droplets (with falling speeds of only about 1 cm, or 0.4 inch, per second) is therefore necessary if they are to fall through the cloud, survive evaporation in the unsaturated air below, and reach the ground as drizzle or rain. The production of a few large particles from a large population of much smaller ones may be achieved in one of two ways.

The first of these depends on the fact that cloud droplets are seldom of uniform size: droplets form on nuclei of various sizes and grow under slightly different conditions and for different lengths of time in different parts of the cloud. A droplet appreciably larger than average will fall faster than the smaller ones and so will collide and fuse (coalesce) with some of those that it overtakes. Calculations show that, in a deep cloud containing strong upward air currents and high concentrations of liquid water, such a droplet will have a sufficiently long journey among its smaller neighbours to grow to raindrop size. This coalescence mechanism is responsible for the showers that fall in tropical and subtropical regions from clouds whose tops do not reach altitudes where air temperatures are below 0°C (32°F) and therefore cannot contain ice crystals. Radar evidence also suggests that showers in temperate latitudes may sometimes be initiated by the coalescence of water drops, although the clouds may later reach heights at which ice crystals may form in their upper parts.

The second method of releasing precipitation can operate only if the cloud top reaches elevations at which air temperatures are below 0°C and the droplets in the upper cloud regions become supercooled. At temperatures below -40°C (-40°F), the droplets freeze automatically or spontaneously. At higher temperatures, they can freeze only if they are infected with special minute particles called ice nuclei. The origin and nature of these nuclei are not known with certainty, but the most likely source is clay-silicate particles carried up from the ground by the wind. As the temperature falls below 0°C, more and more ice nuclei become active, and ice crystals appear in increasing numbers among the supercooled droplets. Such a mixture of supercooled droplets and ice crystals is unstable, however. The cloudy air is usually only slightly supersaturated with water vapour with respect to the droplets and is strongly oversaturated with respect to ice crystals; the latter thus grow more rapidly than the droplets. After several minutes, the growing ice crystals acquire falling speeds of tens of centimetres per second, and several of them may become joined to form a snowflake. In falling into the warmer regions of the cloud, this flake may melt and hit the ground as a raindrop. The extensive multilayer cloud systems are generally formed in cyclonic depressions (lows) and near fronts (the interface between two air masses of different density and temperature). Cloud systems of this type are associated with feeble upcurrents of only a few centimetres per second that last for at least several hours. Although the structure of these great raincloud systems is being explored by aircraft and radar, it is not yet well understood. That such systems rarely produce rain, as distinct from drizzle, unless their tops are colder than about -12°C (10°F) suggests that ice crystals are mainly responsible. This view is supported by the fact that the radar signals from these clouds usually take a characteristic form that has been clearly identified with the melting of snowflakes.

Showers, thunderstorms, and hail

Precipitation from shower clouds and thunderstorms, whether in the form of raindrops, pellets of soft hail, or true hailstones, is generally of greater intensity and shorter duration than that from layer clouds and is usually composed of larger particles. The clouds are characterized by their large vertical depth, strong vertical air currents, and high concentrations of liquid water – all factors favouring the rapid growth of precipitation elements by the accretion of cloud droplets.

Structure of a thunderstorm.

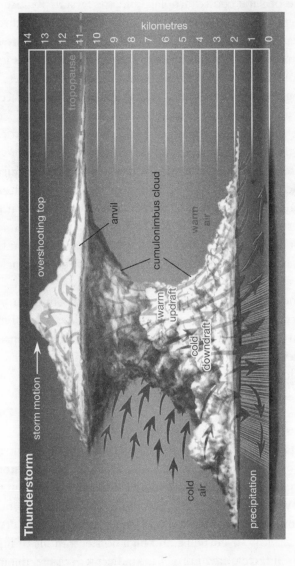

Figure 2.4 © Encyclopædia Britannica, Inc.

The hailstones that fall from deep, vigorous clouds in warm weather consist of a core surrounded by several alternate layers of clear and opaque ice. When the growing particle traverses a region of relatively high air temperature or high concentration of liquid water, or both, the transfer of heat from the hailstone to the air cannot occur rapidly enough to allow all of the deposited water to freeze immediately. This results in the formation of a wet coating of slushy ice, which may later freeze to form a layer of compact, relatively trans- parent ice. If the hailstone then enters a region of lower temperature and lower water content, the impacting droplets may freeze individually to produce ice of relatively low density with air spaces between the droplets. The alternate layers are formed as the stone passes through regions in which the combination of air temperature, liquid-water content, and updraft speed allows alternately wet and dry growth.

Some authorities consider lightning to be closely associated with the appearance of precipitation, especially in the form of soft hail, and that the charge and the strong electric fields are produced by ice crystals or cloud droplets striking and boun- cing off the undersurfaces of the hail pellets.

World distribution of precipitation

Regional and latitudinal distribution
The yearly precipitation averaged over the whole Earth is about 100 cm (39 inches), but this is distributed very unevenly. The regions of highest rainfall are found in the equatorial zone and the monsoon area of Southeast Asia. Middle latitudes receive moderate amounts of precipitation, but little falls in the desert regions of the subtropics or around the poles.

If the surface of the Earth were perfectly uniform, the long- term average rainfall would be distributed in distinct latitudinal bands. But the situation is complicated by the pattern of the global winds, the distribution of land and sea, and the presence of mountains. Because rainfall results from the ascent and cooling of moist air, the areas of heavy rain indicate regions of rising air, whereas the deserts occur in regions in which the air is warmed and dried during descent. In the subtropics, the trade winds bring plentiful rain to the east coasts of the continents, but the west coasts tend to be dry. Conversely, in high latitudes the west coasts are generally wetter than the east coasts. Rain tends to be abundant on the windward slopes of mountain ranges but sparse on the lee sides.

In the equatorial belt, the trade winds from both hemispheres converge and give rise to a general upward motion of air, which becomes intensified locally in tropical storms that produce very heavy rains in the Caribbean, the Indian and southwest Pacific Oceans, and the China Sea, and in thunderstorms that are especially frequent and active over the land areas. During the annual cycle the doldrums (regions of light ocean currents within the intertropical convergence zone [ITCZ]) move toward the summer hemisphere, so outside a central region near the equator, which has abundant rain at all seasons, there is a zone that receives much rain in summer but a good deal less in winter.

The dry areas of the subtropics – such as the desert regions of North Africa, the Arabian Peninsula, South Africa, Australia, and central South America – are due to the presence of semi-permanent subtropical anticyclones in which the air subsides and becomes warm and dry. These high-pressure belts tend to migrate with the seasons and cause summer dryness on the poleward side and winter dryness on the equatorward side of their mean positions (see "Cyclones and anticyclones", Chapter 6, page 290). The easterly trade winds, having made a long passage over the warm oceans, bring plentiful rains to the east coasts of the subtropical landmasses, but the west coasts and the interiors of the continents, which are often sheltered by mountain ranges, are very dry.

In middle latitudes, weather and rainfall are dominated by travelling depressions and fronts that yield a good deal of rain in all seasons and in most places except the far interiors of the Asian and North American continents. Generally, rainfall is more abundant in summer, except on the western coasts of North America, Europe, and North Africa, where it is higher during the winter.

At high latitudes and especially in the polar regions, the low precipitation is caused partly by subsidence of air in the high-pressure belts and partly by the low temperatures. Snow or rain occur at times, but evaporation from the cold sea and land surfaces is slow, and the cold air has little capacity for moisture.

The influence of oceans and continents on rainfall is particularly striking in the case of the Indian Ocean monsoon (see Chapter 6, page 305). During the northern hemisphere winter, cool dry air from the interior of the continent flows southward and produces little rain over the land areas. After the air has travelled some distance over the warm tropical ocean, however, it releases heavy shower rains over the East Indies. During the northern summer, when the monsoon blows from the south-west, rainfall is heavy over India and Southeast Asia. These rains are intensified where the air is forced to ascend over the windward slopes of the Western Ghats and the Himalayas.

The combined effects of land, sea, mountains, and prevailing winds are manifest in South America. There the desert in southern Argentina is sheltered by the Andes from the westerly winds (i.e. from west to east) blowing from the Pacific Ocean, and the west-coast desert is not only situated under the South Pacific subtropical anticyclone but is also protected by the Andes against rain-bearing winds from the Atlantic.

Amounts and variability

The long-term average amounts of precipitation for a season or a year give little information on the regularity with which rain may be expected, particularly for regions where the average amounts are small. For example, at Iquique, a city in northern Chile, four years once passed without rain, whereas the fifth year gave 15 mm (0.6 inch); the five-year average was therefore 3 mm (0.1 inch). Clearly, such averages are of little practical value, and the frequency distribution or the variability of precipitation must also be known.

The variability of the annual rainfall is closely related to the average amounts. For example, over the British Isles, which have a very dependable rainfall, the annual amount varies by less than 10 per cent relative to the long-term average value. A variability of less than 15 per cent is typical of the mid-latitude cyclonic belts of the Pacific and Atlantic Oceans and of much of the wet equatorial regions. In the interiors of the desert areas of Africa, Arabia, and Central Asia, however, the rainfall in a particular year may deviate from the normal long-term average by more than 40 per cent. The variability for individual seasons or months may differ considerably from that for the year as a whole, but again the variability tends to be higher where the average amounts are low.

The heaviest annual rainfall in the world was recorded at the village of Cherrapunji, India, where 26,470 mm (1,042 inches) fell between August 1860 and July 1861. The heaviest rainfall in a period of 24 hours was 1,870 mm (74 inches), recorded at the village of Cilaos, Réunion, in the Indian Ocean on March 15–16 1952. The lowest recorded rainfall in the world occurred at Arica, a port city in northern Chile. The annual average, taken over a 43-year period, was only 0.5 mm (0.02 inch).

Although past records give some guide, it is not possible to estimate very precisely the maximum possible precipitation that may fall in a given locality during a specified interval of time. Much will depend on a favourable combination of several factors, including the properties of the storm and the effects of local topography. Therefore it is possible only to make estimates that are based on analyses of past storms or on theoretical calculations that attempt to maximize statistically the various factors or the most effective combination of factors that are known to control the duration and intensity of the precipitation. For some purposes, however – such as important planning and design problems – estimates of the greatest precipitation to be expected at a given location within a specified number of years are required. For example, in the designing of a dam the highest 24-hour rainfall to be expected once in 30 years over the whole catchment area might be relevant. For dealing with such problems, a great deal of work has been devoted to determining from past records the frequency with which rainfalls of a given intensity and total amount may be expected to recur at particular locations, and also to determining the statistics of rainfall for a specific area from measurements made at only a few points.

Effects of precipitation

Raindrop impact and soil erosion

Large raindrops, up to 6 mm (0.2 inch) in diameter, have terminal velocities of about 10 metres (30 feet) per second and so may cause considerable compaction and erosion of the soil by the force of their impact. The formation of a compacted crust makes it more difficult for air and water to reach the roots of plants and encourages the water to run off the surface and carry away the topsoil with it. In hilly and mountainous areas, heavy rain may turn the soil into mud and slurry, which may produce enormous erosion by mudflow generation.

Rainwater running off hard impervious surfaces or water-logged soil may cause local flooding. (For more about soil erosion see Chapter 4, page 196.)

Surface runoff

The rainwater that is not evaporated or stored in the soil eventually runs off the surface and finds its way into rivers, streams, and lakes or percolates through the rocks and becomes stored in natural underground reservoirs. The runoff may be determined by measuring the flow of water in the rivers with stream gauges, and the precipitation may be measured by a network of rain gauges, but storage and evaporation are more difficult to estimate.

Of all the water that falls on the Earth's surface, the relative amounts that run off, evaporate, or seep into the ground vary so much for different areas that no firm figures can be given for Earth as a whole. It has been estimated, however, that in the United States 10–50 per cent of the rainfall runs off at once, 10–30 per cent evaporates, and 40–60 per cent is absorbed by the soil. Of the entire rainfall, 15–30 per cent is estimated to be used by plants, either to form plant tissue or in transpiration. The wide range in these average figures is indicative of the huge geographical variation within the United States.

Climatic classification

General considerations

The climate of an area is the synthesis of the environmental conditions (soils, vegetation, weather, etc.) that have prevailed there over a long period of time. This synthesis involves both averages of the climatic elements and measurements of variability (such as extreme values and probabilities). Climate is a complex, abstract concept involving data on all aspects of Earth's environment. As such, no two localities on Earth may be said to have exactly the same climate.

Nevertheless, it is clear that, over restricted areas of the planet, climates vary within a limited range and that climatic regions are discernible, within which some uniformity is apparent in the patterns of climatic elements. Moreover, widely separated areas of the world possess similar climates when the set of geographic relationships occurring in one area parallels that of another. This symmetry and organization of the climatic environment suggests an underlying worldwide regularity and order in the phenomena causing climate (e.g. patterns of incoming solar radiation, atmospheric pressure, wind, soils, temperature, precipitation, and vegetation).

Climate classification is an attempt to formalize this process of recognizing climatic similarity – of organizing, simplifying, and clarifying the vast amount of environmental data, and of systematizing the long-term effects of interacting climatic processes to enhance scientific understanding of climates. Users of climate classifications should be aware of the limitations of the procedure, however.

First, climate is a multidimensional concept, and it is not an obvious decision as to which of the many observed environmental variables should be selected as the basis of the classification. This choice must be made on a number of grounds, both practical and theoretical. For example, using too many different elements runs the risk that the classification will have too many categories to be readily interpreted and that many of the categories will not correspond to real climates.

Moreover, measurements of many of the elements of climate are not available for large areas of the world or have been collected for only a short time. The major exceptions are soil, vegetation, temperature, and precipitation data, which are more extensively available and have been recorded for extended periods of time.

The choice of variables is also determined by the purpose of the classification (e.g. to account for the distribution of natural vegetation, to explain soil formation processes, or to classify climates in terms of human comfort). The variables relevant in the classification will be determined by this purpose, as will the threshold values of the variables chosen to differentiate climatic zones.

A second difficulty results from the generally gradual nature of changes in the climatic elements over Earth's surface. Except in unusual situations caused by mountain ranges or coastlines, temperature, precipitation, and other climatic variables tend to change only slowly over distance. As a result, climate types generally change imperceptibly as one moves from one locale on Earth's surface to another. Choosing a set of criteria to distinguish one climatic type from another is thus equivalent to drawing a line on a map to distinguish the climatic region possessing one type from that having the other. While this is in no way different from many other classification decisions that one makes routinely in daily life, it must always be remembered that boundaries between adjacent climatic regions are placed somewhat arbitrarily through regions of continuous, gradual change and that the areas defined within these boundaries are far from homogeneous in terms of their climatic characteristics.

Most classification schemes are intended for global- or continental-scale application and define regions that are major subdivisions of continents, hundreds to thousands of kilometres across. These may be termed macroclimates. Not only will there be slow changes (from wet to dry, hot to cold, etc.) across such a region as a result of the geographic gradients of climatic elements, but there will exist mesoclimates (the climate of a small area) within these regions, associated with climatic processes occurring at a scale of tens to hundreds of kilometres – created by elevation differences, slope aspect, bodies of water, differences in vegetation cover, urban areas, and the like. Mesoclimates, in turn, may be resolved into numerous microclimates, which occur at scales of less than 0.1 km (0.06 mile) – as in the climatic differences between forests, crops, and bare soil, at various depths in a plant canopy, at different depths in the soil, on different sides of a building, and so on.

These limitations notwithstanding, climate classification plays a key role as a means of generalizing the geographic distribution and interactions among climatic elements, of identifying mixes of climatic influences important to various climatically dependent phenomena, of stimulating the search to identify the controlling processes of climate, and, as an educational tool, in showing some of the ways in which distant areas of the world are both different from and similar to one's home region.

Approaches to climatic classification

The earliest known climatic classifications were those of classical Greek times. Such schemes generally divided Earth into latitudinal zones based on the significant parallels of 0°, 23.5°, and 66.5° of latitude and on the length of day. Modern climate classification has its origins in the mid-nineteenth century, with the first published maps of temperature and precipitation over Earth's surface, which permitted the development of methods of climate grouping that used both variables simultaneously.

Many different schemes of classifying climate have been devised (more than 100), but all of them may be broadly differentiated as either empirical or genetic methods. This distinction is based on the nature of the data used for classification. Empirical methods make use of observed environmental data, such as temperature, humidity, and precipitation, or simple quantities derived from them (e.g. evaporation). A genetic method classifies climate on the basis of its causal elements, or climatic processes – the activity and characteristics of all factors (air masses, circulation systems, fronts, jet streams, solar radiation, topographic effects, and so forth) that give rise to the spatial and temporal patterns of climatic data.

Hence, while empirical classifications are largely descriptive of climate, genetic methods are (or should be) explanatory. Unfortunately, genetic schemes, while scientifically more desirable, are inherently more difficult to implement because they do not use simple observations. As a result, such schemes are both less common and less successful overall. Moreover, the regions defined by the two types of classification scheme do not necessarily correspond: in particular, it is not uncommon for similar climatic forms resulting from different climatic processes to be grouped together by many common empirical schemes.

Genetic classifications

Genetic classification methods group climates by their causes. Among such methods, three types may be distinguished: (1) those based on the geographic determinants of climate, (2) those based on the surface energy budget, and (3) those derived from air-mass analysis. In the first class are a number of schemes (largely the work of German climatologists) that categorize climates according to such factors as latitudinal control of temperature, continentality versus ocean-influenced features, location with respect to pressure and wind belts, and the effects of mountains. These classifications share a common shortcoming: they are qualitative, so that climatic regions are designated in a subjective manner rather than as a result of the application of some rigorous differentiating formula.

An interesting example of a method based on the energy budget (inputs and outputs of solar energy) of Earth's surface is the 1970 classification of Werner H. Terjung, an American geographer. His method utilizes data for more than 1,000 locations worldwide on the net solar radiation received at the surface, the available energy for evaporating water, and the available energy for heating the air and subsurface. The annual patterns are classified according to the maximum energy input, the annual range in input, the shape of the annual curve, and the number of months with negative magnitudes (energy deficits). The combination of characteristics for a location is represented by a label consisting of several letters with defined meanings, and regions having similar net radiation climates are mapped.

Probably the most extensively used genetic systems, however, are those that employ air-mass concepts. Air masses are large bodies of air that, in principle, possess relatively homogeneous properties of temperature, humidity, etc. from one side of the air mass to the other. Weather on individual days may be interpreted in terms of these features and their contrasts at fronts.

Two American geographer-climatologists have been most influential in classifications based on air mass. In 1951, Arthur N. Strahler described a qualitative classification based on the combination of air masses present at a given location throughout the year. Some years later (in 1968 and 1970), John E. Oliver placed this type of classification on a firmer footing by providing a quantitative framework that designated particular air masses and air-mass combinations as "dominant", "subdominant", or "seasonal" at particular locations. He also provided a means of identifying air masses from diagrams of mean monthly temperature and precipitation plotted on a "thermohyet diagram", a procedure that obviates the need for less common upper-air data to make the classification.

Empirical classifications

Most empirical classifications are those that seek to group climates based on one or more measureable aspects of the climate system. While many such phenomena have been used in this way, natural vegetation stands out as one of prime importance. The view held by many climatologists is that natural vegetation functions as a long-term integrator of the climate in a region – the vegetation, in effect, is an instrument for measuring climate in a similar way to that in which a thermometer measures temperature. The pre-eminence of this feature is apparent in the fact that many textbooks and other sources refer to climates using the names of vegetation: as, for example, rainforest, taiga, or tundra.

Wladimir Köppen, a German botanist-climatologist, developed the most popular (but not the first) of these vegetation-based classifications. His aim was to devise formulae that would define climatic boundaries in such a way as to correspond to those of the vegetation zones that were being mapped for the first time during his lifetime. Köppen published his first scheme in 1900 and a revised version in 1918. He continued to revise his system of classification until his death in 1940. Other climatologists have modified portions of Köppen's procedure on the basis of their experience in various parts of the world.

Köppen's classification is based on a subdivision of terrestrial climates into five major types, which are represented by the capital letters A, B, C, D, and E. Each of these climate types except for B is defined by temperature criteria.

Type B designates climates in which the controlling factor on vegetation is dryness (rather than coldness). Aridity is not a matter of precipitation alone but is defined by the relationship between the precipitation input to, and the evaporative losses from, the soil in which the plants grow. Since evaporation is difficult to evaluate and is not a conventional measurement at meteorological stations, Köppen was forced to substitute a formula that identifies aridity in terms of a temperature-precipitation index (i.e. evaporation is assumed to be controlled by temperature). Dry climates are divided into arid (BW) and semi-arid (BS) subtypes, and each may be differentiated further by adding a third code: h for warm and k for cold.

As noted above, the other four major climate types are defined by temperature. These are also subdivided, with additional letters again used to designate the various subtypes. Type A climates (the warmest) are differentiated on the basis of the seasonality of precipitation: Af (no dry season), Am (short dry season), or Aw (winter dry season). Type E climates (the coldest) are conventionally separated into tundra (ET) and snow-and-ice climates (EF). The mid-latitude C and D climates are given a second letter, f (no dry season), w (winter dry), or s (summer dry), and a third symbol (a, b, c, or d [the last subclass exists only for D climates]), indicating the warmth of the summer or the coldness of the winter. Table 2.2 gives the specific criteria for the Köppen-Geiger-Pohl system of 1953.

The Köppen classification has been criticized on many grounds. It has been argued that extreme events, such as a periodic drought or an unusual cold spell, are just as significant in controlling vegetation distributions as the mean conditions upon which Köppen's scheme is based. It also has been pointed out that factors other than those used in the classification, such as sunshine and wind, are important to vegetation. Moreover, it has been contended that natural vegetation can respond only slowly to environmental change, so that the vegetation zones observable today are in part adjusted to past

Table 2.2 Classification of major climatic types according to the Köppen-Geiger-Pohl scheme

1st	2nd	3rd	Criterion
Letter symbol			
A			Temperature of coolest month 18 °C or higher
	f		Precipitation in driest month at least 60 mm
	m		Precipitation in driest month less than 60 mm but equal to or greater than 100 – $(r/25)$[1]
	w		Precipitation in driest month less than 60 mm and less than 100 – $(r/25)$
B[2]			70 per cent or more of annual precipitation falls in the summer half of the year and r less than $20t + 280$, or 70 per cent or more of annual precipitation falls in the winter half of the year and r less than $20t$, or neither half of the year has 70 per cent or more of annual precipitation and r less than $20t + 140$[3]
	W		r is less than one-half of the upper limit for classification as a B type (see above)
	S		r is less than the upper limit for classification as a B type but is more than one-half of that amount
		h	t equal to or greater than 18 °C
		k	t less than than 18 °C
C			Temperature of warmest month greater than or equal to 10 °C, and temperature of coldest month less than 18 °C but greater than -3 °C
	s		Precipitation in driest month of summer half of the year is less than 30 mm and less than one-third of the wettest month of the winter half
	w		Precipitation in driest month of the winter half of the year less than one-tenth of the amount in the wettest month of the summer half
	f		Precipitation more evenly distributed throughout year; criteria for neither s nor w satisfied
		a	Temperature of warmest month 22 °C or above
		b	Temperature of each of four warmest months 10 °C or above but warmest month less than 22 °C
		c	Temperature of one to three months 10 °C or above but warmest month less than 22 °C
D			Temperature of warmest month greater than or equal to 10 °C, and temperature of coldest month -3 °C or lower
	s		Same as for type C
	w		Same as for type C
	f		Same as for type C
		a	Same as for type C
		b	Same as for type C
		c	Same as for type C
		d	Temperature of coldest month less than -38 °C (d designation then used instead of a, b, or c)
E			Temperature of warmest month less than 10 °C
	T		Temperature of warmest month greater than 0 °C but less than 10 °C
	F		Temperature of warmest month 0 °C or below
H[4]			Temperature and precipitation characteristics highly dependent on traits of adjacent zones and overall elevation – highland climates may occur at any latitude

[1] In the formulas above, r is average annual precipitation total (mm) and t is average annual temperature (°C). All temperatures are monthly means (°C), and all other precipitation amounts are mean monthly totals (mm)

[2] Any climate that satisfies the criteria for designation as a B type is classified as such, irrespective of its other characteristics

[3] The summer half of the year is defined as the months April–September for the Northern Hemisphere and October–March for the Southern Hemisphere

[4] Most modern climate schemes consider the role of altitude. The highland zone has been taken from Trewartha (1968).

climates. Many critics have drawn attention to the rather poor correspondence between the Köppen zones and the observed vegetation distribution in many areas of the world. However, in spite of these and other limitations, the Köppen system remains the most popular climatic classification in use today.

A major contribution to climate grouping was made by the American geographer-climatologist C. Warren Thornthwaite in 1931 and 1948. He first used a vegetation-based approach that made use of the derived concepts of temperature efficiency and precipitation effectiveness as a means of specifying atmospheric effects on vegetation. His second classification retained these concepts in the form of a moisture index and a thermal efficiency index, but radically changed the classification criteria and rejected the idea of using vegetation as the climatic integrator, attempting instead to classify "rationally" on the basis of the numerical values of these indices. Thornthwaite's 1948 scheme is encountered in many climatology texts, but it has not gained as large a following among a wide audience as the Köppen classification system, perhaps because of its complexity and the large number of climatic regions it defines.

One drawback of vegetation-based climate classifications is that, while they could be regarded as having relevance to human activity through what they may indicate about agricultural potential and natural environment, they cannot give any sense of how human beings would feel within the various climate types. Terjung's 1966 scheme was an attempt to group climates on the basis of their effects on human comfort. The classification makes use of four physiologically relevant parameters: temperature, relative humidity, wind speed, and solar radiation. The first two are combined in a "comfort index" to express atmospheric conditions in the following terms: extremely hot, hot, oppressive, warm, comfortable, cool, keen, cold, very cold, extremely cold, and ultra cold. Temperature, wind speed, and solar radiation are combined in a "wind effect index" expressing the net effect of wind chill (the cooling power of wind on exposed surfaces) and the addition of heat to the human body by solar radiation. These indices are combined for different seasons in different ways to express how humans feel in various geographic areas on a yearly basis. Terjung visualized that his classification would find applicability in medical geography, climatological education, tourism, housing, clothing, and as a general analytical tool.

Many other specialized empirical classifications have been devised. For example, there are those that differentiate between types of desert and coastal climates, those that account for different rates of rock weathering or soil formation, and those based on the identification of similar agricultural climates.

World distribution of major climatic types

The following discussion of the climates of the world is based on groupings of Köppen's climatic types. It should be read in conjunction with Table 2.2 (for the specific criteria defining each type) and Köppen's climate classification map (for the geographic extent of each type; Figure 2.5).

Type A climates

Köppen's type A climates are found in a nearly unbroken belt around Earth at low latitudes, mostly within 15°N and 15°S. Their location within a region in which the available net solar radiation is large and relatively constant from month to month ensures both high temperatures (generally in excess of 18°C [64°F]) and a virtual absence of thermal seasons.

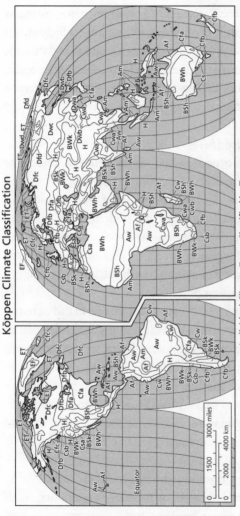

Köppen Climate Classification

Individual Climate Zones Classified by Type

Type A
Tropical humid
Af, Am, Aw

Type B
Dry
BSk, BSh,
BWh, BWk

Type C
Humid subtropical
Cf, Cfa, Cw, Cwa, Cwb
Mediterranean
Cs, Csa, Csb

Type C
Marine west coast
Cfb, Csc

Type D
Humid continental
Dfa, Dfb, Dwa, Dwb

Type D
Continental subarctic
Dfc, Dfd, Dwc, Dwd

Type E, H
Polar
ET-Tundra,
EF-Snow and Ice
Highland
H

Figure 2.5
Adapted from Arthur N. Strahler, *Physical Geography*, 3rd edition, John Wiley & Sons, Inc.

Typically, the temperature difference between day and night is greater than that between the warmest and the coolest month – the opposite of the situation in mid-latitudes. The terms "winter" and "summer" have little meaning, but in many locations annual rhythm is provided by the occurrence of wet and dry seasons. Type A climates are controlled mainly by the seasonal fluctuations of the trade winds, the intertropical convergence zone (ITCZ), and the Indian Ocean monsoon (see Chapter 6, pages 303 and 305 respectively).

Wet equatorial climate (Af)
Within about 12° latitude of the equator lies a region of consistently high temperatures (around 30°C [86°F]), with plentiful precipitation (1,500–10,000 mm [59–394 inches]), heavy cloud cover, and high humidity, with very little annual temperature variation. Such regions lie within the influence of the ITCZ in all months; the converging, ascending air spawns convectional thunderstorm activity, with much of the rainfall occurring in late afternoon or early evening. Other zones of Af climate are found beyond the usual margins of ITCZ activity: in coastal Madagascar, south-east coastal Brazil, and much of Central America and western Colombia, where trade winds blow onshore all year to encounter coastlines backed by mountain barriers that stimulate the formation of precipitation as warm, moist tropical air is forced to ascend and cool. Some of these regions also may receive precipitation from tropical disturbances, including hurricanes.

Tropical monsoon and trade-wind littoral climates (Am)
These climates resemble the Af in most characteristics, with small annual temperature ranges, high temperatures, and plentiful precipitation (often more than Af climates in annual total). They differ from the latter, however, in that they exhibit

a short dry season, usually in the low-sun ("winter") season. Two distinct processes can give rise to Am climate types. The largest areas, mostly in southern and southeastern Asia, result from the Asian monsoon circulation that brings convective and orographic precipitation in the summer when warm, moist, maritime tropical air moves over land to converge into the low-pressure zone north of the Himalayas. In winter, by contrast, cool dry air diverges out of the Siberian anticyclone to the north, bringing a cooler, drier, and clearer period of variable length. In the Americas and in Africa, Am climates are of the trade-wind variety. These areas receive precipitation on narrow coastal strips through orographic effects as the moist air of the trade winds ascends mountain chains.

Tropical wet-dry climate (Aw)

This climate has distinct wet and dry seasons, with most of the precipitation occurring in the high-sun ("summer") season. Total amounts of rainfall are less than in the previous two climate types (500–1,750 mm [20–69 inches]), most of which occur in convectional thunderstorms. The dry season is longer than in the Am climates and becomes progressively longer as one moves poleward through the region. Temperatures are high throughout the year but show a greater range than Af and Am climates. Throughout most of the region, the cause of the seasonal cycle is the shift in the tropical circulation throughout the year. During the high-sun season, the ITCZ moves poleward and brings convergent and ascending air to these locations, which stimulates convective rainfall. During the low-sun season, the convergence zone moves off to the winter hemisphere and is replaced by the periphery or core of the subtropical anticyclone, with its subsiding, stable air resulting in a period of dry, clear weather, the intensity and length of which depend on latitude.

Type B climates

Arid and semi-arid climates cover about a quarter of the Earth's land surface, mostly between 50°N and 50°S, but they are mainly found in the 15–30° latitude belt in both hemispheres. They exhibit low precipitation, great variability in precipitation from year to year, low relative humidity, high evaporation rates (when water is available), clear skies, and intense solar radiation.

Tropical and subtropical desert climate (BWh and BWk)

Most of Earth's tropical, true desert (BW) climates occur between 15° and 30° latitude. These regions are dominated in all months by the subtropical anticyclone, with its descending air, atmospheric inversions, and clear skies: an atmospheric environment that inhibits precipitation. The most extreme arid areas are also far removed from sources of moisture-bearing winds in the interiors of continents and are best developed on the western sides of continents, where the subtropical anticyclone shows its most intense development. An exception to the general tendency for aridity to be associated with subsidence is in the so-called Horn of Africa region, where the dryness of Somalia is caused more by the orientation of the landmass in relation to the atmospheric circulation.

In most low-latitude deserts, cloud cover is uncommon (in some areas fewer than 30 days per year have clouds). Precipitation amounts are mostly in the range 0–250 mm (0–10 inches), although the unreliability of precipitation is more significant than the small totals. Temperatures are high, with monthly means in the range 21–32°C (70–90°F). Daily temperature variations are extreme – ranges of 35°C (63°F) are not unknown. The highest air temperatures recorded on Earth have been in the BWh regions: for example, in shaded, well-ventilated parts Death Valley in the western United States has reached 57°C (135°F), while al-'Aziziyah in Libya has had a recorded high of 58°C (136°F).

An interesting variant of tropical and subtropical deserts are the so-called West Coast Desert areas found on the western coastal margins of regions such as the Sonoran Desert of North America, the Peru and Atacama deserts of South America, and the Sahara (Moroccan part) and Namib deserts of Africa. These areas are much cooler than their latitude would suggest, with monthly mean temperatures of only 15–21°C (59–70°F), and parts are classified as BWk in Köppen's scheme. The cooling results from airflow off adjacent coastal waters where upwelling of the ocean gives rise to cold currents. Deserts of this sort are subject to frequent fog and low-level clouds, yet they are extremely arid. Some parts of the Atacama Desert, for example, have recorded no precipitation for 20 years.

Tropical and subtropical steppe climate (BSh)

The low-latitude semi-arid (or steppe) climate occurs primarily on the periphery of the true deserts treated above. It is transitional to the Aw climate on the equatorward side (showing a summer rainfall maximum associated with the ITCZ and a small annual temperature range) and to the Mediterranean climate on its poleward margin (with a cooler, wetter winter resulting from the higher latitude and mid-latitude frontal cyclone activity). Annual precipitation totals are greater than in BW climates (380–630 mm [15–25 inches]). Yearly variations in amount are not as extreme as in the true deserts but are nevertheless large.

Mid-latitude steppe and desert climate (BSk and BWk)

Although these climates are contiguous with the tropical dry climates of North and South America and of Central Asia, they have different origins. Cool true deserts extend to 50° latitude and cool steppes reach nearly 60°N in the Canadian Prairies, well beyond the limits of the subtropical anticyclone.

These climates owe their origins to locations deep within continental interiors, far from the windward coasts and sources of moist, maritime air. Temperature conditions are extremely variable. In the higher latitudes, winters are severely cold, with meagre precipitation associated with polar and arctic air masses. Summer precipitation is more often convective, arriving in the form of scattered thunderstorm activity brought about by irregular incursions of moist air.

Both BSk and BWk climates in the mid-latitudes owe their origins to these mechanisms, but the steppe type tends to be located peripheral to the true desert, either adjacent to the moister C and D climates or at the poleward extent of the range, where reduced evaporation under cooler conditions makes more of the scarce precipitation available as soil moisture for plant growth.

Types C and D climates

Through a major portion of the middle and high latitudes (mostly from 25° to 70° N and S) lies a group of climates classified within the Köppen scheme as C and D types. Most of these regions lie beneath the upper-level, mid-latitude westerlies throughout the year, and their climatic character derives from the seasonal variations in location and intensity of these winds and their associated features. During summer, the polar front and its jet stream move poleward, and air masses of tropical origin are able to extend to high latitudes. During winter, as the circulation moves equatorward, tropical air retreats and cold polar outbreaks influence weather, even within the subtropical zone.

The relative frequency of these air masses of different origins varies gradually from low to high latitude and is largely responsible for the observed temperature change across the belt, which is most marked in winter. The air masses interact in the frontal systems commonly found embedded within the travelling cyclones that lie beneath the polar-front jet stream. Ascent induced by convergence into these low-pressure cells and by uplift at fronts induces precipitation, the main location of which shifts with the seasonal circulation cycle.

Other important sources of precipitation are convection, mainly in tropical air, and forced uplift at mountain barriers. Monsoon effects modify this general pattern, while the sub-tropical anticyclone plays a role in the climate on the western sides of the continents in the subtropics.

Humid subtropical climate (Cfa and Cwa)

These climates are found on the eastern sides of the continents between 20° and 35° N and S latitude. Most (the Cfa types) show a relatively uniform distribution of precipitation throughout the year, with totals in the range 750–1,500 mm (30–59 inches). In summer, these regions are largely under the influence of moist, maritime airflow from the western side of the subtropical anticyclonic cells over low-latitude ocean waters. Temperatures are high: the warmest months generally average about 27°C (81°F), with mean daily maxima from 30°C to 38°C (86°F to 100°F) and warm, oppressive nights. Summers are usually somewhat wetter than winters, with much of the rainfall coming from convectional thunderstorm activity; tropical cyclones also enhance warm-season rainfall in some regions. The coldest month is usually quite mild, although frosts are not uncommon.

In North America, the spring and early summer seasons – when the front begins its northward return – are notorious for

the outbreak of tornadoes associated with frontal thunderstorms along the zone of interaction between tropical and
polar air.

In eastern and southern Asia, the monsoon influence results
in a modified humid subtropical climate (Cwa) that has a
clearly defined dry winter, when air diverges from the Siberian
anticyclone, and the polar front and cyclone paths are deflected around the region. These areas generally lie on the
poleward side of Am and Aw climates and exhibit a somewhat
larger annual temperature range than Cfa types. Winters are
sunny and rather cool. Annual precipitation averages 1,000
mm (39 inches) but may vary from 750 cm to 2,000 mm (30–
79 inches).

Mediterranean climate (Csa and Csb)

Between about 30° and 45° latitude on the western sides of the
continents are a series of climates with the unusual combination of hot, dry summers and cool, wet winters. Poleward
extension and expansion of the subtropical anticyclonic cells
over the oceans bring subsiding air to the region in summer,
with clear skies and high temperatures. When the anticyclone
moves equatorward in winter, it is replaced by travelling,
frontal cyclones with their attendant precipitation.

Annual temperature ranges are generally smaller than those
found in the Cfa climates, since locations on the western sides
of continents are not well positioned to receive the coldest
polar air, which develops over land rather than over the ocean.
Mediterranean climates also tend to be drier than humid
subtropical ones, with precipitation totals ranging from 350
mm to 900 mm (14 to 35 inches); the lowest amounts occur in
interior regions adjacent to the semi-arid steppe climates.

Some coastal locations (e.g. southern California in the
western United States) exhibit relatively cool summer

conditions and frequent fogs where cold offshore currents prevail. Only in Europe, where the latitude for this climate type fortuitously corresponds to an ocean basin (that of the Mediterranean, from which this climate derives its name), does this climate type extend eastward away from the coast for any significant distance.

Marine west coast climate (Cfb and Cfc)

Poleward of the Mediterranean climate region on the western sides of the continents, between 35° and 60° N and S latitude, are regions that exhibit ample precipitation in all months. Unlike their equatorward neighbours, these areas are located beyond the furthest poleward extent of the subtropical anticyclone, and they experience the mid-latitude westerlies and travelling frontal cyclones all year. Annual precipitation ranges from 500 to 2,500 mm (20 to 98 inches), with local totals exceeding 5,000 mm (197 inches) where onshore winds encounter mountain ranges. Precipitation is not only plentiful but also reliable and frequent. Fog is common in autumn and winter, but thunderstorms are rare. Strong gales with high winds may be encountered in winter.

These are equable climates with few extremes of temperature. Annual ranges are rather small (10–15°C or 50–59°F) – about half those encountered further to the east in the continental interior at the same latitude. Mean annual temperatures are usually 7–13°C (45–55°F) in lowland areas, the winters are mild, and the summers are relatively moderate, rarely having monthly temperatures above 20°C (68°F).

In North and South America, Australia, and New Zealand, north–south mountain ranges backing the west coasts of the landmasses at these latitudes confine the marine west coast climate to relatively narrow coastal strips (but enhance precipitation). By contrast, in Europe the major mountain chains (the Alps and Pyrenees) run east–west, permitting Cfb and Cfc climates to extend inland some 2,000 km (1,250 miles) into eastern Germany and Poland.

Humid continental climate (Dfa, Dfb, Dwa, and Dwb)

The type D climates are primarily northern hemispheric phenomena, since landmasses are absent at the significant latitudes in the southern hemisphere. The humid continental subgroup occupies a region between 30°N and 60°N in central and eastern North America and Asia in the major zone of conflict between polar and tropical air masses. These regions exhibit large seasonal temperature contrasts, with hot summers and cold winters. Precipitation tends to be ample throughout the year in the Df section, being derived both from frontal cyclones and, in summer months, from convectional showers when maritime tropical air pushes northward behind the retreating polar front. Severe thunderstorms and tornadoes are an early summer occurrence when the polar front is in the southern margin of the Dfa region. Winter precipitation often occurs in the form of snow, and a continuous snow cover is established for between one and four months in many parts of the region, especially in the north.

The changeable nature of weather in all seasons is a characteristic feature of these climates, especially in such areas as the eastern United States and Canada, where there are few topographic barriers to limit the exchange of air masses between high and low latitudes. Mean temperatures are typically below freezing for up to several months. Annual precipitation totals range from 500 to 1,250 mm (20–50 inches), with higher amounts in the south of the region and in the uplands.

In eastern Asia (Manchuria and Korea), a monsoonal variant of the humid continental climate (Dwa and Dwb) occurs.

This climate type has a pronounced summer precipitation maximum and a cold, dry winter dominated by continental polar air diverging out of the nearby Siberian anticyclone.

Continental subarctic climate (Dfc, Dfd, Dwc and Dwd)

North of the humid continental climate, from about 50°N to 70°N, in a broad swathe extending from Alaska to Newfoundland in North America and from northern Scandinavia to Siberia in Eurasia, lie the continental subarctic climates. These are regions dominated by the winter season – a long, bitterly cold period with short, clear days, relatively little precipitation (mostly in the form of snow), and low humidity. In Asia the Siberian anticyclone, the source of continental polar air, dominates the interior, and mean temperatures of -40°C to -50°C (-40°F to -58°F) are not unusual. The North American version of this climate is not as severe but is still profoundly cold: mean monthly temperatures are below freezing for six to eight months, and mean temperatures in summer only rarely exceed 16°C (61°F). As a result of these temperature extremes, annual temperature ranges are larger in continental subarctic climates than in any other climate type on Earth, up to 30°C (54°F) through much of the area and more than 60°C (108°F) in central Siberia, although coastal areas are more moderate.

Annual precipitation totals are mostly less than 500 mm (20 inches), with higher levels occurring in marine areas near warm ocean currents. Such areas also are generally somewhat more equable and may be designated marine subarctic climates. Areas with a distinct dry season in winter, which results in the Köppen climate types Dwc and Dwd, occur in eastern Siberia, both in the region where the wintertime anticyclone is established and in the peripheral areas subject to dry, divergent airflow from it.

Type E climates

Köppen's type E climates are controlled by the polar and arctic air masses of high latitudes (60°N and S and higher). These climates are characterized by low temperatures and precipitation and by a surprisingly great diversity of subtypes.

Tundra climate (ET)

Tundra climates occur between 60° and 75° of latitude, mostly along the Arctic coast of North America and Eurasia and on the coastal margins of Greenland. Mean annual temperatures are below freezing and annual ranges are large (but not as large as in the adjacent continental subarctic climate). Summers are generally mild, with daily maxima of 15–18°C (59–64°F), although the mean temperature of the warmest month must be less than 10°C (50°F). Summer days are long (a result of the high latitude) but often cloudy. Although the snow cover of winter melts in the warmer season, in places the ground at depth remains permanently frozen as permafrost. However, frosts and snow are possible in any month. Winters are long and cold: temperatures may be below 0°C (32°F) for 6 to 10 months, especially in the region north of the Arctic Circle. Winter precipitation generally consists of dry snow, with typical annual totals of less than 350 mm (about 14 inches).

Snow-and-ice climate (EF)

This climate occurs poleward of 65° N and S latitude, over the ice caps of Greenland and Antarctica and over the permanently frozen portion of the Arctic Ocean. Temperatures are below freezing throughout the year, and annual temperature ranges are large but again not as large as in the continental subarctic climates. Winters are frigid, with mean monthly temperatures from -20°C to -65°C (-4°F to -85°F). The lowest recorded temperatures at Earth's surface are recorded in this climate: the Vostok II research station in Antarctica holds the record for the lowest extreme temperature (-89.2°C [-129°F]), while the Plateau Station has the lowest annual mean (-56°C [-69°F]). Daily temperature variations are very small, because the presence of snow and ice at the surface refrigerates the air.

Precipitation is meagre in the cold, stable air: in most cases 50–500 mm (2–20 inches), with the largest amounts occurring on the coastal margins. Most of this precipitation results from the periodic penetration of a cyclone into the region, which brings snow and ice pellets and, with strong winds, blizzards. High winds also occur in the outer portions of the Greenland and Antarctic EF climates, where cold, dense air drains off the higher, central sections of the ice caps as katabatic winds (see "Local wind systems", Chapter 6, page 299).

Type H climates

The major highland regions of the world – including the Cascades, Sierra Nevadas, and Rockies of North America, and the Andes, the Himalayas and the Tibetan Plateau – cannot be classified realistically on the Köppen system, since the effects of altitude and relief give rise to myriad mesoclimates and microclimates. As such, modified classifications have been developed over time to account for this and other variations – Type H climates one such form modified group. This diversity over short horizontal distances is unmappable at the continental scale. Very little of a universal nature can be written about such mountain areas except to note that, as a rough approximation, they tend to resemble cooler, wetter versions of the climates of nearby lowlands. Otherwise, only the most general characteristics may be noted.

With increasing height, the temperature, pressure, atmo-

spheric humidity, and dust content decrease. The reduced
amount of air overhead results in high atmospheric transpar-
ency and enhanced receipt of solar radiation (especially of
ultraviolet wavelength) at elevation. Altitude also tends to
increase precipitation, at least for the first 4,000 metres
(13,100 feet). The orientation of mountain slopes has a major
impact on solar radiation receipt and temperature and also
governs exposure to wind. Mountains can have other effects
on the wind climate: valleys can increase wind speeds by
"funnelling" regional flows and may also generate mesoscale
mountain- and valley-wind circulations. Cold air also may
drain from higher elevations to create "frost pockets" in low-
lying valleys. Furthermore, mountains can act as barriers to
the movement of air masses, can cause differences in precipitation amounts between windward and leeward slopes (the
reduced precipitation on and downwind from lee slopes is
called a rain shadow), and, if high enough, can collect permanent snow and ice on their peaks and ridges. The snow line
varies in elevation from sea level in the subarctic to about
5,500 metres (18,000 feet) at 15–25° N and S latitudes.

Climate and life

The connection between climate and life arises from a two-way
exchange of mass and energy between the atmosphere and the
biosphere (the life-supporting part of Earth's surface, comprising living organisms and non-living things from which they
derive energy and nutrients). In Earth's early history, before
life evolved, only geochemical and geophysical processes determined the composition, structure, and dynamics of the
atmosphere. But since life evolved, biochemical and biophysical processes have played a role in these atmospheric

characteristics. Human beings (*Homo sapiens*) are increasingly shouldering this role by mediating interactions between the biosphere and the atmosphere.

The living organisms of the biosphere use gases from, and return "waste" gases to, the atmosphere, and the composition of the atmosphere is a product of this gas exchange. It is very likely that, prior to the evolution of life on Earth, 95 per cent of the atmosphere was made up of carbon dioxide and that water vapour was the second most abundant gas. Other gases were present in trace amounts. This atmosphere was the product of geochemical and geophysical processes in the interior of the Earth and was mediated by volcanic outgassing. It is estimated that the great mass of carbon dioxide in this early atmosphere gave rise to an atmospheric pressure 60 times that of modern times.

Today only about 0.035 per cent of Earth's atmosphere is carbon dioxide. Much of the carbon dioxide present in our planet's first atmosphere has been removed by photosynthesis, chemosynthesis (the synthesis of organic compounds by way of energy derived from inorganic chemical reactions, instead of sunlight), and weathering. Currently, most of the carbon dioxide now resides in Earth's limestone sedimentary rocks, in coral reefs, in fossil fuels, and in the living components of the present-day biosphere.

In this transformation, the atmosphere and the biosphere co-evolved through continuous exchanges of mass and energy. "Biogenic gases" are gases critical for, and produced by, living organisms. In our current atmosphere they include oxygen (O_2), nitrogen (N_2), water vapour (H_2O), carbon dioxide (CO_2), carbon monoxide (CO), methane (CH_4), ozone (O_3), nitrogen dioxide(NO_2), nitric acid (HNO_3), ammonia (NH_3) and ammonium ions (NH_4^+), nitrous oxide (N_2O), sulphur dioxide (SO_2), hydrogen sulphide (H_2S), carbonyl sulphide

(COS), dimethyl sulphide (DMS; C_2H_6S), and a complex array of non-methane hydrocarbons. Of these gases, only nitrogen and oxygen are not "greenhouse gases".

Added to this roster of biogenic gases is a much longer list of human-generated gases from industrial, commercial, and cultural activities that reflect the diversity and extent of the human enterprise on Earth.

Evolution of life and the atmosphere

Life on Earth began at least as early as 3.5 billion years ago, during the middle of the Archean Eon (about 4 billion to 2.5 billion years ago). It was during this period that life first began to exercise certain controls on the atmosphere. Although some nitrogen was present in the atmosphere's prebiological state, little if any oxygen was available. Chemical reactions with hydrogen sulphide, hydrogen, and reduced compounds of nitrogen and sulphur precluded any but the shortest lifetime for free oxygen in the atmosphere. As a result, life evolved in an atmosphere that was "reducing" (high hydrogen content) rather than "oxidizing" (high oxygen content).

In addition to their chemically reducing character, and with the exception of nitrogen, the predominant gases of this prebiotic atmosphere were largely transparent to incoming sunlight but opaque to outgoing terrestrial infrared radiation. These gases are therefore called (perhaps improperly) "greenhouse gases", because they are able to slow the release of outgoing radiation back into space. In the Archean Eon the sun produced as much as 25 per cent less light than it does today but, despite this, Earth's temperature was much like that of today. This is because the greenhouse-gas-rich Archean atmosphere was effective in retarding the loss of terrestrial radiation to space. The resulting long residence time of energy within Earth's atmospheric system resulted in a warmer atmosphere than would have been possible otherwise.

The average temperature of Earth's surface in the early Archean Eon was higher than the modern global average. It was, according to some sources, probably similar to temperatures found in today's tropics. Depending on the amount of nitrogen present during the Archean Eon, it has been suggested that the atmosphere may have held more than 1,000 times as much carbon dioxide than it does today.

Archean organisms included photosynthetic and chemosynthetic bacteria, methane-producing bacteria, and a more primitive group of organisms now called the "Archaea" (a group of prokaryotes more related to eukaryotes than to bacteria and found in extreme environments). Through their metabolic processes, organisms of the Archean Eon slowly changed the atmosphere. Hydrogen rose from trace amounts to about 1 part per million (ppm) of dry air. Methane concentrations increased from near zero to about 100 ppm. Oxygen increased from near zero to 1 ppm, whereas nitrogen concentrations rose to encompass 99 per cent of all atmospheric molecules excluding water vapour. Carbon dioxide concentrations decreased to only 0.3 per cent of the total; however, this was nearly 10 times the prevailing concentration at the beginning of the Archean Eon.

The composition of the atmosphere, its radiation budget, its thermodynamics, and its fluid dynamics were transformed by life from the Archean Eon. The American geochemist Robert Garrels has calculated that, in the absence of life and given the burial rate of carbon in rocks, oxygen would be unavailable to form water, and free hydrogen would be lost to space. Without the presence of life and compounded by this loss of hydrogen there would be no oceans, and Earth would have become merely a dusty planet by the middle of the Archean Eon.

By the end of the Archean Eon, 2.5 billion years ago, both the pigment chlorophyll and photosynthetic organisms had evolved such that the production of oxygen increased rapidly. The atmosphere became transformed from a reducing atmosphere with carbon dioxide, limited oxygen, and anaerobic organisms (life forms that do not require oxygen for respiration) in control to one with an oxidizing atmosphere that was rich in oxygen, poor in carbon dioxide, and dominated by aerobic organisms (life forms requiring oxygen for respiration). With the decline in carbon dioxide and a rise in oxygen, the "greenhouse" warming capacity of Earth's atmosphere was sharply reduced. However, this happened over a period of time when the energy produced by the sun increased systematically. These compensating changes resulted in a relatively constant planetary temperature over much of Earth's history.

The role of the biosphere in the Earth–atmosphere system

The biosphere and Earth's energy budget

Biogenic gases in the atmosphere play a role in the dynamics of Earth's planetary radiation budget, the thermodynamics of the planet's moist atmosphere, and, indirectly, the mechanics of the fluid flows that are Earth's planetary wind systems. In addition, human cultural and economic activities have added a new dimension to the relationship between the biosphere and the atmosphere.

While humans are biologically trivial compared with bacteria in the exchange of gases with the atmosphere, chemical compounds produced from human industrial activities and other economic enterprises are changing the gaseous composition of the atmosphere in climatically significant ways. The largest changes involve the harvesting of ancient carbon stores. This organic material has been transformed into fossil fuels (coal, petroleum, natural gas, and others) by geological processes acting upon the remains of plants and animals over many millions of years. Different forms of carbon may be burned and thus used as energy sources. In so doing, organic carbon is converted into carbon dioxide. Additionally, humans are also burning trees, grasses, and other biomass for cooking purposes, and clearing the land for agriculture and other activities (see Chapter 4). The combination of burning both fossil fuels and biomass is enriching the atmosphere with carbon dioxide and adding to the essential reservoir of greenhouse gases.

Earth's atmosphere today is largely transparent to sunlight. Although oxygen and nitrogen make up nearly 99 per cent of the atmosphere, these diatomic molecules (consisting of two atoms) do not vibrate in a way that permits them to absorb terrestrial radiation. They are largely transparent to outgoing terrestrial radiation as well as to incoming solar radiation. Of the sunlight absorbed by the entire Earth–atmosphere system, about one-third is absorbed by the atmosphere and two-thirds by Earth's surface. Sunlight is absorbed by other molecules of the atmosphere, by cloud droplets, and by dust and debris.

Over the continents, the surface cover of vegetation is the principal absorbing medium of Earth's surface, although other surfaces such as bare rock, sand, and water also absorb solar radiation. At night, absorption at the surface (that is, between the ground and 1.2 metres [4 feet] above the actual ground–atmosphere interface) is reradiated, in the form of long-wave infrared radiation, back toward space. Most of this infrared radiation is absorbed by the principal biogenic trace gases of the atmosphere – the so-called greenhouse gases: water vapour, carbon dioxide, nitrous oxide, and methane. Without these gases, Earth would be 33°C (59°F) colder on average than it is. A moderate-emission scenario from the 2007 Inter-governmental Panel on Climate Change (IPCC) report (see "Causes of global warming", Chapter 1, page 11) predicts that the continued addition of greenhouse gases from fossil fuels will increase the average global temperature by between 1.8°C and 4.0°C (3.2–7.2°F) over the next century. Other scenarios, predicting greater greenhouse gas emissions, forecast even greater global warming.

Cycling of biogenic atmospheric gases

The cycling of oxygen, nitrogen, water vapour, and carbon dioxide, as well as the trace gases – methane, ammonia, various oxides of nitrogen and sulphur, and non-methane hydrocarbons – between the atmosphere and the biosphere results in relatively constant proportions of these compounds in the atmosphere over time. Without the continuous genera-tion of these gases by the biosphere, they would quickly disappear from the atmosphere.

The carbon cycle, as it relates to the biosphere, is simple in its essence. Inorganic carbon (carbon dioxide, CO_2) is con-verted to organic carbon (the molecules of life), which is in turn converted back to inorganic carbon. Ultimately, the carbon cycle is powered by sunlight, as green plants and cyanobacteria (blue-green algae) use sunlight to split water into oxygen and hydrogen and to "fix" carbon dioxide from the atmosphere into organic carbon. So, carbon dioxide is removed from the atmosphere and oxygen is added. Animals engage in aerobic respiration, in which oxygen is consumed and organic carbon (from the molecules in their bodies) is oxidized to manufacture inorganic carbon dioxide – so oxygen is removed from the atmosphere and carbon dioxide is added.

It should be noted that chemosynthetic bacteria, which are found in deep-ocean and cave ecosystems, also fix carbon dioxide and produce organic carbon. Instead of using sunlight as an energy source, these bacteria rely on the oxidation of either ammonia or sulphur.

The carbon cycle is fully coupled to the oxygen cycle. Each year, photosynthesis fixes carbon dioxide and releases 100,000 megatons of oxygen (O_2) to the atmosphere. Respiration by animals and living organisms consumes about the same amount of oxygen and produces carbon dioxide in return. Oxygen and carbon dioxide are thus coupled in two linked cycles. On a seasonal basis, an enrichment of atmospheric carbon dioxide occurs in the winter half of the year, whereas a drawdown of atmospheric carbon dioxide takes place during the summer.

The nitrogen cycle begins with the "fixing" of inorganic atmospheric nitrogen (N_2), by nitrogen-fixing bacteria and blue-green algae, into compounds that can be used by organisms. Through the process of denitrification (by denitrifying bacteria), these nitrogen-containing compounds are converted back to inorganic atmospheric nitrogen. Ammonia and ammonium ions are the products of nitrogen fixation and may be incorporated into some organisms as organic nitrogen-containing molecules. In addition, ammonium ions may be oxidized (by nitrifying bacteria) to form nitrites, which can be further oxidized into nitrates. Although both nitrites and nitrates may be used to make organic nitrogen-containing molecules, nitrates are especially useful for plant growth and are key compounds that support both terrestrial and aquatic food chains. Nitrates may also be denitrified by bacteria to produce nitrogen gas. This process completes the nitrogen cycle.

The nitrogen cycle is coupled to both the carbon and oxygen cycles. Seventy-eight per cent of the gases of Earth's atmosphere by volume is diatomic nitrogen (N_2). Diatomic nitrogen is the most stable of the nitrogen-containing gases of the atmosphere – only 300 megatons of nitrogen need to be produced each year by denitrifying bacteria to account for losses. These losses mostly occur during lightning discharges and during nitrogen-fixation activities by algae and bacteria. Nitrogen-fixing bacteria, which are found in the root nodules of certain plants called legumes, also produce oxides of nitrogen from atmospheric nitrogen. Denitrifying bacteria convert nitrates in soils and wetlands to nitrogen gas.

Nitrous oxide (N_2O) occurs in trace amounts (0.3 ppm) in the atmosphere. Between 100 and 300 megatons of nitrous oxide are produced by soil and marine bacteria each year to maintain this concentration. In the atmosphere, nitrous oxide is short lived because it is quickly broken down by ultraviolet light. Nitric oxide (NO), a minor contributor in the breakdown of stratospheric ozone, is the by-product of this reaction.

Like nitrous oxide, ammonia (NH_3) is also produced in soils and marine waters by bacteria and escapes into the atmosphere. Nearly 1,000 megatons are added to the atmosphere each year. Ammonia decreases the acidity of precipitation and serves as a nutrient for plants when returned to the land via precipitation.

There are six major sulphur-containing biogenic atmospheric gases that are part of the sulphur cycle. They include hydrogen sulphide (H_2S), carbon disulphide (CS_2), carbonyl sulphide (COS), dimethyl sulphide (DMS, C_2H_6S), dimethyl disulphide ($[CH_3S]_2$), and methyl mercaptan (CH_3SH). Sulphur dioxide (SO_2) is an oxidation product of these sulphur gases, and it is also added to the atmosphere by volcanoes, burning biomass, and anthropogenic sources (i.e. smelting metals and coal ignition). SO_2 is removed from the atmosphere and returned to the biosphere in rainfall. The increased acidity of rain and snow from anthropogenic additions of SO_2 and oxides of nitrogen is often referred to as "acid precipitation" or "acid rain" (see Chapter 6, page 330). The acidity of this precipitation and other phenomena, such as "acid fog", is partly offset by the release of ammonia in the atmosphere.

The concentration of methane (CH_4) in the atmosphere at any one time is only about 1.7 ppm. Though only a trace gas, methane is highly reactive and plays a key role in the chemical reactions that control the composition of the atmosphere. Methanogenic bacteria in wetland sediments decompose organic matter and release 1,000 megatons of gaseous methane to the atmosphere per year. In the lower atmosphere, methane reacts with oxygen to produce water and carbon dioxide. Each year 2,000 megatons of oxygen are removed from the atmosphere by this mechanism. This loss of oxygen must be replaced by photosynthesis.

Some methane reaches the upper stratosphere, where its interaction with oxygen is a major source of upper stratospheric moisture. Within wetlands, bacteria produce methyl halide compounds (methyl chloride [CH_3Cl] and methyl iodide [CH_3I] gases), while in forests these same methyl halides are produced by fungi. These gases, upon reaching the stratosphere, regulate the production of stratospheric ozone by contributing to its natural breakdown. Without the continual production of methane by methanogenic bacteria, the oxygen concentration of the atmosphere would increase by 1 per cent every 12,000 years. Dangerously high levels of oxygen in the atmosphere would greatly increase the incidence of wildfires. If the oxygen concentration of Earth's atmosphere rose from its current concentration of 21 per cent to 25 per cent, even damp twigs and grass would easily ignite. Non-methane hydrocarbons of terrestrial origin are generally well mixed in the free atmosphere above the planetary boundary layer (PBL; see

"Biosphere controls on the planetary boundary layer", below). These organic particles weaken incoming solar radiation as it passes through the atmosphere, and reductions of 1 per cent have been recorded.

The influence of the biosphere on atmosphere and temperature

Biosphere controls on the structure of the atmosphere

Because the biosphere plays a key role in the flux of energy from the surface to the atmosphere, it also contributes to the thermal structure of the atmosphere. Three major fluxes are important: the direct transfer of heat from the surface to the atmosphere by conduction and convection (sensible heating), the energy flux to the atmosphere carried by water vapour via evaporation and transpiration from the surface (latent heat energy), and the flux of radiant energy from the surface to the atmosphere (infrared terrestrial radiation). These fluxes differ in the altitude at which the heating of the air takes place.

Sensible heating primarily warms the planetary boundary layer (PBL) of the atmosphere. In contrast, the latent heat of the atmosphere is released when the water vapour is converted into cloud droplets by condensation. Heating by latent energy release generally occurs above the PBL. The availability of water to evaporate from the surface limits the sensible heating of the air near the surface and so limits the maximum daytime surface air temperature (see "Biosphere controls on maximum temperatures by evaporation and transpiration", below).

Heating of the atmosphere by radiation from the surface depends on the density of the atmosphere and its water vapour content. Radiative heating from the surface declines with increasing altitude.

Biosphere controls on the planetary boundary layer

The planetary boundary layer (PBL) is the layer of the atmosphere closest to the Earth's surface – the sublayer of the troposphere. The top of the PBL can be visually marked by the elevation of the base of the clouds. The PBL can also be denoted by a thin layer of haze, often seen by passengers aboard airplanes during take-off from airports. During the day, the air within the PBL is thoroughly mixed by convection induced by the heating of Earth's surface. The thickness of the PBL depends on the intensity of this surface heating and the amount of water evaporated into the air from the biosphere. In general, the greater the heating of the surface, the deeper the PBL; the wetter the air advected (transported within a fluid) into the region and the greater the additional water added by evaporation and transpiration, the shallower the PBL. Over deserts, the PBL may extend up to 4–5 km (13,100–16,400 feet) in altitude. In contrast, over ocean areas it is less than 1 km (3,300 feet) thick, since little surface heating takes place there because of the vertical mixing of water. In heavily vegetated areas, it is 1–2 km (3,300–6,600 feet) thick. For every 1°C (1.8°F) increase in daily maximum surface temperature for a well-mixed PBL, the top of the PBL is elevated 100 metres (about 325 feet). In New England forests during the days following the spring leafing, it has been shown that the top of the PBL is lowered to 200–400 metres (650–1,300 feet). By contrast, during the months before the leafing out, the PBL thickens from solar heating as the sun rises higher in the sky and day length increases.

At the top of the PBL, a small inversion, where temperatures increase with height, develops. If convective mixing of the air in the PBL is vigorous, convection currents may penetrate through this temperature inversion. The cooling of the lifting air initiates the condensation of water vapour and the development of minuscule particles of liquid water called cloud droplets. The small clouds just above the PBL are known as planetary boundary layer clouds. These clouds scatter direct sunlight. As the ratio of diffuse sunlight to direct-beam sunlight increases, the photosynthetic productivity the biosphere below increases. The result is a dynamic synergy between the atmosphere and biosphere.

The landscapes of most human-dominated ecosystems are decidedly "patchy" in their geography. Cities, suburbs, fields, forests, lakes, and shopping centres both heat and evaporate water into the air of the PBL according to the nature of the surfaces involved. Convection and the prospect of air currents breaking through the top of the PBL vary markedly across such heterogeneous landscapes. These upward and downward currents or vertical eddies within the PBL transfer mass and energy upward from the surface. The frequency, timing, and strength of convective weather elements, including thunderstorms, vary according to the patchiness of the land use and land-cover pattern of the area. In general, the greater the patchiness of the landscape and the earlier the hour in the day, the more frequent and more intense these rain-producing systems become.

In the absence of a storm in the region, the air above the PBL sinks gently and the air below lifts. The temperature inversion at the top of the PBL is essentially a stable layer in the atmosphere. Emissions from the biosphere below are thus contained within the PBL and may build up below this layer over time. Consequently, the PBL may become quite turbid, hazy, or filled with smog.

When the sinking from above is vigorous, the PBL inversion grows in thickness. This has the effect of hindering the development of thunderstorms, which depend on rapidly rising air. This situation often occurs over southern California, and thus the chance of thunderstorms forming there is small. Emissions from both the biosphere and from anthropogenic activities accumulate in this part of the atmosphere, and pollution may build up to such an extent that health warnings are required. In locations free of strong temperature inversions, convection processes are strong enough, particularly during the summer months, that emissions are scavenged and quickly lifted by thunderstorms to regions high above the PBL. Often, acidic compounds from these emissions are returned to the surface in the precipitation that falls.

Biosphere controls on maximum temperatures by evaporation and transpiration

Solar radiation is converted to sensible and latent heat at the Earth's surface. A change in sensible heat results in a change in the temperature of a medium, whereas energy stored as latent heat is used to drive a process, such as a phase change in a substance from its liquid to its gaseous state, and does not produce a change in temperature. Thus, the daily maximum surface temperature at a given location is dependent on the amount of radiant energy converted to sensible heat. Water available for evaporation increases latent heating by adding water vapour to the atmosphere. As a result, relatively little energy remains to heat the air, and thus the sensible heating of the air near the ground is minimized. Daily maximum temperatures are therefore not as high in locations with strong latent heating.

As day length increases from winter to summer, sensible heating and maximum surface temperatures rise. In the US Midwest, prior to the leafing out of vegetation in the springtime and the resulting rise in transpiration, sensible heating causes an average increase in maximum surface temperatures of about 0.3°C (0.5°F) per day. The process of leaf production creates a surge in transpiration and results in increased latent heating and reduced sensible heating. After leafing, since most of the available thermal energy is used to convert liquid water to water vapour rather than to heat the air, the average day-to-day rise in daily maximum temperatures is reduced to about 0.1°C (0.2°F) per day.

This effect extends upward through the atmosphere. Prior to leafing out, the one-kilometre-thick layer occurring between the 850-to-750-millibar pressure level (which typically occurs between 1,650 and 2,750 metres [5,400–9,000 feet]) in the Midwest was found to warm at the rate of 0.1°C (0.2°F) per day. Following leafing out, the warming rate fell to 0.02°C (0.04°F) per day. Scientists have used computer models of the atmosphere to study the effect of transpiration from vegetation on maximum surface air temperatures. In these models, the variable controlling transpiration by vegetation was "turned off" and the character of the resulting modelled climate was studied. By subtracting the effect of transpiration, temperatures in central North America and on the other continents were predicted to equilibrate at a very hot 45°C (113°F). Such warming is nearly realized in desert areas where moisture is unavailable for transpiration.

Biosphere controls on minimum temperatures
During the late 1860s, the British experimental physicist John Tyndall, based on his studies of the infrared radiation absorption by atmospheric gases, concluded that night-time minimum temperatures were dependent on the concentration of trace gases in the atmosphere. Of these gases, water vapour had the greatest impact. To emphasize the significance of water vapour on decreases in air temperature during the night, he wrote that if all the water vapour in the air over England was removed even for a single night, it would be

"attended by the destruction of every plant which a freezing temperature could kill." It follows that the greater the water content of the atmosphere, the lower the radiative loss of energy to the sky and the less the surface atmosphere is cooled. Locations with substantial amounts of water vapour experience reduced nocturnal cooling.

Water vapour in the atmosphere also influences the limit to which temperatures fall at night. This limiting temperature is known as the dew point, which is defined as the temperature at which condensation begins. Over North America east of the 100th meridian (a line of longitude traditionally dividing the moist eastern part of North America from the drier western areas), average night-time minimum temperatures are within a degree or two of the dew point temperature. Upon nocturnal cooling, the dew point is reached, condensation begins, and latent energy is converted to heat. Additional temperature falls are retarded by this release of heat to the atmosphere. A significant fraction of the water in the atmosphere over the continents comes from the evaporation of water from soils and the transpiration from vegetation. Transpired water directly moderates temperature by increasing humidity and thus raising the dew point. As a consequence, the amount of outgoing terrestrial radiation released to space is reduced. This results in the elevation of the minimum temperature of the air.

The effect of spring leafing on the build-up of humidity in the lower atmosphere has received the attention of researchers in recent years. In the late 1980s, the American climatologists M.D. Schwartz and T.R. Karl used the superimposed epoch method to study the climate before and after the leafing out of lilac plants in the spring in the US Midwest. (This method uses time-series data from multiple locations, which can be compared with one another by adjusting each data set around the respective onset date of, in this case, lilac blooming.) Prior to the average date of leafing, the atmospheric humidity (vapour pressure) was found to be relatively constant and minimum temperatures to hover near freezing. At leafing, there was an abrupt increase in atmospheric humidity. Following leafing, daily minimum temperatures also increased abruptly. Although frosts are possible until 10 June in many parts of the Midwest, the chances of frost decline as the atmosphere is humidified.

Climate and changes in the albedo of the Earth's surface

The amount of solar energy available at the Earth's surface for sensible and latent heating of the atmosphere depends on the albedo, or the reflectivity, of the surface. Surface albedos vary by location, season, and land-cover type. The albedo of unvegetated ground devoid of snow ranges from 0.1 to 0.6 (10–60 per cent), while the albedo of fully forested lands ranges from 0.08 to 0.15. An increase of 0.1 in regional albedo has been associated with a 20 per cent decline in rainfall events connected with thunderstorms. Equivalent reductions in both evaporation and transpiration have also been reported in areas with sudden increases in albedo.

The greatest changes in albedo occur in regions undergoing desertification and deforestation. Depending on the albedo of the underlying soil, reductions in vegetative land cover may give rise to albedo increases of as much as 0.2. Model studies of the vegetative zone known as the Sahel in Africa reveal that albedo increased from 0.14 to 0.35 due to desertification occurring during the twentieth century. This coincided with a 40 per cent decrease in rainfall. In addition, it is likely that the clearing of forests and prairies for agricultural crops over the past several hundred years has altered the albedo of extensive regions of the middle latitudes.

Contemporary agricultural practices give rise to large variations in albedo from season to season as the land passes through the cycle of tilling, planting, crop growth, and harvest. At larger scales, an agricultural mosaic often emerges as each different plot of ground is covered by plantings of a single species. Viewed from the air, landscapes in the middle latitudes appear as a heterogeneous mix of forests, grasslands, meadows, water bodies, farmlands, wetlands, and urban types. The resultant patchiness in the landscape produces a patchiness in surface albedo, and thereby a mix in the fluxes of sensible and latent heat to the atmosphere. Such changes to the heat flux have been shown to cause changes in the timing, intensity, and frequency of summer thunderstorms.

Biosphere controls on surface friction and localized winds

Averaged annually over Earth's entire surface, the sun provides about 342 watts of energy per square metre. About 30 per cent of this energy is reflected away to space and is never used in the Earth–atmosphere system. Of that which remains, a little less than 1 per cent (3.1 watts per square metre) accelerates the air by generating winds. An equal amount of energy must eventually be lost, or else wind speeds would perpetually increase.

Earth as a thermodynamic system is dissipative – the mechanical energy of the winds is eventually converted to heat through friction. Over the continents, it is the combination of terrain and the veneer of vegetation that offers the frictional roughness to dissipate the surface winds and convert this kinetic energy (the energy it has by virtue of being in motion) into heat. Marine winds approaching the British Isles average about 12 metres per second (27 miles per hour), but they are decelerated to 6 metres per second (13 miles per hour) because of the friction of the landscape's surface shortly after the winds make landfall. Without vegetation cover, the continents would offer much less friction to the wind – wind speeds in unvegetated landscapes are nearly twice as high as those in vegetated landscapes.

The correct specification of Earth's surface roughness due to vegetation, for use in computer models of the atmosphere, is critical to the accurate modelling of the patterns of Earth's winds, global geography, and rainfall. When modelling newly desertified areas, such as the Sahel, it is important to understand that desertification creates vegetation of lower stature and thus lower surface-roughness values. Both wind velocities and wind direction could change from previous patterns over landscapes with taller vegetation.

The extent of this impact of the biosphere on the atmosphere is revealed in climate model studies. One such study modelled the influence of reduced vegetation on surface roughness over the Indian subcontinent and provided evidence for a weaker monsoon and reduced rainfall. Given that much of the northwest third of India underwent a severe desertification and cultural collapse near the beginning of recorded history, the role cultures play in vegetation reduction and climate change should not be ignored.

The vegetation cover of the continents is not passive in response to the winds. Greenhouse-grown trees subjected to mechanical forces designed to mimic the winds lay down new woody tissue called "reaction wood", which results in a stiffer tree over time. This material helps trees become more resilient and offer more frictional resistance to wind. This negative feedback, where increased winds result in stiffer vegetation and thereby subsequently reduced wind speeds, might well apply at the global scale, by balancing an increase in energy used to heat and accelerate the air with an increase in the surface friction needed to dissipate it.

Biosphere impacts on precipitation processes

Cloud condensation nuclei

The formation and subsequent freezing of cloud droplets depend on the presence of cloud condensation nuclei and ice nuclei, respectively. Significantly, the biosphere is a major source of both of these kinds of nuclei. Over the continents, condensation nuclei are readily available and are of biogenic as well as anthropogenic origin. Examples of condensation nuclei include sea salt, small soil particles, and dust.

As atmospheric convection increases with the heating of the day, cloud condensation nuclei are mixed into and above the planetary boundary layer and into the troposphere. In the bottom 0.5 km (1,600 feet) of the atmosphere, nuclei typically number 2.2×10^{10} per cubic metre. In the next 0.5 km (1,600–3,300 feet), half as many nuclei are found. The number of condensation nuclei continues to decline with increased altitude. Furthermore, in general, the number of nuclei in the air over land is ten times higher than that over the oceans.

Cloud condensation nuclei are generally abundant. They do not limit cloud formation over the continents, although low numbers of condensation nuclei over the oceans may limit cloud formation there. In addition to natural sources, particulates from fuel combustion and sulphur dioxide gas resulting from high-sulphur fuels also contribute to the load of condensation nuclei over the continents. Both the number and kind of condensation nuclei present in the atmosphere affect the cloudiness and the brightness of clouds in a given region. In this way, condensation nuclei play a significant role in determining both regional and global albedo.

There is a type of condensation nuclei that forms in the marine air over the margins of continents. Though these nuclei are often few in number, they play a large role in cloud formation near the coasts of continents and may contribute significantly to both planetary albedo and global average temperature. Typically, sources of condensation nuclei in marine air are sulphate aerosols formed from the biogenic production of dimethyl sulphide (DMS; C_2H_6S) by marine algae. Given that DMS production increases with sea-surface temperatures, a negative feedback may result. That is, warmer waters result in the increased production of condensation nuclei by phytoplankton and thus produce more clouds. Increased cloudiness shades the ocean surface and results in lower temperatures that limit condensation nuclei production. It is estimated that a 30 per cent increase in marine condensation nuclei would increase planetary albedo by 0.005 (0.5 per cent), or produce a 0.7 per cent reduction in solar radiation and a planetary average temperature decrease of 1.3°C (2.3°F). The extent to which this negative feedback may influence planetary temperatures remains in active debate.

Biogenic ice nuclei

As water vapour condenses onto condensation nuclei, the droplets grow in size. Growth proceeds at relative humidity as low as 70 per cent, but the rate of growth is very slow. Growth by condensation is most rapid where the air is slightly supersaturated with water vapour. At this point, cloud droplets typical of the size of fog droplets arise. Should temperatures fall to the level where freezing begins, the temperature difference between the droplet and the surrounding air (the vapour pressure deficit) strongly favours rapid condensation into the crystalline lattice of an ice particle. Ice particles that grow rapidly soon reach sizes where they begin to fall. As they fall, they collide and merge with smaller droplets and thereby grow larger.

The formation of ice is of critical importance. Upon ice formation, heat energy in the order of 80 calories per gram of water frozen is released. This energy increases the sensible heat of the air and causes the air to become more buoyant. The process of ice formation encourages convection, cloudiness, and precipitation from clouds.

A droplet of pure water, such as distilled water, will automatically freeze in the atmosphere at a temperature of -40°C (-40°F). Freezing at warmer temperatures requires a substance upon which ice crystallization can take place. The common clay mineral kaolinite, a contaminant of the droplet, raises this freezing point to around -25°C (-13°F). Silver iodide (often used in cloud seeding to encourage rainfall) and sea salts also cause ice to form at -25°C. Freezing at still warmer temperatures is most common with biogenic ice nuclei, a major source of which is the decomposition of organic matter. Ice-crystal formation has been shown to occur at temperatures as warm as -2°C to -3°C (28.5–26.6°F) when biogenic ice nuclei are involved. The common freezing temperature for biogenic nuclei varies systematically according to biome (the largest geographic biotic unit) and latitude – the coldest freezing-temperature nuclei occur above the tropics, whereas the warmest occur above the Arctic. The production of biogenic nuclei from organic matter decomposition is greatest during the warm months when bacterial decomposition is greatest.

Freezing produces greater buoyancy of the particles and helps them to reach higher vertical velocities within the clouds.

The vertical motions and the larger droplet size that occur with biogenic materials favour the charge separation needed to produce lightning. As a result, oceanic areas with few biogenic ice nuclei are also areas of low lightning frequency.

Recycled rainfall

The water that is transpired into the atmosphere from the biosphere is eventually returned to the surface as precipitation. This vegetation-transpiration component of the hydrologic cycle is referred to as "recycled rainfall". While the oceans are the major source of atmospheric water vapour and rainfall, water from plant transpiration is also significant. For example, in the 1970s and 1980s, analyses performed by the American meteorologist Michael Garstang on the city of Manaus, Brazil, in the Amazon basin revealed that around 20 per cent of the precipitation came from water transpired by vegetation; the remaining 80 per cent of this precipitation (an estimate made by the German American meteorologist Heinz Lettau in the 1970s) was generated by the Atlantic Ocean. Isotopic studies of rainwater collected at various points in the Amazon basin indicated that nearly half of the total rain came from water originating in the ocean and half transpired through the vegetation. Evidence of the proportion of transpired water in rainfall reaching as high as 88 per cent has been reported for the Amazon foothills of the Andes. General climate circulation models indicate that, without transpired water from plants, rainfall in the central regions of the continents would be greatly reduced. As a general rule, the further the distance from oceanic water sources, the higher the fraction of rainwater originating from transpiration.

Urban climate

Urban climates are distinguished from those of less built-up areas by differences of air temperature, humidity, wind speed and direction, and amount of precipitation. These differences are attributable largely to the altering of the natural terrain through the construction of artificial structures and surfaces. For example, tall buildings, paved streets, and parking lots affect wind flow, precipitation run-off, and the energy balance of a locale.

Also characteristic of the atmosphere over urban centres are substantially higher concentrations of pollutants such as carbon monoxide, the oxides of sulphur and nitrogen, hydrocarbons, oxidants, and particulate matter. Foreign matter of this kind is introduced into the air by industrial processes (e.g. chemical discharges by oil refineries), fuel combustion (for the operation of motor vehicles and for the heating of offices and factories), and the burning of solid wastes. Urban pollution concentrations depend on the magnitude of local emissions' sources and the prevailing meteorological ventilation of the area – i.e. the height of the atmospheric layer through which the pollutants are being mixed and the average wind speed through that layer. Heavy concentrations of air pollutants have a considerable impact on temperature, visibility, and precipitation in and around cities. Moreover, there occasionally arise weather conditions that allow the accumulation of pollutants over an urban area for several days. Such conditions strongly inhibit atmospheric mixing and can cause acute distress in the population and even, under extremely severe conditions, loss of life. Atmospheric inversion caused an air-pollution disaster in London in December 1952 in which about 3,500 people died from respiratory diseases.

The centre of a city is warmer than are outlying areas. Daily minimum temperature readings at related urban and rural sites frequently show that the urban site is 6–11°C (10–20°F) warmer than the rural site. Two primary processes influence the formation of this "heat island". Firstly, during summer, masonry and asphalt absorb, store, and reradiate more solar energy per unit area than do vegetation and soil. Furthermore, less of this energy can be used for evaporation in urban areas, which characteristically exhibit greater precipitation run-off from streets and buildings. At night, radiative losses from urban building and street materials keep the city's air warmer than that of rural areas. Secondly, during winter the urban atmosphere is warmed slightly, but significantly, by energy from fuel combustion for home heating, power generation, industry, and transportation. Also contributing to the warmer urban atmosphere is the blanket of pollutants and water vapour that absorbs a portion of thermal radiation emitted by the Earth's surface. Part of the absorbed radiation warms the surrounding air – a process that tends to stabilize the air over a city, which in turn increases the probability of higher pollutant concentrations.

The average relative humidity in cities is usually several per cent lower than that of adjacent rural areas, primarily because of increased run-off of precipitation and the lack of transpiration from vegetation in urban areas. Some moisture, however, is added to urban atmospheres by the many combustion sources.

The flow of wind through a city is characterized by mean speeds that are 20–30 per cent lower than those of winds blowing across the adjacent countryside. This difference occurs as a result of the increased frictional drag on air flowing over built-up urban terrain, which is rougher than rural areas. Another difference between urban and rural wind flow is the convergence of low-level wind over a city (i.e. air tends to flow into a city from all directions). This is caused primarily by the horizontal thermal gradients of the urban heat island.

The amount of solar radiation received by cities is reduced by the blanket of particulates in the overlying atmosphere. The higher particulate concentrations in urban atmospheres reduce visibility by both scattering and absorbing light. In addition, some particles provide opportunities for the condensation of water vapour to form water droplets, the ingredients of fog.

A city also influences precipitation patterns in its vicinity. Such city-generated or city-modified weather factors as wind turbulence, thermal convection, and high concentrations of condensation nuclei might be expected to increase precipitation. Although appropriate continuous, quantitative measurements have not been made for a sufficient length of time, there is some data to suggest that the amount of precipitation over many large cities is about 5–10 per cent greater than that over nearby rural areas, with the greatest increases occurring downwind of the city centre.

Bioclimatology and palaeoclimatology

Bioclimatology is the branch of climatology that deals with the effects of the physical environment on living organisms over an extended period of time. Although Hippocrates touched on these matters 2,000 years ago in his treatise on *Air, Waters, and Places*, the science of bioclimatology is relatively new. It developed into a significant field of study during the 1960s owing largely to a growing concern over the deteriorating environment.

Because almost every aspect of climate and weather has some effect on living organisms, the scope of bioclimatology is almost limitless. Certain areas are emphasized more than others, however – among them studies of the influence of weather and climate on small plant organisms and insects responsible for the development of plant, animal, and human diseases; the influence of weather and climate on physiological processes in normal healthy humans and their diseases; the influence of microclimate in dwellings and urban centres on human health; and the influence of past climatic conditions on the development and distribution of plants, animals, and humans.

The scientific study of the climatic conditions of past geological ages is called palaeoclimatology (also spelled paleoclimatology). Palaeoclimatologists seek to explain climate variations for all parts of Earth during any given geological period, beginning with the time of Earth's formation. Many related fields contribute to the field of palaeoclimatology, but the basic research data are drawn mainly from geology and palaeobotany. Speculative attempts at explanation have come largely from astronomy, atmospheric physics, meteorology, and geophysics.

Two major factors in the study of both ancient and present-day climatic conditions of Earth are the changes in the relationship between Earth and the sun (e.g. the slight alteration in the configuration of Earth's orbit) and the changes in the surface of the planet itself (such phenomena as volcanic eruptions, mountain-building events, the transformations of plant communities, and the dispersal of the continents after the break-up of the supercontinent Pangea). Some of the questions that were studied in the past have been largely explained. Palaeoclimatologists have found, for example, that the warmth of the northern hemispheric landmasses during at least 90 per cent of the past 570 million years is mainly due to the drift of the continents across the latitudes: until about 150 million years ago, both North America and Europe were much closer to the equator than they are today. Other questions, such as the reasons behind the irregular advances and retreats of the ice sheets (i.e. glacial and interglacial episodes), are much more difficult to explain, and no completely satisfactory theory has been presented.

All historical sciences share a problem: as they probe further back in time, they become more reliant on fragmentary and indirect evidence. Earth-system history is no exception. High-quality instrumental records spanning the past century exist for most parts of the world, but the records become sparse in the nineteenth century, and, although some very rare records date back over 1,000 years, few predate the late eighteenth century. Other historical documents, including ships' logs, diaries, court and church records, and tax rolls, can sometimes be used. Within strict geographic contexts, these sources can provide information on frosts, droughts, floods, sea ice, the dates of monsoons, and other climatic features, perhaps up to several hundred years ago.

Fortunately, climatic change also leaves a variety of signatures in the natural world. Researchers studying climatic changes predating the instrumental record rely increasingly on these natural archives – biological or geological processes that record some aspect of past climate. Natural archives, often referred to as proxy evidence, are extraordinarily diverse: they include, but are not limited to, fossil records of past plant and animal distributions, sedimentary and geochemical indicators of former conditions of oceans and continents, and land-surface features characteristic of past climates. Palaeoclimatologists study these natural archives by collecting cores, or cylindrical samples, of sediments from lakes, bogs, and oceans; by studying surface features and geological strata; by examining tree-ring patterns from cores or sections of living and dead trees; by drilling into marine corals and cave stalagmites; by drilling into the ice sheets of Antarctica and Greenland and into the Plateau of Tibet, the

Andes, and other montane regions; and by a wide variety of other means.

Most of the evidence of past climatic change is circumstantial, so palaeoclimatology involves a great deal of investigative work. Wherever possible, palaeoclimatologists try to use multiple lines of evidence to cross-check their conclusions. They are frequently confronted with conflicting evidence, but this, as in other sciences, usually leads to an enhanced understanding of the Earth system and its complex history. New sources of data, analytical tools, and instruments are becoming available, and the field is moving quickly. Ongoing climatic changes are being monitored by networks of sensors in space, on the land surface, and both on and below the surface of the world's oceans. Revolutionary changes in the understanding of Earth's climate history have occurred since the 1990s, and coming decades are expected to bring many new insights and interpretations.

The following sections explore the patterns of Earth's climatic variation over various timescales – from the human lifespan to the scale of human history and over geologic time. Climatic changes have constantly influenced human cultures, their survival and geographical expansion, and their agricultural and ecological practices.

Climatic variation and change within a human lifespan

Regardless of their locations on the planet, all humans experience climate variability and change within their lifetimes. The most familiar and predictable phenomena are the seasonal cycles, to which people adjust their clothing, outdoor activities, thermostats, and agricultural practices. However, no two summers or winters are exactly alike in the same place: some are warmer, wetter, or stormier than others. This interannual variation in climate is partly responsible for year-to-year variations in fuel prices, crop yields, road maintenance budgets, and wildfire hazards. Single-year, precipitation-driven floods can cause severe economic damage, such as those of the upper Mississippi River drainage basin during the summer of 1993; and loss of life, such as those that devastated much of Bangladesh in the summer of 1998. Similar damage and loss of life can also occur as the result of wildfires, severe storms, hurricanes, heat waves, and other climate-related events.

Climate variation and change may also occur over longer periods, e.g. over decades. Some locations experience multiple years of drought, floods, or other harsh conditions. Such decadal variation of climate poses challenges to human activities and planning. For example, multiyear droughts can disrupt water supplies, induce crop failures, and cause economic and social dislocation, as in the case of the Dust Bowl droughts in the midcontinent of North America during the 1930s. Multiyear droughts may even cause widespread starvation, as in the Sahel drought that occurred in northern Africa during the 1970s and 1980s.

Seasonal variation

Every place on Earth experiences seasonal variation in climate, although the shift can be slight in some tropical regions. This cyclic variation is driven by seasonal changes in the supply of solar radiation to Earth's atmosphere and surface. Earth's orbit around the sun is elliptical: it is closer to the sun (147 million km [about 91 million miles]) near the winter solstice and further from the sun (152 million km [about 94 million miles]) near the summer solstice in the northern hemisphere.

Furthermore, Earth's axis of rotation occurs at an oblique angle (23.5°) with respect to its orbit. Thus, each hemisphere is tilted away from the sun during its winter period and toward the sun in its summer period. When a hemisphere is tilted away from the sun, it receives less solar radiation than the opposite hemisphere, which at that time is pointed toward the sun. Thus, despite the closer proximity of the sun at the winter solstice, the northern hemisphere receives less solar radiation during the winter than it does during the summer.

Earth's climate system is driven by solar radiation: seasonal differences in climate ultimately result from the seasonal changes in Earth's orbit. The circulation of air in the atmosphere and water in the oceans responds to seasonal variations of available energy from the sun. Specific seasonal changes in climate occurring at any given location on Earth's surface largely result from the transfer of energy from atmospheric and oceanic circulation. Differences in surface heating taking place between summer and winter cause storm tracks and pressure centres to shift position and strength. These heating differences also drive seasonal changes in cloudiness, precipitation, and wind.

Seasonal responses of the biosphere (especially vegetation) and cryosphere (glaciers, sea ice, and snowfields) also feed into atmospheric circulation and climate. Leaf fall by deciduous trees as they go into winter dormancy increases the albedo (reflectivity) of Earth's surface and may lead to greater local and regional cooling (see "Climate and changes in the albedo of the Earth's surface", page 114). Similarly, snow accumulation also increases the albedo of land surfaces and often amplifies winter's effects.

Interannual variation

Interannual climate variations, including droughts, floods, and other events, are caused by a complex array of factors and Earth-system interactions. One important feature that plays a role in these variations is the periodic change of atmospheric and oceanic circulation patterns in the tropical Indo-Pacific region, collectively known as the El Niño–Southern Oscillation (ENSO) variation (see Chapter 5, page 264). Although its primary climatic effects are concentrated in the tropical Pacific, ENSO has cascading effects that often extend to the Atlantic Ocean region, the interior of Europe and Asia, and the polar regions. These effects, called teleconnections, occur because alterations in low-latitude atmospheric circulation patterns in the Pacific region influence atmospheric circulation in adjacent and downstream systems. As a result, storm tracks are diverted and atmospheric pressure ridges (areas of high pressure) and troughs (areas of low pressure) are displaced from their usual patterns.

As an example, El Niño events occur when the easterly trade winds in the tropical Pacific weaken or reverse direction. This shuts down the upwelling of deep, cold waters off the west coast of South America, warms the eastern Pacific, and reverses the atmospheric pressure gradient in the western Pacific. As a result, air at the surface moves eastward from Australia and Indonesia toward the central Pacific and the Americas. These changes produce high rainfall and flash floods along the normally arid coast of Peru, and severe drought in the normally wet regions of northern Australia and Indonesia. Particularly severe El Niño events lead to monsoon failure in the Indian Ocean region, resulting in intense drought in India and East Africa. At the same time, the westerlies and storm tracks are displaced toward the equator, providing California and the desert Southwest of the United States with wet, stormy winter weather and causing winter conditions in the Pacific Northwest, which are typically wet, to become warmer and drier. Displacement of the westerlies also results in drought in northern China and from north-eastern Brazil through sections of Venezuela. Long-term records of ENSO variation from historical documents, tree rings, and reef corals indicate that El Niño events occur, on average, every two to seven years. However, the frequency and intensity of these events vary through time.

The North Atlantic Oscillation (NAO) is another example of an interannual oscillation that produces important climatic effects within the Earth system and can influence climate throughout the northern hemisphere. This phenomenon results from variation in the pressure gradient, or the difference in atmospheric pressure between the subtropical high (usually situated between the Azores and Gibraltar) and the Icelandic low (centred between Iceland and Greenland). When the pressure gradient is steep due to a strong subtropical high and a deep Icelandic low (positive phase), northern Europe and northern Asia experience warm, wet winters with frequent strong winter storms. At the same time, southern Europe is dry. The eastern United States also experiences warmer, less snowy winters during positive NAO phases, although the effect is not as great as in Europe. The pressure gradient is dampened when NAO is in a negative mode – that is, when a weaker pressure gradient exists from the presence of a weak subtropical high and Icelandic low. When this happens, the Mediterranean region receives abundant winter rainfall, while northern Europe is cold and dry. The eastern United States is typically colder and snowier during a negative NAO phase.

The ENSO and NAO cycles are driven by feedbacks and interactions between the oceans and atmosphere. Interannual climate variation is driven by these and other cycles, interactions among cycles, and perturbations in the Earth system, such as those resulting from large injections of aerosols from volcanic eruptions. One example of a perturbation due to volcanism is the 1991 eruption of Mount Pinatubo in the Philippines, which led to a decrease in the average global temperature of approximately 0.5°C (0.9°F) the following summer.

Decadal variation

Climate also varies on decadal timescales, with multiyear clusters of wet, dry, cool, or warm conditions. These clusters can have dramatic effects on human activities and welfare. For instance, a severe three-year drought in the late sixteenth century probably contributed to the destruction of Sir Walter Raleigh's "Lost Colony" at Roanoke Island in what is now North Carolina, and a subsequent seven-year drought (1606–12) led to high mortality at the Jamestown Colony in Virginia. Some scholars have implicated persistent and severe droughts as the main reason for the collapse of the Maya civilization in Mesoamerica between AD 750 and 950 (although discoveries in the early twenty-first century suggest that war-related trade disruptions played a role, possibly interacting with famines and other drought-related stresses).

Although decadal-scale climate variation is well documented, the causes are not entirely clear. Much decadal variation in climate is related to interannual variations. For example, the frequency and magnitude of ENSO change through time. The early 1990s were characterized by repeated El Niño events, and several such clusters have been identified as having taken place during the twentieth century. The steepness of the NAO gradient also changes at decadal timescales; it has been particularly steep since the 1970s, for example.

Recent research has revealed that decadal-scale variations in climate result from interactions between the ocean and the atmosphere. One such variation is the Pacific Decadal Oscillation (PDO), also referred to as the Pacific Decadal Variability

(PDV), which involves changing sea-surface temperatures in the North Pacific Ocean. These temperatures influence the strength and position of the Aleutian Low, which in turn strongly affects precipitation patterns along the Pacific Coast of North America. PDO variation consists of an alternation between "cool-phase" periods, when coastal Alaska is relatively dry and the Pacific Northwest relatively wet (e.g. 1947–76), and "warm-phase" periods, characterized by relatively high precipitation in coastal Alaska and low precipitation in the Pacific Northwest (e.g. 1925–46; 1977–98). Tree-ring and coral records, which span at least the last four centuries, document this variation.

A similar oscillation, the Atlantic Multidecadal Oscillation (AMO), occurs in the North Atlantic and strongly influences precipitation patterns in eastern and central North America. A warm-phase oscillation (relatively warm North Atlantic sea-surface temperatures) is associated with relatively high rainfall in Florida and low rainfall in much of the Ohio Valley. However, it interacts with PDO, and both interact with interannual variations, such as ENSO and NAO, in complex ways. Such interactions may lead to the amplification of droughts, floods, or other climatic anomalies. For example, severe droughts over much of the 48 states of the continental United States in the first few years of the twenty-first century were associated with warm-phase Atlantic Multidecadal combined with cool-phase Pacific Decadal Oscillations.

The mechanisms underlying decadal variations such as these are poorly understood, but they are probably related to ocean–atmosphere interactions with larger time constants than interannual variations. Decadal climatic variations are the subject of intense study by climatologists and palaeoclimatologists.

Climatic variation and change
since the emergence of civilization

Human societies have experienced climate change since the
development of agriculture some 10,000 years ago, and these
changes have often had profound effects on cultures and societies.
They include annual and decadal climate fluctuations such as
those described above, as well as large-magnitude changes that
occur over centennial to multimillennial timescales. Such changes
are believed to have influenced and even stimulated the initial
cultivation and domestication of crop plants, as well as the
domestication and pastoralization of animals. Human societies
have changed adaptively in response to climate variations,
although evidence abounds that certain societies and civilizations
have collapsed in the face of rapid and severe climatic changes.

Centennial-scale variation

Historical records as well as proxy records (particularly tree
rings, corals, and ice cores) indicate that climate has changed
during the past 1,000 years at centennial timescales – that is,
no two centuries have been exactly alike. During the past 150
years, the Earth system has emerged from a period called the
Little Ice Age, which was characterized in the North Atlantic
region and elsewhere by relatively cool temperatures. The
twentieth century in particular saw a substantial pattern of
warming in many regions. Some of this warming may be
attributable to the transition from the Little Ice Age or other
natural causes. However, many climate scientists believe that
much of the twentieth-century warming, especially in the later
decades, has resulted from atmospheric accumulation of
greenhouse gases, in particular carbon dioxide.

The Little Ice Age was most noticeable in Europe and the
North Atlantic region, which experienced relatively cool

conditions between the early fourteenth and mid-nineteenth centuries. This was not a period of uniformly cool climate, since interannual and decadal variability brought many warm years. Furthermore, the coldest periods did not always coincide among different regions – some regions experienced relatively warm conditions at the same time that others were subjected to severely cold conditions. Alpine glaciers advanced far below their previous (and present) limits, obliterating farms, churches, and villages in Switzerland, France, and elsewhere. Frequent cold winters and cool, wet summers ruined wine harvests and led to crop failures and famines over much of northern and central Europe. The North Atlantic cod fisheries declined as ocean temperatures fell in the seventeenth century. The Norse colonies on the coast of Greenland were cut off from the rest of Norse civilization during the early fifteenth century as pack ice and storminess increased in the North Atlantic. The western colony of Greenland collapsed through starvation, and the eastern colony was abandoned. In addition, Iceland became increasingly isolated from Scandinavia.

The Little Ice Age was preceded by a period of relatively mild conditions in northern and central Europe. This interval, known as the Medieval Warm Period, occurred from approximately AD 1000 to the first half of the thirteenth century. Mild summers and winters led to good harvests in much of Europe. Wheat cultivation and vineyards flourished at far higher latitudes and elevations than today. Norse colonies in Iceland and Greenland prospered, and Norse parties fished, hunted, and explored the coasts of Labrador and Newfoundland. The Medieval Warm Period is well documented in much of the North Atlantic region, with evidence including ice cores from Greenland. Like the Little Ice Age, this time was neither a climatically uniform period nor a period of uniformly warm temperatures everywhere in the world. Other regions of the globe lack evidence for high temperatures during this period.

Much scientific attention continues to be devoted to a series of severe droughts that occurred between the eleventh and fourteenth centuries. These droughts, each spanning several decades, are well documented in tree-ring records across western North America and in the peatland records of the Great Lakes region. The records appear to be related to ocean temperature anomalies in the Pacific and Atlantic basins, but they are still inadequately understood. The information suggests that much of the United States is susceptible to persistent droughts that would be devastating for water resources and agriculture.

Millennial and multimillennial variation

The climatic changes of the past thousand years are superimposed upon variations and trends at both millennial and greater timescales. Numerous indicators from eastern North America and Europe show trends of increased cooling and increased effective moisture during the past 3,000 years. For example, in the Great Lakes–St Lawrence regions along the US–Canadian border, water levels of the lakes rose, peatlands developed and expanded, moisture-loving trees such as beech and hemlock expanded their ranges westward, and populations of boreal trees, such as spruce and tamarack, increased and expanded southward. These patterns all indicate a trend of increased effective moisture, which may be a sign of increased precipitation, decreased transpiration due to cooling, or both. The patterns do not necessarily indicate a monolithic cooling event: more complex climatic changes probably occurred. For example, beech expanded northward and spruce southward during the past 3,000 years in both eastern North America and western Europe. The beech expansions may indicate milder winters or longer growing seasons, whereas the spruce expansions appear related to cooler, moister summers. Palaeoclimatologists are applying a variety of approaches and proxies to help identify such changes in seasonal temperature and moisture during the Holocene Epoch (the interglacial period that began around 10,000 years ago).

Just as the Little Ice Age was not associated with cool conditions everywhere, so the cooling and moistening trend of the past 3,000 years was not universal. Some regions became warmer and drier during the same time period. For example, northern Mexico and the Yucatan experienced decreasing moisture in the past 3,000 years. Heterogeneity of this type is characteristic of climatic change, which involves changing patterns of atmospheric circulation. As circulation patterns change, the transport of heat and moisture in the atmosphere also changes, resulting in the apparent paradox of opposing temperature and moisture trends in different regions.

The trends of the past 3,000 years are just the latest in a series of climatic changes that occurred over the Holocene Epoch. At the start of the Holocene, remnants of continental glaciers from the last glaciation still covered much of eastern and central Canada and parts of Scandinavia. These ice sheets had largely disappeared by 6,000 years ago. Their absence – along with increasing sea-surface temperatures, rising sea levels (as glacial meltwater flowed into the world's oceans), and especially changes in the radiation budget of Earth's surface owing to Milankovitch variations (changes in the seasons resulting from periodic adjustments of Earth's orbit around the sun) – affected atmospheric circulation. The diverse changes of the past 10,000 years across the globe are difficult to summarize, but some general highlights and large-scale patterns are worthy of note. These include the presence of early-to-mid-Holocene thermal maxima in various locations, variation in ENSO patterns, and an early-to-mid-Holocene amplification of the Indian Ocean monsoon.

Thermal maxima

Many parts of the globe experienced higher temperatures than today some time during the early-to-mid-Holocene. In some cases the increased temperatures were accompanied by decreased moisture availability. Although the thermal maximum has been referred to in North America and elsewhere as a single widespread event (variously termed the "altithermal", "xerothermic interval", "climatic optimum", or "thermal optimum"), it is now recognized that the periods of maximum temperatures varied among regions. For example, north-western Canada experienced its highest temperatures several thousand years earlier than central or eastern North America. Similar heterogeneity is seen in moisture records. For instance, the record of the prairie-forest boundary in the Midwestern region of the United States shows eastward expansion of prairie in Iowa and Illinois 6,000 years ago (indicating increasingly dry conditions), whereas Minnesota's forests expanded westward into prairie regions at the same time (indicating increasing moisture). The Atacama Desert, located primarily in present-day Chile and Bolivia, on the western side of South America, is one of the driest places on Earth today, but it was much wetter during the early Holocene when many other regions were at their driest.

The primary driver of changes in temperature and moisture during the Holocene was orbital variation, which slowly changed the latitudinal and seasonal distribution of solar radiation on Earth's surface and atmosphere. However, the heterogeneity of these changes was caused by changing patterns of atmospheric circulation and ocean currents.

ENSO variation in the Holocene

Because of the global importance of El Niño–Southern Oscillation (ENSO) variation today (see Chapter 5, page 264), Holocene variation in ENSO patterns and intensity is under serious study by palaeoclimatologists. The record is still fragmentary, but accumulating evidence from fossil corals, tree rings, lake records, climate modelling, and other approaches suggests that (1) ENSO variation was relatively weak in the early Holocene, (2) ENSO has undergone centennial-to-millennial variations in strength during the past 10,000 years, and (3) ENSO patterns and strength similar to those currently in place developed within the past 5,000 years. This evidence is particularly clear when comparing ENSO variation over the past 3,000 years to today's patterns. The causes of long-term ENSO variation are still being explored, but changes in solar radiation owing to Milankovitch variations are strongly implicated by modelling studies.

Amplification of the Indian Ocean monsoon

Much of Africa, the Middle East, and the Indian subcontinent are under the strong influence of an annual climatic cycle known as the Indian Ocean monsoon. The climate of this region is highly seasonal, alternating between clear skies with dry air (winter) and cloudy skies with abundant rainfall (summer). Monsoon intensity, like other aspects of climate, is subject to interannual, decadal, and centennial variations, at least some of which are related to ENSO and other cycles. Abundant evidence exists for large variations in monsoon intensity during the Holocene Epoch. Palaeontological and palaeoecological studies (studies of the geologic past from fossils, and of prehistoric ecosystems from fossils, respectively) show that large portions of the region experienced much greater precipitation during the early Holocene

(10,000–6,000 years ago) than today. Lake and wetland sediments dating to this period have been found under the sands of parts of the Sahara Desert. These sediments contain fossils of elephants, crocodiles, hippopotamuses, and giraffes, together with pollen evidence of forest and woodland vegetation. In arid and semi-arid parts of Africa, Arabia, and India, large and deep freshwater lakes occurred in basins that are now dry or are occupied by shallow, saline lakes. Civilizations based on plant cultivation and grazing animals, such as the Harappan civilization of north-western India and adjacent Pakistan, flourished in these regions, which have since become arid.

These and similar lines of evidence, together with palaeontological and geochemical data from marine sediments and climate-modelling studies, indicate that the Indian Ocean monsoon was greatly amplified during the early Holocene, supplying abundant moisture far inland into the African and Asian continents. This amplification was driven by high solar radiation in summer, which was approximately 7 per cent higher 10,000 years ago than today and resulted from orbital forcing (changes in Earth's eccentricity, equinoctial precession [the orientation of its rotational axis], and axial tilt). High summer insolation resulted in warmer summer air temperatures and lower surface pressure over continental regions and, hence, increased inflow of moisture-laden air from the Indian Ocean to the continental interiors. Modelling studies indicate that the monsoonal flow was further amplified by feedbacks involving the atmosphere, vegetation, and soils. Increased moisture led to wetter soils and lusher vegetation, which in turn led to increased precipitation and greater penetration of moist air into continental interiors. Decreasing summer insolation during the past 4,000–6,000 years led to the weakening of the Indian Ocean monsoon.

Climatic variation and change since the advent of humans

The history of humanity – from the initial appearance of genus *Homo* over 2 million years ago to the advent and expansion of the modern human species (*Homo sapiens*), beginning some 150,000 years ago – is integrally linked to climate variation and change. *H. sapiens* has experienced nearly two full glacial-interglacial cycles, but its global geographical expansion, massive population increase, cultural diversification, and worldwide ecological domination began only during the last glacial period and accelerated during the last glacial-interglacial transition. The first bipedal apes appeared in a time of climatic transition and variation, and *H. erectus*, an extinct species possibly ancestral to modern humans, originated during the colder Pleistocene Epoch and survived both the transition period and multiple glacial-interglacial cycles. Thus, it can be said that climate variation has been the midwife of humanity and its various cultures and civilizations.

Recent glacial and interglacial periods
Most recent glacial phase
With glacial ice restricted to high latitudes and altitudes, the Earth 125,000 years ago was in an interglacial period similar to the one occurring today. (The current interglacial period, known as the Holocene Epoch, started 10,000–11,000 years ago.) During the past 125,000 years, however, the Earth system went through an entire glacial-interglacial cycle – just the most recent of many that have place over the last million years. The most recent period of cooling and glaciation began approximately 120,000 years ago. Significant ice sheets developed and persisted over much of Canada and northern Eurasia.

After the initial development of glacial conditions, the Earth system alternated between two modes, one of cold temperatures and growing glaciers, and the other of relatively warm temperatures (although much cooler than today) and retreating glaciers. These Dansgaard-Oeschger cycles, recorded both in ice cores and marine sediments, occurred approximately every 1,500 years. A lower-frequency cycle, called the Bond cycle, is superimposed on the pattern of Dansgaard-Oeschger cycles; Bond cycles occurred every 3,000–8,000 years. Each Bond cycle is characterized by unusually cold conditions that take place during the cold phase of a Dansgaard-Oeschger cycle, the subsequent Heinrich event (which is a brief dry and cold phase), and the rapid warming phase that follows each Heinrich event. During each Heinrich event, massive fleets of icebergs are released into the North Atlantic, carrying rocks picked up by the glaciers far out to sea. Heinrich events are marked in marine sediments by conspicuous layers of iceberg-transported rock fragments.

Many of the transitions in the Dansgaard-Oeschger and Bond cycles were rapid and abrupt, and they are being studied intensely by palaeoclimatologists and Earth-system scientists to understand the driving mechanisms of such dramatic climatic variations. These cycles now appear to result from interactions between the atmosphere, oceans, ice sheets, and continental rivers that influence thermohaline circulation (the pattern of ocean currents driven by differences in water density, salinity, and temperature, rather than wind). Thermohaline circulation, in turn, controls ocean heat transport such as the Gulf Stream.

Last Glacial Maximum

During the past 25,000 years, the Earth system has undergone a series of dramatic transitions. The most recent glacial period peaked 21,500 years ago during the Last Glacial Maximum, or LGM. At that time, the northern third of North America was covered by the Laurentide Ice Sheet, which extended as far south as Des Moines, Iowa; Cincinnati, Ohio; and New York City. The Cordilleran Ice Sheet covered much of western Canada as well as northern Washington, Idaho, and Montana in the United States. In Europe the Scandinavian Ice Sheet sat atop the British Isles, Scandinavia, north-eastern Europe, and north-central Siberia. Montane glaciers were extensive in other regions, even at low latitudes in Africa and South America. The global sea level was 125 metres (410 feet) below modern levels, because of the long-term net transfer of water from the oceans to the ice sheets. Temperatures near Earth's surface in unglaciated regions were about 5°C (9°F) cooler than today. Many northern hemisphere plant and animal species inhabited areas far south of their present ranges. For example, jack pine and white spruce trees grew in north-western Georgia, 1,000 km (600 miles) south of their modern range limits in the Great Lakes region of North America.

Last deglaciation

The continental ice sheets began to melt back about 20,000 years ago. Drilling and dating of submerged fossil coral reefs provide a clear record of increasing sea levels as the ice melted. The most rapid melting began 15,000 years ago. For example, the southern boundary of the Laurentide Ice Sheet in North America was north of the Great Lakes and St Lawrence regions by 10,000 years ago, and had completely disappeared by 6,000 years ago.

The warming trend was punctuated by transient cooling events, most notably the Younger Dryas climate interval of 12,800–11,600 years ago (see "Abrupt climate changes in Earth's history", page 155). The climatic regimes that developed during the deglaciation period in many areas, including much of North America, have no modern analogue (i.e. no regions exist with comparable seasonal regimes of temperature and moisture). For example, in the interior of North America, climates were much more continental (characterized by warm summers and cold winters) than they are today. Also, palaeontological studies indicate assemblages of plant, insect, and vertebrate species that do not occur anywhere today. Spruce trees grew with temperate hardwoods (ash, hornbeam, oak, and elm) in the upper Mississippi River and Ohio River regions. In Alaska, birch and poplar grew in woodlands, and there were very few of the spruce trees that dominate the present-day Alaskan landscape. Boreal and temperate mammals, whose geographic ranges are widely separated today, co-existed in central North America and Russia during this period of deglaciation. These unparalleled climatic conditions probably resulted from the combination of a unique orbital pattern that increased summer insolation and reduced winter insolation in the northern hemisphere, and the continued presence of northern hemisphere ice sheets, which themselves altered atmospheric circulation patterns.

Climatic change and the emergence of agriculture

The first known examples of animal domestication occurred in western Asia between 11,000 and 9,500 years ago, when goats and sheep were first herded. Examples of plant domestication date to 9,000 years ago, when wheat, lentils, rye, and barley were first cultivated. This phase of technological increase occurred during a time of climatic transition that followed the last glacial period. A number of scientists have suggested that, although climate change imposed stresses on hunter-gatherer-forager societies by causing rapid shifts in resources, it also provided opportunities as new plant and animal resources appeared.

Glacial and interglacial cycles of the Pleistocene

The glacial period that peaked 21,500 years ago was only the most recent of five glacial periods in the last 450,000 years. In fact, the Earth system has alternated between glacial and interglacial regimes for more than two million years – a period of time known as the Pleistocene. The duration and severity of the glacial periods increased during this period, with a particularly sharp change occurring between 900,000 and 600,000 years ago.

The continental glaciations of the Pleistocene left signatures on the landscape in the form of glacial deposits and landforms, but the best knowledge of the magnitude and timing of the various glacial and interglacial periods comes from oxygen isotope records in ocean sediments. These records provide both a direct measure of sea level and an indirect measure of global ice volume. Water molecules composed of a lighter isotope of oxygen, ^{16}O, are evaporated more readily than molecules bearing a heavier isotope, ^{18}O. Glacial periods are characterized by high ^{18}O concentrations and represent a net transfer of water, especially with ^{16}O, from the oceans to the ice sheets. Oxygen isotope records indicate that interglacial periods have typically lasted 10,000–15,000 years, and that maximum glacial periods were of similar length. Most of the past 500,000 years – approximately 80 per cent – have been spent within various intermediate glacial states that were warmer than glacial maxima but cooler than interglacial periods. During these intermediate times, substantial glaciers occurred over much of Canada and probably covered Scandinavia as well. These intermediate states were not constant; they were characterized by continual, millennial-scale climate variation. There has been no average or typical state for global climate during Pleistocene and Holocene times – the Earth system has been in continual flux between interglacial and glacial patterns.

The cycling of the Earth system between glacial and interglacial modes has been ultimately driven by orbital variations. However, orbital forcing is by itself insufficient to explain all of this variation, and Earth-system scientists are focusing their attention on the interactions and feedbacks between the myriad components of the system. For example, the initial development of a continental ice sheet increases albedo over a portion of Earth, reducing surface absorption of sunlight and leading to further cooling. Similarly, changes in terrestrial vegetation, such as the replacement of forests by tundra, feed back into the atmosphere via changes in both albedo and latent heat flux from evapotranspiration. Forests – particularly those of tropical and temperate areas, with their large leaf area – release great amounts of water vapour through transpiration. Tundra plants, which are much smaller, possess tiny leaves designed to slow water loss; they release only a small fraction of the water vapour that forests do.

The discovery in ice-core records that atmospheric concentrations of two potent greenhouse gases, carbon dioxide and methane, have decreased during past glacial periods and peaked during interglacials indicates important feedback processes in the Earth system. A reduction in greenhouse gas concentrations during the transition to a glacial phase would reinforce and amplify cooling already under way. The reverse is true for the transition to interglacial periods. The glacial carbon sink remains a topic of considerable research activity. A full understanding of glacial-interglacial carbon dynamics requires knowledge of the complex interplay between ocean chemistry and circulation, the ecology of marine and terrestrial organisms, ice sheet dynamics, and atmospheric chemistry and circulation.

Last Great Cooling

The Earth system has undergone a general cooling trend for the past 50 million years, culminating in the development of permanent ice sheets in the northern hemisphere about 2.75 million years ago. These ice sheets expanded and contracted in a regular rhythm, with each glacial maximum separated from adjacent ones by 41,000 years (based on the cycle of Earth's axial tilt). As the ice sheets waxed and waned, global climate drifted steadily toward cooler conditions, characterized by increasingly severe glaciations and increasingly cool interglacial phases. Beginning around 900,000 years ago, the glacial-interglacial cycles shifted frequency. Ever since, the glacial peaks have been 100,000 years apart, and the Earth system has spent more time in cool phases than before. The 41,000-year periodicity has continued, with smaller fluctuations superimposed on the 100,000-year cycle. In addition, a smaller, 23,000-year, cycle has occurred through both the 41,000-year and 100,000-year cycles.

The 23,000-year and 41,000-year cycles are driven ultimately by two components of Earth's orbital geometry: the equinoctial precession cycle (23,000 years) and the axial-tilt cycle (41,000 years). Although the third parameter of Earth's orbit, eccentricity, varies on a 100,000-year cycle, its magnitude is insufficient to explain the 100,000-year cycles of glacial and interglacial periods of the past 900,000 years. The origin of the periodicity present in Earth's eccentricity is an important question in current palaeoclimate research.

Climatic variation and change through geologic time

The Earth system has undergone dramatic changes throughout its 4.5-billion-year history. These have included climatic changes diverse in mechanisms, magnitudes, rates, and consequences. Many of these past changes are obscure and controversial, and some have been discovered only recently. Nevertheless, the history of life has been strongly influenced by these changes, some of which radically altered the course of evolution. Life itself is implicated as a causative agent of some of these changes, as the processes of photosynthesis and respiration have largely shaped the chemistry of Earth's atmosphere, oceans, and sediments (see "Climate and life", page 98).

Cenozoic climates

The Cenozoic Era, encompassing the 65 million years that have elapsed since the mass extinction event marking the end of the Cretaceous Period, has shown a broad range of climatic variation characterized by alternating intervals of global warming and cooling. Earth has experienced both extreme warmth and extreme cold during this period. These changes have been driven by tectonic forces, which have altered the positions and elevations of the continents as well as ocean passages and bathymetry (measurement of ocean depth). Feedbacks between different components of the Earth system (atmosphere, biosphere, lithosphere [Earth's crust], cryosphere [glaciers, sea ice, and snow fields], and hydrosphere) are being increasingly recognized as influences of global and regional climate. In particular, the atmospheric concentration of carbon dioxide has varied substantially during the Cenozoic for reasons that are poorly understood, though its fluctuation must have involved feedbacks between the planet's different spheres.

Orbital forcing is also evident in the Cenozoic, although, when compared on such a vast era-level timescale, orbital variations can be seen as oscillations against a slowly changing backdrop of lower-frequency climatic trends. Descriptions of the orbital variations have evolved according to the growing understanding of tectonic and biogeochemical changes. A pattern emerging from recent palaeoclimatologic studies suggests that the climatic effects of eccentricity, equinoctial precession, and axial tilt have been amplified during cool phases of the Cenozoic, whereas they have been dampened during warm phases.

The meteor impact that occurred at or very close to the end of the Cretaceous Period came at a time of global warming, which continued into the early Cenozoic. Tropical and subtropical flora and fauna occurred at high latitudes until at least 40 million years ago, and geochemical records of marine sediments have indicated the presence of warm oceans. The interval of maximum temperature occurred during the late Palaeocene and early Eocene epochs (48 million to 59 million years ago). The highest global temperatures of the Cenozoic occurred during the Palaeocene-Eocene Thermal Maximum (PETM), a short interval lasting approximately 100,000 years. Although the underlying causes are unclear, the onset of the PETM about 56 million years ago was rapid, occurring within a few thousand years, and ecological consequences were large, with widespread extinctions in both marine and terrestrial ecosystems. Sea surface and continental air temperatures increased by more than 5°C (9°F) during the transition into the PETM. Sea-surface temperatures in the high-latitude Arctic may have been as warm as 23°C (73°F), comparable to modern subtropical and warm-temperate seas. Following the PETM, global temperatures declined to pre-PETM levels, but they gradually increased again to near-PETM levels over the next few million years, during a period known as the Eocene Optimum. This temperature maximum was followed by a steady decline in global temperatures toward the Eocene-Oligocene boundary, which occurred about 34 million years ago. These changes are well represented in marine sediments and in palaeontological records from the continents, where vegetation zones moved equatorward.

Mechanisms underlying the cooling trend are under study, but it is most likely that tectonic movements played an important role. This period saw the gradual opening of the sea passage between Tasmania and Antarctica, followed by the opening of the Drake Passage between South America and Antarctica. The latter, which isolated Antarctica within a cold polar sea, produced global effects on atmospheric and oceanic circulation. Recent evidence suggests that decreasing atmospheric concentrations of carbon dioxide during this period may have initiated a steady and irreversible cooling trend over the next few million years.

A continental ice sheet developed in Antarctica during the Oligocene Epoch, persisting until a rapid warming event took place 27 million years ago. The late Oligocene and early-to-mid-Miocene epochs (26 million to 15 million years ago) were relatively warm, though not nearly as warm as the Eocene. Cooling resumed 15 million years ago, and the Antarctic Ice Sheet expanded again to cover much of the continent. The cooling trend continued through the late Miocene and accelerated into the early Pliocene Epoch, 5.3 million years ago. During this period the northern hemisphere remained ice free, and palaeobotanical studies show cool-temperate Pliocene flora at high latitudes on Greenland and the Arctic Archipelago. The northern hemisphere glaciation, which began 3.2 million years ago, was driven by tectonic events, such as the closing of the Panama seaway and the uplift of the Andes, the Tibetan Plateau, and western parts of North America. These tectonic events led to changes in the circulation of the oceans and the atmosphere, which in turn fostered the development of persistent ice at high northern latitudes. Small-magnitude variations in carbon dioxide concentrations, which had been relatively low since at least the mid-Oligocene Epoch (about 28 million years ago), are also thought to have contributed to this glaciation.

Phanerozoic climates

The Phanerozoic Eon (542 million years ago to the present), which includes the entire span of complex, multicellular life on Earth, has witnessed an extraordinary array of climatic states and transitions. The sheer antiquity of many of these regimes and events renders them difficult to understand in detail. However, a number of periods and transitions are well known, owing to good geological records and intense study by scientists. Furthermore, a coherent pattern of low-frequency climatic variation is emerging, in which the Earth system alternates between warm ("greenhouse") phases and cool ("icehouse") phases. The warm phases are characterized by high temperatures, high sea levels, and an absence of continental glaciers. Cool phases are marked by low temperatures, low sea levels, and the presence of continental ice sheets, at least at high latitudes. Superimposed on these alternations are higher-frequency variations, where cool periods are embedded within greenhouse phases and warm periods are embedded within icehouse phases. For example, glaciers developed for a brief period (between 1 million and 10 million years) during the late Ordovician and early Silurian (about 430 million years ago), in the middle of the early Palaeozoic greenhouse phase (540 million to 350 million years ago). Similarly, warm periods with glacial retreat occurred within the late Cenozoic cool period during the late Oligocene and early Miocene epochs.

The Earth system has been in an icehouse phase for the past 30 million to 35 million years, ever since the development of ice sheets on Antarctica. The previous major icehouse phase occurred between about 350 million and 250 million years ago, during the Carboniferous and Permian periods of the late Palaeozoic Era. Glacial sediments dating to this period have been identified in much of Africa as well as in the Arabian Peninsula, South America, Australia, India, and Antarctica. At the time, all these regions were part of Gondwana, a high-latitude supercontinent in the southern hemisphere. The glaciers atop Gondwana extended to at least 45°S latitude, similar to the latitude reached by northern hemisphere ice sheets during the Pleistocene. Some late Palaeozoic glaciers extended even further equatorward, to 35°S. One of the most striking features of this time period are cyclothems, repeating sedimentary beds of alternating sandstone, shale, coal, and limestone. The great coal deposits of North America's Appalachian region, the American Midwest, and northern Europe are interbedded in these cyclothems, which may represent repeated transgressions (producing limestone) and retreats (producing shales and coals) of ocean shorelines in response to orbital variations.

The two most prominent warm phases in Earth's history occurred during the Mesozoic and early Cenozoic eras (approximately 250 million to 35 million years ago) and the early and mid-Palaeozoic (approximately 500 million to 350 million years ago). Climates of each of these greenhouse periods were distinct; continental positions and ocean bathymetry were very different, and terrestrial vegetation was absent from the continents until relatively late in the Palaeozoic warm period. Both of these periods experienced substantial long-term climate variation and change; increasing evidence indicates brief glacial episodes during the mid-Mesozoic.

Understanding the mechanisms underlying icehouse-greenhouse dynamics is an important area of research, involving an interchange between geologic records and the modelling of the Earth system and its components. Two processes have been implicated as drivers of Phanerozoic climate change. First, tectonic forces caused changes in the positions and elevations of continents and the bathymetry of oceans and seas. Second, variations in greenhouse gases were also important, though at these long timescales they were largely controlled by tectonic processes, during which sinks and sources of greenhouse gases varied.

Climates of early Earth

The pre-Phanerozoic interval, also known as Precambrian time, comprises some 88 per cent of the time elapsed since the origin of Earth. The pre-Phanerozoic is a poorly understood phase of Earth system history. Much of the sedimentary record of the atmosphere, oceans, biota, and crust of the early Earth has been obliterated by erosion, metamorphism, and subduction (where a continental and an oceanic plate meet and the former slides over the latter). However, a number of pre-Phanerozoic records have been found in various parts of the world, mainly from the later portions of the period. Pre-Phanerozoic Earth system history is an extremely active area of research, in part because of its importance in understanding the origin and early evolution of life on Earth. Furthermore, the chemical composition of Earth's atmosphere and oceans largely developed during this period, with living organisms playing an active role. Geologists, palaeontologists, microbiologists, planetary geologists, atmospheric scientists, and geochemists are focusing intense efforts on understanding this period. Three areas of particular interest and debate are the "faint young sun paradox", the role of organisms in shaping Earth's atmosphere, and the possibility that Earth went through one or more "snowball" phases of global glaciation.

Faint young sun paradox

Astrophysical studies indicate that the luminosity of the sun was much lower during Earth's early history than it has been in the Phanerozoic. In fact, radiative output was low enough to suggest that all surface water on Earth should have been frozen solid during its early history, but evidence shows that it was not. The solution to this "faint young sun paradox" appears to lie in the presence of unusually high concentrations of greenhouse gases at the time, particularly methane and carbon dioxide. As solar luminosity gradually increased over time, if these concentrations had remained at their early high levels, Earth may have heated up beyond a life-sustaining threshold. Therefore, greenhouse gas concentrations must have decreased proportionally with increasing solar radiation, implying a feedback mechanism to regulate greenhouse gases. One of these mechanisms might have been rock weathering, which is temperature dependent and serves as an important sink for, rather than source of, carbon dioxide by removing sizable amounts of this gas from the atmosphere. Scientists are also looking to biological processes (many of which also serve as carbon dioxide sinks) as complementary or alternative regulating mechanisms of greenhouse gases on the young Earth. (See also "The Gaia hypothesis", Chapter 8.)

Photosynthesis and atmospheric chemistry

The evolution by photosynthetic bacteria of a new photosynthetic pathway, substituting water (H_2O) for hydrogen sulphide (H_2S) as a reducing agent for carbon dioxide, had dramatic consequences for Earth-system geochemistry.

Molecular oxygen (O_2) is given off as a by-product of photo-synthesis using the H_2O pathway, which is energetically more efficient than the more primitive H_2S pathway. Using H_2O as a reducing agent in this process led to the large-scale deposition of banded-iron formations, or BIFs, a source of 90 per cent of present-day iron ores. Oxygen present in ancient oceans oxidized dissolved iron, which precipitated out of solution onto the ocean floors. This deposition process, in which oxygen was used up as fast as it was produced, continued for millions of years until most of the iron dissolved in the oceans was precipitated. By approximately 2 billion years ago, oxygen was able to accumulate in dissolved form in seawater and to outgas to the atmosphere. Although oxygen does not have greenhouse gas properties, it plays important indirect roles in Earth's climate, particularly in phases of the carbon cycle. Scientists are studying the role of oxygen and other contributions of early life to the development of the Earth system.

Snowball Earth hypothesis

Geochemical and sedimentary evidence indicates that Earth experienced as many as four extreme cooling events between 750 million and 580 million years ago. Geologists have proposed that Earth's oceans and land surfaces were covered by ice from the poles to the equator during these events. This "Snowball Earth" hypothesis is a subject of intense study and discussion. Two important questions arise from the hypoth-esis. First, how, once frozen, could Earth thaw? Second, how could life survive periods of global freezing? A proposed solution to the first question involves the outgassing of massive amounts of carbon dioxide by volcanoes, which could have warmed the planetary surface rapidly, especially given that major carbon dioxide sinks (rock weathering and photosynthesis) would have been dampened by a frozen Earth. A possible answer to the second question may lie in the existence of present-day life forms within hot springs and deep-sea vents, which would have persisted long ago despite the frozen state of Earth's surface.

A counter-premise known as the "Slushball Earth" hypothesis contends that Earth was not completely frozen over. Rather, in addition to massive ice sheets covering the continents, parts of the planet (especially ocean areas near the equator) could have been draped only by a thin, watery layer of ice amid areas of open sea. Under this scenario, photosynthetic organisms in low-ice or ice-free regions could continue to capture sunlight efficiently and survive these periods of extreme cold.

Abrupt climate changes in Earth's history

An important new area of research, abrupt climate change, has developed since the 1980s. This research has been inspired by the discovery, in the ice-core records of Greenland and Antarctica, of evidence for abrupt shifts in regional and global climates of the past. These events, which have also been documented in ocean and continental records, involve sudden shifts of Earth's climate system from one equilibrium state to another. Such shifts are of considerable scientific concern because they can reveal something about the controls and sensitivity of the climate system. In particular, they point out nonlinearities, the so-called "tipping points", where small, gradual changes in one component of the system can lead to a large change in the entire system. Such nonlinearities arise from the complex feedbacks between components of the Earth system. Abrupt climate shifts are of great societal concern, for any such shifts in the future might be so rapid and radical as to outstrip the capacity of agricultural, ecological, industrial, and economic systems to respond and adapt. Climate scientists are working with social scientists, ecologists, and economists to assess society's vulnerability to such "climate surprises".

The most intensely studied and best-understood example of abrupt climate change is the Younger Dryas event, 12,800 to 11,600 years ago. This event took place during the last deglaciation, a period of global warming when the Earth system was in transition from a glacial mode to an interglacial one. The Younger Dryas was marked by a sharp drop in temperatures in the North Atlantic region – cooling in northern Europe and eastern North America is estimated at 4–8°C (7.2–14.4°F). Terrestrial and marine records indicate that the Younger Dryas had detectable effects of lesser magnitude over most other regions of Earth. The termination of the Younger Dryas was very rapid, occurring within a decade. The event resulted from an abrupt shutdown of the thermohaline circulation in the North Atlantic, which is critical for the transport of heat from equatorial regions northward (today the Gulf Stream is a part of that circulation). The cause of this shutdown is under study; an influx of large volumes of freshwater from melting glaciers into the North Atlantic has been implicated, although other factors probably played a role.

Palaeoclimatologists are devoting increasing attention to identifying and studying other abrupt changes. The Dansgaard-Oeschger cycles of the last glacial period are now recognized as representing alternation between two climate states, with rapid transitions from one state to the other. A 200-year-long cooling event in the northern hemisphere approximately 8,200 years ago resulted from the rapid draining of the glacial Lake Agassiz into the North Atlantic via the Great Lakes and St Lawrence drainage. This event, characterized as a miniature version of the Younger Dryas, had ecological impacts in Europe and North America that included a rapid decline of hemlock populations in New England forests. In addition, evidence of another such transition, marked by a rapid drop in the water levels of lakes and bogs in eastern North America, occurred 5,200 years ago. It is recorded in ice cores from glaciers at high altitudes in tropical regions as well as tree-ring, lake-level, and peatland samples from temperate regions.

Abrupt climatic changes occurring before the Pleistocene have also been documented. A transient thermal maximum has been documented near the Palaeocene–Eocene boundary (56 million years ago), and evidence of rapid cooling events are observed near the boundaries between both the Eocene and Oligocene epochs (34 million years ago) and the Oligocene and Miocene epochs (23 million years ago). All three of these events had global ecological, climatic, and biogeochemical consequences. Geochemical evidence indicates that the warm event occurring at the Palaeocene–Eocene boundary was associated with a rapid increase in atmospheric carbon dioxide concentrations, possibly resulting from the massive outgassing and oxidation of methane hydrates (a compound whose chemical structure traps methane within a lattice of ice) from the ocean floor. The two cooling events appear to have resulted from a transient series of positive feedbacks among the atmosphere, oceans, ice sheets, and biosphere, similar to those observed in the Pleistocene. Other abrupt changes, such as the Palaeocene–Eocene Thermal Maximum, are recorded at various points in the Phanerozoic.

Abrupt climate changes can evidently be caused by a variety of processes. Rapid changes in an external factor can push the climate system into a new mode. Examples of such external forcing are the outgassing of methane hydrates and the sudden influx of glacial meltwater into the ocean. Alternatively, gradual changes in external factors can lead to the crossing of a threshold: the climate system is unable to return to the former equilibrium and passes rapidly to a new one. Such nonlinear system behaviour is a potential concern as human activities, such as fossil-fuel combustion and land-use change, alter important components of Earth's climate system.

Humans and other species have survived countless climatic changes in the past, and humans are a notably adaptable species. Adjustment to climatic changes, whether biological (as in the case of other species) or cultural (for humans), is easiest and least catastrophic when the changes are gradual and can be largely anticipated. Rapid changes are more difficult to adapt to and incur more disruption and risk. Abrupt changes, especially unanticipated "climate surprises", put human cultures and societies, as well as both the populations of other species and the ecosystems they inhabit, at considerable risk of extreme disruption. Such changes may well be within humanity's capacity to adapt to, but not without paying severe penalties in the form of economic, ecological, agricultural, human health, and other costs. Knowledge of past climate variability provides guidelines on the natural variability and sensitivity of the Earth system. This knowledge also helps in the identification of the risks associated with altering the Earth system with greenhouse-gas emissions and regional-to-global-scale changes in land cover.

3

A HISTORY OF WEATHER FORECASTING

Weather forecasting is the prediction of the weather through the application of the principles of physics, supplemented by a variety of statistical and empirical techniques. In addition to predictions of atmospheric phenomena themselves, weather forecasting includes predictions of changes on the Earth's surface caused by atmospheric conditions, e.g. snow and ice cover, storm tides, and floods.

General considerations

Measurements as the basis for weather prediction

The observations of few other scientific enterprises are as vital or affect as many people as those related to weather forecasting. From the days when early humans ventured from caves and other natural shelters, perceptive individuals in all likelihood became leaders by being able to detect nature's signs of impending snow, rain, or wind; indeed, of any change in weather. With such information they must have enjoyed greater success in the search for food and safety – the major objectives of that time.

In a sense, weather forecasting is still carried out in basically the same way as it was by the earliest humans – by making observations and predicting changes. The results of the modern tools used to measure temperature, pressure, wind, and humidity in the twenty-first century are obviously better. But even the most sophisticated numerically calculated forecast made on a supercomputer requires a set of measurements of the condition of the atmosphere – an initial picture of temperature, wind, and other basic elements, somewhat comparable to that formed by our forebears when they looked out of their cave dwellings. The primeval approach entailed insights based on the accumulated experience of the perceptive observer, while the modern technique consists of solving equations. In each case the forecaster asks "What is?" in the sense of "What kind of weather prevails today?" and then seeks to determine how it will change in order to extrapolate what it will be.

Because observations are so critical to weather prediction, an account of meteorological measurements and weather forecasting is a story in which ideas and technology are closely intertwined, with creative thinkers drawing new insights from available observations and pointing to the need for new or better measurements, and technology providing the means for making new observations and for processing the data derived from measurements. The basis for weather prediction started with the theories of the ancient Greek philosophers and continued with Renaissance scientists, the scientific revolution of the seventeenth and eighteenth centuries, and the theoretical models of twentieth- and twenty-first-century atmospheric scientists and meteorologists. Likewise, it tells of the development of the "synoptic" idea – that of characterizing the weather over a large region at exactly the same time in order to organize information about prevailing conditions. In synoptic meteorology, simultaneous observations for a specific time are plotted on a map for a broad area, from which a general view of the weather in that region is gained. (The term synoptic is derived from the Greek word meaning "general or comprehensive view".) The so-called synoptic weather map came to be the principal tool of nineteenth-century meteorologists and continues to be used today in weather stations and on television weather reports around the world.

Since the mid-twentieth century, digital computers have made it possible to calculate changes in atmospheric conditions mathematically and objectively – i.e. in such a way that anyone can obtain the same result from the same initial conditions. The widespread adoption of numerical weather-prediction models brought a whole new group of players – computer specialists and experts in numerical processing and statistics – to the scene to work with atmospheric scientists and meteorologists. Moreover, the enhanced capability to process and analyze weather data stimulated the long-standing interest of meteorologists in securing more observations of greater accuracy. Technological advances since the 1960s have led to a growing reliance on remote sensing, particularly the gathering of data with specially instrumented Earth-orbiting satellites. By the late 1980s, forecasts of weather were largely based on the determinations of numerical models integrated by high-speed supercomputers. The exceptions to this were some shorter-range predictions, particularly those related to local thunderstorm activity, which were made by specialists directly interpreting radar and satellite measurements.

Practical applications of weather forecasting

Systematic weather records were kept after instruments for measuring atmospheric conditions became available during the seventeenth century. Undoubtedly these early records were employed mainly by those engaged in agriculture. Planting and harvesting obviously can be planned better and carried out more efficiently if long-term weather patterns can be estimated. In the United States, national weather services were first provided by the Army Signal Corps beginning in 1870. These operations were taken over by the Department of Agriculture in 1891. By the early 1900s a free mail service and the telephone were providing forecasts daily to millions of American farmers. The US Weather Bureau established a Fruit-Frost (forecasting) Service during World War I, and by the 1920s radio broadcasts to agricultural interests were being made in most states.

Weather forecasting became an important tool for aviation during the 1920s and 1930s. Its application in this area gained in importance after Francis W. Reichelderfer was appointed chief of the US Weather Bureau in 1939. Reichelderfer had previously modernized the Navy's meteorological service and made it a model of support for naval aviation. During World War II the discovery of very strong wind currents at high altitudes (the jet streams, which can affect aircraft speed) and the general susceptibility of military operations in Europe to weather led to a special interest in weather forecasting.

One of the most famous wartime forecasting problems was for Operation Overlord, the invasion of the European mainland at Normandy by Allied forces. An unusually intense June storm brought high seas and gales to the French coast, but a moderation of the weather that was successfully predicted by Colonel J.M. Stagg of the British forces (after consultation with both British and American forecasters) enabled General Dwight D. Eisenhower, supreme commander of the Allied Expeditionary Forces, to make his critical decision to invade on June 6, 1944.

The second half of the twentieth century has seen unprecedented growth of commercial weather-forecasting firms in the United States and elsewhere. Marketing organizations and stores commonly hire weather-forecasting consultants to help with the timing of sales and promotions of products from snow tyres and roofing materials to summer clothes and holiday resorts. Many ocean-going shipping vessels as well as military ships use optimum ship-routing forecasts to plan their routes in order to minimize lost time, potential damage, and fuel consumption in heavy seas. Similarly, airlines carefully consider atmospheric conditions when planning long-distance flights so as to avoid the strongest headwinds and to ride with the strongest tailwinds.

International trading of foodstuffs such as wheat, corn (maize), beans, sugar, cocoa, and coffee can be severely affected by weather news. For example, in 1975 a severe freeze in Brazil caused the price of coffee to increase substantially within just a few weeks, and in 1977 a freeze in Florida nearly doubled the price of frozen concentrated orange juice in a matter of days. Weather-forecasting organizations are thus frequently called upon by banks, commodity traders, and food companies to give them advance knowledge of the possibility of such sudden changes. The expenditure on all sorts of commodities and services, whether they are tents for outdoor events or plastic covers for the daily newspapers, can be reduced or eliminated if reliable information about possible precipitation can be obtained in advance.

Forecasts must be quite precise for applications that are tailored to specific industries. Gas and electric utilities, for example, may require forecasts of temperature to within one or two degrees a day ahead of time, or ski-resort operators may need predictions of night-time relative humidity on the slopes of a margin of accuracy within 5–10 per cent in order to schedule snow making.

Early measurements and ideas

The Greek philosophers had much to say about meteorology, and many who subsequently engaged in weather forecasting no doubt made use of their ideas. Unfortunately, they probably made many bad forecasts, because Aristotle, who was the most influential, did not believe that wind is air in motion. He did believe, however, that west winds are cold because they blow from the sunset.

The scientific study of meteorology did not develop until measuring instruments became available. Its beginning is commonly associated with the invention of the mercury barometer by Evangelista Torricelli, an Italian physicist-mathematician, in the mid-seventeenth century, and the nearly concurrent development of a reliable thermometer. (Galileo had constructed an elementary form of gas thermometer in 1607, but it was defective; the efforts of many others finally resulted in a reasonably accurate liquid-in-glass device.)

A succession of notable achievements by chemists and physicists of the seventeenth and eighteenth centuries contributed significantly to meteorological research. The formulation of the laws of gas pressure, temperature, and density by Robert Boyle and Jacques-Alexandre-César Charles, the development of calculus by Sir Isaac Newton and Gottfried Wilhelm Leibniz, the development of the law of partial pressures of mixed gases by John Dalton, and the formulation of the doctrine of latent heat (the release of heat by condensation or freezing) by Joseph Black are just a few of the major scientific break-throughs of the period that made it possible to measure and better understand previously unknown aspects of the atmosphere and its behaviour. During the nineteenth century, all of these illuminating ideas began to produce results in terms of useful weather forecasts.

The emergence of synoptic forecasting methods

Analysis of synoptic weather reports

An observant person who has learned nature's signs can interpret the appearance of the sky, the wind, and other local effects and "foretell the weather". A scientist can use instruments at one location to do so even more effectively. The modern approach to weather forecasting, however, can be realized only when many such observations are exchanged quickly by experts at various weather stations and entered on a synoptic weather map to depict the patterns of pressure, wind, temperature, clouds, and precipitation at a specific time. Such a rapid exchange of weather data became feasible with the development of the electric telegraph in 1837 by Samuel F.B. Morse of the United States. By 1849 Joseph Henry of the Smithsonian Institution in Washington, DC, was plotting daily weather maps based on telegraphic reports, and in 1869 Cleveland Abbe at the Cincinnati Observatory began to provide regular weather forecasts using data received telegraphically.

Synoptic weather maps resolved one of the great controversies of meteorology: the rotary storm dispute. By the early decades of the nineteenth century, it was known that storms were associated with low barometric readings, but the relation of the winds to low-pressure systems, called cyclones, remained unrecognized. William Redfield, a self-taught meteorologist from Middletown, Connecticut, noticed the pattern of fallen trees after a New England hurricane and suggested in 1831 that the wind flow was a rotary counter-clockwise circulation around the centre of lowest pressure. The American meteorologist James P. Espy subsequently proposed in his *Philosophy of Storms* (1841) that air would flow toward the regions of lowest pressure and then would be forced upward, causing clouds and precipitation. Both Redfield and Espy proved to be right. The air does spin around the cyclone, as Redfield believed, while the layers close to the ground flow inward and upward as well. The net result is a rotational wind circulation that is slightly modified at Earth's surface to produce inflow toward the storm centre, just as Espy had proposed. Furthermore, the inflow is associated with clouds and precipitation in regions of low pressure, though that is not the only cause of clouds there.

In Europe the writings of Heinrich Dove, a Polish scientist who directed the Prussian Meteorological Institute, greatly influenced views concerning wind behaviour in storms. Unlike his American contemporaries, Dove did not focus on the pattern of the winds around the storm but rather on how the wind should change at one place as a storm passed. It was many years before his followers understood the complexity of the possible changes.

Establishment of weather-station networks and services

Routine production of synoptic weather maps became possible after networks of stations were organized to take measurements and report them to some type of central observatory. As early as 1814, US Army Medical Corps personnel were ordered to record weather data at their posts; this activity was subsequently expanded and made more systematic. Actual weather-station networks were established in the United States by New York

University, the Franklin Institute, and the Smithsonian Institution during the early decades of the nineteenth century.

In Britain, James Glaisher organized a similar network, as did Christophorus H.D. Buys Ballot in The Netherlands. Other such networks of weather stations were developed near Vienna, Paris, and St Petersburg in Russia.

It was not long before national meteorological services were established on the Continent and in the United Kingdom. The UK's Met Office was founded in 1854 to provide forecasting for the marine community.

The original purpose of the service was to provide storm warnings for the Atlantic and Gulf coasts and for the Great Lakes. Within the next few decades, national meteorological services were set up in such countries as Japan, India, and Brazil. The importance of international cooperation in weather prognostication was recognized by the directors of these services, and by 1880 they had formed the International Meteorological Organization (IMO).

The proliferation of weather-station networks linked by telegraphy made synoptic forecasting a reality by the close of the nineteenth century. Yet the daily weather forecasts generated left much to be desired. Many errors occurred as predictions were largely based on the experience that each individual forecaster had accumulated over several years of practice, vaguely formulated rules of thumb (e.g. of how pressure systems move from one region to another), and associations that were poorly understood, if at all.

Progress during the early twentieth century

An important aspect of weather prediction is to calculate the atmospheric pressure pattern – the positions of the highs and lows and their changes. Modern research has shown that sea-level pressure patterns respond to the motions of the upper-atmospheric winds, with their narrow, fast-moving jet streams and waves that propagate through the air and pass air through themselves.

Frequent surprises and errors in estimating surface atmospheric pressure patterns undoubtedly caused nineteenth-century forecasters to seek information about the upper atmosphere for possible explanations. The British meteorologist Glaisher made a series of ascents by balloon during the 1860s, reaching an unprecedented height of 9 km (5.6 miles). At about this time investigators on the Continent began using unmanned balloons to carry recording barographs, thermographs, and hygrographs to high altitudes. During the late 1890s meteorologists in both the United States and Europe used kites equipped with instruments to probe the atmosphere up to altitudes of about 3 km. Notwithstanding these efforts, knowledge about the upper atmosphere remained very limited at the turn of the century. The situation was aggravated by the confusion created by observations from weather stations located on mountains or hilltops. Such observations often did not show what was expected, partly because so little was known about the upper atmosphere and partly because the mountains themselves affect measurements, producing results that were not representative of what would be found in the free atmosphere at the same altitude.

Fortunately, enough scientists had already advanced ideas that would make it possible for weather forecasters to think three-dimensionally, even if sufficient meteorological measurements were lacking. Henrik Mohn, the first of a long line of highly creative Norwegian meteorologists; Wladimir Köppen, the noted German climatologist; and Max Margules, an influential Russian-born meteorologist, all contributed to the view that mechanisms of the upper air generate the energy of storms.

In 1911 William H. Dines, a British meteorologist, published data that showed how the upper atmosphere compensates for the fact that the low-level winds carry air toward low-pressure centres. Dines recognized that the inflow near the ground is more or less balanced by a circulation upward and outward aloft. Indeed, for a cyclone to intensify, which would require a lowering of central pressure, the outflow must exceed the inflow; the surface winds can converge quite strongly toward the cyclone, but sufficient outflow aloft can produce falling pressure at the centre.

Meteorologists of the time were now aware that vertical circulations and upper-air phenomena were important, but they had still not determined how such knowledge could improve weather forecasting. Then, in 1919, the Norwegian meteorologist Jacob Bjerknes introduced what has been referred to as the Norwegian cyclone model. This theory pulled together many earlier ideas and related the patterns of wind and weather to a low-pressure system that exhibited fronts – sharp sloping boundaries between cold and warm air masses. Bjerknes pointed out the rainfall/snowfall patterns that are characteristically associated with the fronts in cyclones: the rain or snow occurs over large areas on the cold side of an advancing warm front poleward of a low-pressure centre. Here, the winds are from the lower latitudes, and the warm air, being light, glides up over a large region of cold air. Widespread sloping clouds spread ahead of the cyclone, barometers fall as the storm approaches, and precipitation from the rising warm air falls through the cold air below. Where the cold air advances to the rear of the storm, squalls and showers mark the abrupt lifting of the warm air being displaced. Thus, the concept of fronts focused attention on the action at air-mass boundaries. The Norwegian cyclone model could be called the frontal model, for the idea of warm air masses being lifted over cold air along their edges (fronts) became a major forecasting tool. The model not only emphasized the idea but it also showed how and where to apply it.

In later work, Bjerknes and several other members of the so-called Bergen school of meteorology expanded the model to show that cyclones grow from weak disturbances on fronts, pass through a regular life cycle, and ultimately die by the inflow filling them. Both the Norwegian cyclone model and the associated life-cycle concept are still used today by weather forecasters.

While Bjerknes and his Bergen colleagues refined the cyclone model, other Scandinavian meteorologists provided much of the theoretical basis for modern weather prediction. Foremost among them were Vilhelm Bjerknes, Jacob's father; and Carl-Gustaf Rossby. Their ideas helped make it possible to understand and carefully calculate the changes in atmospheric circulation and the motion of the upper-air waves that control the behaviour of cyclones.

Modern trends and developments

Upper-air observations by means of balloon-borne sounding equipment

Developments in technology provided the means to test new scientific ideas and stimulate newer ones. During the late 1920s and 1930s, several groups of investigators (those headed by Yrjö Väisälä of Finland and Pavel Aleksandrovich Malchanov of the Soviet Union, for example) began using small radio transmitters with balloon-borne instruments, eliminating the need to recover the instruments and speeding up access to the upper-air data. These radiosondes, as they came to be called, gave rise to the upper-air observation networks that still exist today. Approximately 75 stations in the United States and more than 500 worldwide release, twice daily, balloons that reach heights of 30,000 metres or more. Observations of temperature and relative humidity at various pressures are radioed back to the station from which the balloons are released as they ascend at a predetermined rate. The balloons are also tracked by radar and global positioning system (GPS) satellites to ascertain the behaviour of winds from their drift.

Forecasters today are able to produce synoptic weather maps of the upper atmosphere twice each day on the basis of radiosonde observations. While new methods of upper-air measurement have been developed, the primary synoptic clock times for producing upper-air maps are still the radiosonde-observation times: 0000 (midnight) and 1200 (noon) Greenwich Mean Time (GMT). Modern computer-based forecasts also use 0000 and 1200 GMT as the starting times from which they calculate the changes that are at the heart of modern forecasts. It is, in effect, the synoptic approach carried out in a different way, intimately linked to the radiosonde networks developed during the 1930s and 1940s.

Application of radar

As in many fields of endeavour, weather prediction experienced several breakthroughs during and immediately after World War II. The British began using microwave radar in the late 1930s to monitor enemy aircraft, but it was soon learned that radar gave excellent returns from raindrops at certain wavelengths (5–10 cm). As a result it became possible to track and study the evolution of individual showers or thunderstorms, as well as to "see" the precipitation structure of larger storms.

Since its initial application in meteorological work, radar has grown as a forecaster's tool. Virtually all tornadoes and severe thunderstorms over the United States and in some other parts of the world are monitored by radar. Radar observation of the growth, motion, and characteristics of such storms provide clues as to their severity. Modern radar systems use the Doppler principle of frequency shift associated with movement toward or away from the radar transmitter/receiver to determine wind speeds as well as storm motions.

Using radar and other observations, the Japanese-American meteorologist Tetsuya T. Fujita discovered many details of severe thunderstorm behaviour and of the structure of the violent local storms common to the Midwest region of the United States. His Doppler-radar analyses of winds revealed "microburst" gusts. These gusts cause the large wind shears (differences in speed and direction) associated with strong rains that have been responsible for some plane crashes.

Other types of radar have been used increasingly for detecting winds continuously, as opposed to twice a day. These wind-profiling radar systems actually pick up signals "reflected" by clear air and so can function even when no clouds or rain are present.

Meteorological measurements from satellites and aircraft

A major breakthrough in meteorological measurement came with the launching of the first meteorological satellite, the TIROS (Television and Infrared Observation Satellite), by the United States on April 1, 1960. The impact of global quantitative views of temperature, cloud, and moisture distributions, as well as of surface properties (e.g. ice cover and soil moisture), has already been substantial. Medium-range forecasts that provide information five to seven days in advance were impossible before satellites began making global observations (particularly over the ocean waters of the southern hemisphere) routinely available in real time. Global forecasting models developed at the US National Center for Atmospheric Research (NCAR), the European Centre for Medium Range Weather Forecasts (ECMWF), and the US National Meteorological Center (NMC) became the standard during the 1980s, making medium-range forecasting a reality. Global weather forecasting models are routinely run by national weather services around the world, including those of Japan, the United Kingdom, and Canada.

Meteorological satellites travel in various orbits and carry a wide variety of sensors. They are of two principal types: the low-flying polar orbiter and the geostationary orbiter. Polar orbiters circle Earth at altitudes of 500–1,000 km and in roughly north–south orbits. They appear overhead at any one locality twice a day and provide very high-resolution data because they fly close to Earth. Such satellites are vital for much of Europe and other high-latitude locations because they orbit near the poles. However, the fact that they can provide a sampling of atmospheric conditions only twice daily is a major limitation.

Geostationary satellites are made to orbit Earth along its equatorial plane at an altitude of about 36,000 kilometres. At that height the eastward motion of the satellite coincides exactly with Earth's rotation, so that the satellite remains in one position above the equator. Satellites of this type are able to provide an almost continuous view of a wide area. Because of this capability, geostationary satellites have yielded new information about the rapid changes that occur in thunderstorms, hurricanes, and certain types of fronts, making them invaluable to weather forecasting as well as meteorological research.

One weakness common to virtually all satellite-borne sensors and to some ground-based radars that use very high frequency (VHF) to ultra-high frequency (UHF) waves is an inability to measure thin layers of the atmosphere. One such layer is the tropopause, the boundary between the relatively dry stratosphere and the more meteorologically active troposphere below. This is often the region of the jet streams. Important information about these kinds of high-speed air currents is obtained with sensors mounted on high-flying commercial aircraft and is routinely included in global weather analyses.

Numerical weather-prediction models

Thinkers frequently advance ideas long before the technology exists to implement them. Few better examples exist than that of numerical weather forecasting. Instead of mental estimates or rules of thumb about the movement of storms, numerical forecasts are objective calculations of changes to the weather map based on sets of physics-based equations called models. Shortly after World War I the British scientist Lewis F. Richardson completed such a forecast that he had been working on for years by tedious and difficult hand calculations. Although the forecast proved to be incorrect, Richardson's general approach was accepted decades later when the electronic computer became available. In fact, it has become the basis for nearly all present-day weather forecasts. Human forecasters may interpret or even modify the results of the computer models, but there are few forecasts that do not begin with numerical-model calculations of pressure, temperature, wind, and humidity for some future time.

The method is closely related to the synoptic approach (see "The emergence of synoptic forecasting methods", page 165). Data are collected rapidly by a Global Telecommunications System for 0000 or 1200 GMT to specify the initial conditions. The model equations are then solved for various segments of the weather map – often a global map – to calculate how much conditions are expected to change in a given time, e.g. 10 minutes. With such changes added to the initial conditions, a new map is generated (in the computer's memory) valid for 0010 or 1210 GMT. This map is treated as a new set of initial conditions, probably not quite as accurate as the measurements for 0000 and 1200 GMT but still very accurate. A new step is undertaken to generate a forecast for 0020 or 1220. This process is repeated step after step. In principle, the process could continue indefinitely. In practice, small errors creep into the calculations, and accumulate. Eventually, the errors become so large by this cumulative process that there is no point in continuing.

Global numerical forecasts are produced regularly (once or twice daily) at the ECMWF, the NMC, and the US military facilities in Omaha, Nebraska, and Monterey, California; as well as in Tokyo, Moscow, London, Melbourne, and elsewhere. In addition, specialized numerical forecasts designed to predict more details of the weather are made for many smaller regions of the world by various national weather services, military organizations, and even a few private companies. Finally, research versions of numerical weather-prediction models are constantly under review, development, and testing at NCAR and at the Goddard Space Flight Center in the United States and at universities in several other nations.

The capacity and complexity of numerical weather-prediction models have increased dramatically since the mid-1940s, when the earliest modelling work was done by the mathematician John von Neumann and the meteorologist Jule Charney at the Institute for Advanced Study in Princeton, New Jersey. Because of their pioneering work and the discovery of important simplifying relationships by other scientists (notably Arnt Eliassen of Norway and Reginald Sutcliffe of Britain), a joint US Weather Bureau, Navy, and Air Force numerical forecasting unit was formed in 1954 in Washington, DC. Referred to as JNWP, this unit was charged with producing operational numerical forecasts on a daily basis.

The era of numerical weather prediction thus really began in the 1950s. As computing power grew, so did the complexity, speed, and capacity for detail of the models. And as new observations became available from such sources as Earth-orbiting satellites, radar systems, and drifting weather balloons, so too did methods sophisticated enough to incorporate the data into the models as improved initial synoptic maps.

Numerical forecasts have improved steadily over the years. The vast Global Weather Experiment, first conceived by Charney, was carried out by many nations in 1979 under the leadership of the World Meteorological Organization to demonstrate what high-quality global observations could do to improve forecasting by numerical prediction models. The results of that effort continue to effect further improvement.

A relatively recent development has been the construction of mesoscale numerical prediction models. These address medium-sized features in the atmosphere, of the scale between large cyclonic storms and individual clouds. Fronts, clusters of thunderstorms, sea breezes, hurricane bands, and jet streams are mesoscale structures, and their evolution and behaviour present crucial forecasting problems that have only recently been dealt with in numerical prediction. An example of such a model is the meso-eta model, developed by the Serbian atmospheric scientist Fedor Mesinger. This model is a finer-scale version of a regional numerical weather prediction model used by the National Weather Service in the United States. The national weather services of several countries produce numerical forecasts of considerable detail by means of such limited-area mesoscale models.

Principles and methodology of weather forecasting

Short-range forecasting

When people wait under a shelter for a downpour to end, they are making a very-short-range weather forecast. They are assuming, based on past experience, that such hard rain will not last very long. In short-term predictions the challenge for the forecaster is to improve on what the layperson can do. For years this type of situation proved particularly vexing for forecasters, but since the mid-1980s they have been developing a method called "nowcasting" to meet precisely this sort of challenge. In this method, radar and satellite observations of local atmospheric conditions are processed and displayed rapidly by computers to project weather only several hours in advance. The US National Oceanic and Atmospheric Administration operates a facility known as PROFS (Program for Regional Observing and Forecasting Services) in Boulder, Colorado, specially equipped for nowcasting.

Another technique for short-range forecasting is called MOS (for Model Output Statistics). Conceived by Harry R. Glahn and D.A. Lowry of the US National Weather Service, this method involves the use of data relating to past weather phenomena and developments to extrapolate the values of certain weather elements, usually for a specific location and time period. It overcomes the weaknesses of numerical models by developing statistical relations between model forecasts and observed weather. These relations are then used to translate the model forecasts directly to specific weather forecasts. For example, a numerical model might not predict the occurrence of surface winds at all, and whatever wind speeds it does predict might always be greater than the speeds actually observed. MOS relations can automatically correct for errors in wind speed and produce quite accurate forecasts of wind occurrence at a specific point, for example at Heathrow Airport near London. As long as numerical weather prediction models are imperfect, there may be many uses for the MOS technique.

Predictive skills and procedures

Short-range weather forecasts generally tend to lose accuracy as forecasters attempt to look further ahead in time. Predictive skill is greatest for periods of about 12 hours and is still quite substantial for 48-hour predictions. An increasingly important group of short-range forecasts are economically motivated. Their value is determined in the marketplace by the economic gains they produce (or the losses they avert).

Weather warnings are a special kind of short-range forecast – the protection of human life is the forecaster's greatest challenge and source of pride. The Weather Bureau in the United States (the predecessor of the National Weather Service) was in fact formed in response to the need for storm warnings on the Great Lakes where hundreds of lives were lost each year. Increase Lapham of Milwaukee urged Congress to take action to reduce the loss and the effectiveness of the warnings and other forecasts assured the future of the American public weather service.

Weather warnings are issued by government and military organizations throughout the world for all kinds of threatening weather events: tropical storms (variously called hurricanes, typhoons, or tropical cyclones, depending on location); great oceanic gales outside the tropics, spanning hundreds of kilometres and at times packing winds comparable to those of tropical storms; and, on land, flash floods, high winds, fog, blizzards, ice, and snowstorms.

A particular effort is made to warn of hail, lightning, and wind gusts associated with severe thunderstorms, sometimes called severe local storms (SELS) or simply "severe weather". Forecasts and warnings are also made for tornadoes, those intense, rotating windstorms that represent the most violent end of the weather scale. Destruction of property and the risk of injury and death are extremely high in the path of a tornado, especially in the case of the largest systems (sometimes called maxi-tornadoes).

Because tornadoes are so uniquely life-threatening and because they are so common in various regions of the United States, the National Weather Service operates a National Severe Storms Forecasting Center in Kansas City, Missouri, where SELS forecasters survey the atmosphere for the conditions that can spawn tornadoes or severe thunderstorms. This group of SELS forecasters, assembled in 1952, monitors temperature and water vapour in an effort to identify the warm, moist regions where thunderstorms may form, and studies maps of pressure and winds to find regions where the storms may organize into mesoscale structures. The group also monitors jet streams and dry air aloft that can combine to distort ordinary thunderstorms into rare rotating ones with tilted chimneys of upward rushing air that, because of the tilt, are unimpeded by heavy falling rain. These high-speed updrafts can quickly transport vast quantities of moisture to the cold upper regions of the storms, thereby promoting the formation of large hailstones. The hail and rain drag down air from aloft to complete a circuit of violent, cooperating updrafts and downdrafts.

By correctly anticipating such conditions, SELS forecasters are able to provide time for the mobilization of special observing networks and personnel. If the storms actually develop, specific warnings are issued based on direct observations. This two-step process consists of the tornado or severe thunderstorm watch, which is the forecast prepared by the SELS forecaster, and the warning, which is usually released by a local observing facility. The watch may be issued when the skies are clear, and it usually covers a number of counties across a given state or states. It alerts the affected area to the threat but does not attempt to pinpoint which communities will be affected. By contrast, the warning is very specific to a locality and calls for immediate action. Radar of various types can be used to detect the large hailstones, the heavy load of raindrops, the relatively clear region of rapid updraft, and even the rotation in a tornado. These indicators, or an actual sighting, often trigger the tornado warning. In effect, a watch is a forecast that warnings may be necessary later in a given region, whereas a warning is a specific statement that danger is imminent.

Long-range forecasting

Techniques
Extended-range, or long-range, weather forecasting has had a different history and a different approach from short- or medium-range forecasting. In most cases, it has not applied the synoptic method of going forward in time from a specific initial map. Instead, long-range forecasters have tended to use the climatological approach, often concerning themselves with the broad weather picture over a period of time rather than attempting to forecast day-to-day details.

There is good reason to believe that the limit of day-to-day forecasts based on the "initial map" approach is about two weeks. Most long-range forecasts thus attempt to predict the departures from normal conditions for a given month or season. Such departures are called anomalies. A forecast might state that "spring temperatures in Minneapolis have a 65 per cent probability of being above normal." It would likely be based on a forecast anomaly map, which shows temperature anomaly patterns. The maps do not attempt to predict the weather for a particular day, but rather forecast trends (e.g. warmer than normal) for an extended amount of time, such as a season.

The US Weather Bureau began making experimental long-range forecasts just before the beginning of World War II, and its successor, the National Weather Service, continues to express such predictions in probabilistic terms, making it clear that they are subject to uncertainty. Verification shows that forecasts of temperature anomalies are more reliable than those of precipitation, that monthly forecasts are better than seasonal ones, and that winter months are predicted somewhat more accurately than other seasons.

Prior to the 1980s the technique commonly used in long-range forecasting relied heavily on the analogue method, in which groups of weather situations (maps) from previous years were compared with those of the current year to determine similarities with the atmosphere's present patterns (or "habits"). An association was then made between what had happened subsequently in those "similar" years and what was going to happen in the current year. Most of the techniques were quite subjective, and there were often disagreements of interpretation and consequently uneven quality and marginal reliability.

The phenomena of persistence (e.g. warm summers follow warm springs) or anti-persistence (e.g. cold springs follow warm winters) were also used, even though, strictly speaking, most forecasters consider persistence forecasts "no-skill" forecasts. Yet they too have had limited success.

Prospects for new procedures

In the last quarter of the twentieth century the approach of and prospects for long-range weather forecasting changed significantly. Stimulated by the work of Jerome Namias, who headed the US Weather Bureau's Long-Range Forecast Division for 30 years, scientists began to look at ocean-surface temperature anomalies as a potential cause for the temperature anomalies of the atmosphere in succeeding seasons and at distant locations. At the same time, other American meteorologists, most notably John M. Wallace, showed how certain repetitive patterns of atmospheric flow were related to each other in different parts of the world. With satellite-based observations available, investigators began to study the El Niño phenomenon. Atmospheric scientists also revived the work of Gilbert Walker, an early twentieth-century British climatologist who had studied the Southern Oscillation, the up-and-down fluctuation of atmospheric pressure over the tropical Indo-Pacific region. Walker had investigated related air circulations (later called the Walker Circulation) that resulted from abnormally high pressures in Australia and low pressures in Argentina or vice versa.

All of this led to new knowledge about how the occurrence of abnormally warm or cold ocean waters and of abnormally high or low atmospheric pressures could be interrelated in vast global connections. Knowledge about these links – the El Niño–Southern Oscillation (ENSO) – and about the behaviour of parts of these vast systems enables forecasters to make better long-range predictions, at least in part, because the ENSO features change slowly and somewhat regularly. This approach of studying interconnections between the atmosphere and the ocean may represent the beginning of a revolutionary stage in long-range forecasting.

Since the mid-1980s, interest has grown in applying numerical weather prediction models to long-range forecasting. In this case, the concern is not with the details of weather predicted 20 or 30 days in advance but rather with objectively predicted anomalies. The reliability of long-range forecasts, like that of short- and medium-range projections, has improved substantially in recent years. Yet many significant problems remain unsolved, posing interesting challenges for all those engaged in the field.

PART 2

THE CHANGING WORLD

4

THE CHANGING PLANET: LAND

In January 2008, the Brazilian government announced that the rate of rainforest destruction in the latter part of 2007 had risen to an unprecedented rate. In December of that year, 948 square km (366 square miles) of forest had been cleared. The increase was attributed to the rise in prices of crops such as soya, which encouraged farmers to clear more land. The Brazilian Amazon comprises nearly 40 per cent of the world's rainforest.

A report by the United Nations University in June 2007 stated that desertification, resulting from climate change and direct human activities, was one of the world's greatest environmental challenges. Due to the large number of people living in affected areas, the resulting economic stress is expected to have a considerable impact on other regions. Drylands, which are the areas most vulnerable to desertification, contain 43 per cent of the world's cultivated land.

The topography and vegetative cover of those parts of Earth's surface not covered by water have varied throughout geological history, as a result of natural changes in temperature, atmospheric conditions and so on. But since the advent of modern civilization, with its population expansion, agricultural techniques and industrial practices, the nature of the land has changed more quickly and dramatically than probably ever before. Human land-use change may bring about climate change in a number of ways. Conversely, the changes in climate that are predicted and are already being experienced in many parts of the world may have profound effects on the ecological and geographical make-up of the land.

In the last few centuries humankind has brought about a widespread conversion of forested land to agricultural pasture and cropland. In certain parts of the world, for example in Europe and North America, virtually all the available productive land has been put to use in this way. In other regions, such as in South America and Indonesia, deforestation for pastureland, the growing of commercial crops, and for revenue from forest products continues apace. Deforestation is of primary concern as a cause of climate change because of the role of forests as a carbon sink: a huge quantity of carbon is stored in organic form in wood and other plant biomass. When these are burned, the carbon returns to the atmosphere in the form of carbon dioxide, one of the major greenhouse gases. (See "Causes of global warming", Chapter 1, page 11.)

Soil is another important repository for carbon, both in its organic component, derived from decomposing plant matter, and in its inorganic mineral content. According to the Intergovernmental Panel on Climate Change, present-day carbon stocks are higher in soils than in plant matter, especially in those regions that are not forested. The degradation and erosion of soil can influence the balance of the systems and cycles in which it plays a part, for example the carbon and hydrologic cycles.

Increasing agricultural pressure from a burgeoning human population, in combination with climatic factors, is resulting in the desertification of many areas on Earth. Often these are marginal ecosystems at the fringes of existing deserts, although deforestation of areas where soils are shallow and vulnerable to erosion by sun and wind may also lead to desertification in a relatively short space of time.

As noted in earlier chapters, deforestation and desertification can affect climate in other ways beyond impacting on the carbon cycle. They increase the Earth's albedo – the degree to which it reflects solar radiation. Reductions in vegetation also influence the rate of heat exchange between the Earth's surface and its atmosphere, since vegetation facilitates the evaporation of water through its leaves and soil (evapotranspiration). The relative importance of these opposing effects (one cooling, the other heating) varies between different regions of the globe and in different seasons, and is difficult to quantify.

This chapter explains how carbon is cycled in soil and vegetation, and the factors to which the system is vulnerable. It describes the human activities that are causing disturbance to the complex ecological balance of the land, and the likely implications of, and for, climate change.

The carbon cycle

Life is built on the conversion of carbon dioxide into the carbon-based organic compounds of living organisms. The central importance of carbon in the biosphere is illustrated by the carbon cycle. It operates along different paths, which recycle carbon at varying rates. The slowest part of the cycle involves carbon that resides in sedimentary rocks, where most of the Earth's carbon is stored. When in contact with water that is acidic (low pH), carbon will dissolve from bedrock; under neutral conditions, carbon will precipitate out as sediment such as calcium carbonate (limestone, $CaCO_3$). This cycling between solution and precipitation is the background against which more rapid parts of the cycle occur.

Short-term cycling of carbon occurs in the continual physical exchange of carbon dioxide (CO_2) between the atmosphere and hydrosphere (the layer of water, in all its forms, on the Earth's surface). Carbon dioxide from the atmosphere becomes dissolved in water (H_2O), with which it reacts to form carbonic acid (H_2CO_3), which dissociates into hydrogen ions (H^+) and bicarbonate ions (HCO_3^-), which further dissociate into hydrogen and carbonate ions (CO_3^{2-}). The more alkaline the water (pH above 7 is alkaline), the more carbon is present in the form of carbonate.

At the same time, carbon dioxide in the water is continually lost to the atmosphere. The exchange of carbon between the atmosphere and hydrosphere links the remaining parts of the cycle, which are the exchanges that occur between the atmosphere and terrestrial organisms and between water and aquatic organisms.

The biological cycling of carbon begins as photosynthetic organisms assimilate carbon dioxide or carbonates from the surrounding environment. In terrestrial communities, plants convert atmospheric carbon dioxide to carbon-based compounds through photosynthesis. During this process, plants cleave the carbon from the two oxygen molecules and release the oxygen back into the surrounding environment. Plants are thus primarily responsible for the presence of atmospheric oxygen. In aquatic communities, plants use dissolved carbon in the form of carbonates or carbon dioxide as the source of carbon for photosynthesis. Once carbon has been assimilated by photosynthetic organisms, as well as by the animals that eat them, it is released again in the form of carbon dioxide as these organisms respire. The release of carbon dioxide into the atmosphere or hydrosphere completes the biological part of the carbon cycle.

The pathways of the global carbon cycle, however, are never completely balanced. That is to say, carbon does not move in and out of all parts of the biosphere at equal rates. Consequently, over time some parts of the biosphere accumulate more carbon than others, thereby serving as major accessible carbon reservoirs. In pre-industrial times the major reservoirs of carbon were the deep and shallow portions of the ocean; the soil, detritus, and flora and fauna of the land; and the atmosphere. The oceans were, and still are, the greatest reservoirs of carbon. Because marine phytoplankton have such short life cycles, the carbon in the ocean cycles rapidly between inorganic and organic states.

In terrestrial environments, forests are the largest carbon reservoirs. Up to 80 per cent of the above-ground carbon in terrestrial communities and about a third of below-ground carbon are contained within forests. Unlike the oceans, much of this carbon is stored directly in the tissues of plants. High-latitude forests include large amounts of carbon, not only in above-ground vegetation but also in peat deposits. It is estimated that half of the carbon in forests occurs in high-latitude forests, and just over a third in low-latitude forests. The two largest forest reservoirs of carbon are the vast expanses in Russia, which hold roughly 25 per cent of the world's forest carbon, and the Amazon basin, which contains about 20 per cent.

Until recent centuries, the equilibrium between the carbon in the world's forests and in the atmosphere remained constant. Samples of carbon dioxide trapped in ice during the past 1,000 years and direct measurements of carbon dioxide in the atmosphere had remained fairly constant until the eighteenth century. However, the cutting of much of the world's forest, along with the increase in consumption of fossil fuels since the Industrial Revolution, has resulted in a disruption of the balance between the amount of carbon dioxide in the forests and in the atmosphere. The concentration of atmospheric carbon dioxide has been increasing steadily: currently the rate of increase is about 4 per cent per decade. If human activities continue to alter the relative sizes of the carbon reservoirs worldwide, they are likely to have a significant effect on the carbon cycle and other biogeochemical cycles. Large-scale deforestation poses a particularly big risk to the equilibrium in global carbon storage and cycling.

The increase in global temperatures as a result of anthropogenic emissions of greenhouse gases is also affecting which ecosystems act as long-term sinks for carbon and which act as sources for carbon dioxide in the atmosphere. For example, the Arctic tundra, with large amounts of carbon stored in its soils, has been a net sink for carbon dioxide during long periods of geologic time. The recent warming of many Arctic regions, however, has accelerated the rate of soil decomposition, transforming these Arctic areas into potential sources of atmospheric carbon dioxide.

The complex cycle of carbon and the varying sizes of different carbon reservoirs illustrate some of the reasons why it has been so difficult to predict the effects that increased atmospheric carbon will have on global change.

Soil

Soil is the biologically active, porous medium that has developed in the uppermost layer of the Earth's crust. It is one of the principal substrata of life on Earth, serving as a reservoir of water and nutrients, as a medium for the filtration and break-down of harmful wastes, and as a participant in the cycling of carbon and other elements through the global ecosystem. It has evolved through weathering processes driven by biological, climatic, geologic, and topographic influences.

Soils provide plants with physical support, water, nutrients, and air for growth. They also sustain an enormous population of microorganisms, such as bacteria and fungi that recycle chemical elements, notably carbon and nitrogen, as well as others that are toxic. The carbon and nitrogen cycles are closely linked to the hydrologic cycle (see Chapter 5), since water functions as the primary medium for chemical transport.

The structure and composition of soil

Physical structure

The bulk of soil consists of mineral particles that are composed of arrays of silicate ions (SiO_4^{4-}) combined with various positively charged metal ions. (An ion is an atom or molecule with a positive or negative charge.)

Soils are classified as clay, silt, or sand, according to their grain size, with clay having the smallest and sand the largest particles. Soil "texture" refers to the relative proportions of sand, silt, and clay particle sizes, irrespective of chemical or mineralogical composition. Sandy soils are called coarse-textured; clay-rich soils called fine-textured. Loam is a textural class representing about one-fifth clay, with the remainder being equal parts of sand and silt. The grain size of soil particles and the aggregate structures they form affect the ability of a soil to transport and retain water, air, and nu-trients.

"Porosity" refers to the capacity of soil to hold air and water, while "permeability" describes the ease of transport of fluids and their dissolved components. The porosity of a soil horizon (a layer of soil with particular characteristics) increases as its texture becomes finer, whereas the permeability decreases as the average pore size becomes smaller. Plant roots open pores between soil aggregates, and cycles of wetting and drying create channels that allow water to pass easily. (However, this structure collapses under waterlogging conditions.) The stability of aggregates increases with humus content (see below), especially humus that originates from grass vegetation. As a general rule, average pore size decreases from certain agricultural practices and other human uses of soil.

Organic content

The second major component of soils is organic matter produced by living organisms. The total organic matter in soil, except for materials identifiable as undecomposed or partially decomposed biomass, is called humus. This solid, dark-coloured component of soil plays a significant role in the control of soil acidity, in the cycling of nutrients, and in the detoxification of hazardous compounds. Humus consists of biological molecules such as proteins and carbohydrates as well as the humic substances (polymeric compounds produced through microbial action). The molecules of humic substances contain hydrogen ions that dissociate in fresh water to form molecules with a negative charge. These can interact with toxic metal ions and effectively remove them from further interaction with the environment.

Much of the molecular framework of soil organic matter, however, is not electrically charged. The uncharged portions of humic substances can react with synthetic organic compounds such as pesticides, fertilizers, solid and liquid waste materials, and their degradation products. Humus, either as a separate solid phase or as a coating on mineral surfaces, can immobilize these compounds and, in some instances, detoxify them significantly.

Soil and climate

Climate influences soil formation primarily through the effects of water and solar energy. Water is the solvent in which chemical reactions take place in the soil, and it is essential to the life-cycles of soil organisms. Water is also the principal medium for the erosive or percolative transport of solid particles. The rates at which these water-mediated processes take place are controlled by the amount of energy available from the sun.

On a global scale, the integrated effects of climate on soil can readily be seen along a transect from pole to equator. As one proceeds from the pole to cool tundra or forested regions, polar desert soils give way to intensively leached soils indicative of humid, boreal climates. As climates become warmer, organic matter accumulates in soils, and eventually, as evapotranspiration increases, lime (calcium carbonate) also begins to accumulate closer to the top of the soil profile. Desert soils, in arid, subtropical climates, are low in organic matter and enriched in soluble salts. Close to the equator, high temperature combines with high precipitation to create red and yellow tropical soils, whose colours reveal the prevalence of residual iron oxide minerals that are resistant to leaching losses because of their low solubility.

In principle, soil profile characteristics that are closely linked to climate can be interpreted as climatic indicators. For example, a soil profile with two well-defined zones of lime accumulation, one shallow and one deep, may signal the existence of a past climate whose greater precipitation drove the lime layer deeper than the present climate is able to do.

Soil erosion

Soil profiles are continually disrupted by the actions of flowing water, wind, or ice and by the force of gravity. These erosive processes remove soil particles from the upper layer (the "A horizon") and expose subsurface horizons to weathering – resulting in the loss of humus, plant nutrients, and beneficial soil organisms. Not only are these losses of paramount importance to agriculture and forestry, but the removal, transport, and subsequent deposition of soil can have significant economic consequences by damaging buildings, bridges, culverts, and other structures.

Erosive processes
Water erosion
Water-induced erosion can take various forms, depending on climate and topography. The impact of raindrops breaks the bonds holding soil aggregates together and catapults the particles into the flowing water from surface run-off. Wholesale removal of soil particles by the sheet flow of water (sheet erosion) or by flow in small channels (rill erosion) accounts for most of the water-induced soil loss from exposed land surfaces. More spectacular but less prevalent types of erosion are gully erosion, in which water concentrates in channels too deep to smooth over by tilling, and streambank erosion, in which the saturated sides of running streams tumble into the moving water below. The same forces at work in streambank erosion are seen in soils on hill slopes that become thoroughly saturated with water. Gravity, able to overcome the cohesive forces that hold soil particles together, can cause the entire soil profile to move downslope – a phenomenon called mass movement. This movement may be either slow (soil creep), rapid (debris flow or mudflow), or catastrophic (landslide).

Wind erosion

The mechanisms involved in wind erosion depend on soil texture and the size of soil particles. Dry soil particles of silt or clay size can be transported over great distance by wind. Larger particles that are the size of fine sand can be vaulted as high as 25 cm (10 inches) into the air, then drop to the ground after a short flight, only to rebound under the continual driving force of the wind. Coarser sand particles are not lifted, but they can tumble along the land surface. The major cause of wind erosion is the jumping motion of the smaller soil particles, a process called saltation. The texture of the windblown surfaces of these soils becomes coarser, making them less chemically reactive and less able to retain plant nutrients or trap pollutants. In arid regions, wind erosion often produces a gravelly land surface known as desert pavement.

Rates of soil erosion

Soil erosion and deposition are natural geomorphic processes that give shape to landforms and provide new parent material for the development of soil profiles. These processes become soil conservation issues when the rate of erosion greatly exceeds the rate expected in the absence of human land use – a situation referred to as accelerated erosion. Rates of normal soil erosion have been estimated from measurements of sediment transport and accumulation, mass movement on hill slopes, and radioactive carbon dating of landforms. The average annual rate of normal soil erosion is nearly 1 tonne per hectare (0.45 tonne per acre), while that of natural soil formation is nearly 0.7 tonne per hectare (0.3 tonne per acre). Broadly speaking, rates of soil loss exceeding 10 tonnes per hectare annually indicate accelerated erosion. It is important to note that this accelerated soil loss is equivalent to less than

1 mm of soil depth, making erosion damage very difficult to observe over short time spans.

When climate and topography are fixed and soil cover is varied, the rate of soil loss by water erosion has a predictable and dramatic dependence on vegetation. Irrespective of location, erosion losses are usually very small from forested land or permanent pastureland, moderate to high from land planted with grain crops, and very high from clean-tilled orchards, vineyards, and land planted with row crops.

Soil resistance to erosion

The ability of soils to resist water and wind erosion depends on their texture and topographic characteristics. Clay-rich soils resist erosion well because of strong cohesive forces between particles and the glue-like characteristics of humus. Both loam and sandy soils are moderately resistant to erosion – the former because they have sufficient clay content to hold the particles together; the latter because their high permeability limits the amount of surface run-off that can wash soil particles away, while their larger particle size makes them too heavy to be easily transported in flowing water. Silty soils, on the other hand, exhibit the least resistance to erosion because their permeability is low (resulting in more surface run-off), and their particle size is neither small enough to promote cohesion nor large enough to prevent entrainment. Soils on steep, long slopes are much more susceptible to erosion than those on shallow, short slopes because the steeper slopes accelerate the flow of surface run-off.

The development of soil conservation strategies requires knowledge of actual and acceptable rates of soil erosion. A practical measure of soil resistance to erosion used by pedologists (soil scientists) in the United States is the soil loss tolerance (T-value, or T-factor). This quantity is defined as the maximum annual rate of soil loss by erosion that will permit high soil productivity for an indefinite period of time. Operationally, the concept is interpreted as the maximum annual loss from the A horizon that does not reduce the thickness of the rooting zone significantly over millennia.

Human land use and climate change

Deforestation

Deforestation, the clearing or thinning of forests usually through human activity, represents one of the largest issues in global land use in the early twenty-first century. Estimates of deforestation traditionally are based on the area of forest cleared for human use, including the removal of the trees for wood products, croplands, and grazing lands. In the practice of clear-cutting, all the trees are removed from the land, which completely destroys the forest. In some cases, however, even partial logging and accidental fires thin out the trees enough to change the forest structure dramatically.

Conversion of forests to land used for other purposes has a long history. The Earth's croplands, which cover about 15 million square km (5.8 million square miles), are mostly deforested land. More than 11 million square km (4.2 million square miles) of present-day croplands receive enough rain and are warm enough to have once supported forests of one kind or another. Of these, only 1 million (390,000 square miles) are in areas that would have been cool boreal forests, as in Scandinavia and northern Canada. Two million square km (770,000 square miles) were once moist tropical forests. The rest were once temperate forests or subtropical forests, including forests in eastern North America, western Europe, and eastern China.

The extent to which forests have become Earth's grazing lands is much more difficult to assess. Cattle or sheep pastures in North America or Europe are easy to identify, and they support large numbers of animals. At least 2 million square km of such forests have been cleared for grazing lands. Less certain are the 5–9 million square km (1.9 to 3.5 million square miles) of humid tropical forests and some drier tropical woodlands that have been cleared for grazing. These often support only very low numbers of cattle, but they may still be considered grazing lands by national authorities. Almost half the world is made up of "drylands" – areas too dry to support large numbers of trees – and most are considered grazing lands. There, goats, sheep, and cattle may harm what few trees are able to grow.

Although most of the areas cleared for crops and grazing represent permanent and continuing deforestation, deforestation can be transient. About half of eastern North America lay deforested in the 1870s, almost all of it having been deforested at least once since European colonization in the early 1600s. Since the 1870s the region's forest cover has increased, though most of the trees are relatively young. Few places exist in eastern North America that retain stands of uncut old-growth forests. In addition, while some forests are being cleared, some are being planted. The United Nations Food and Agriculture Organization (FAO) estimates that there are approximately 1.3 million square km (500,000 square miles) of such plantations on Earth. These are often of eucalyptus or fast-growing pines – and almost always of species that are not native to the places where they are planted.

Elsewhere, forests are shrinking. The FAO estimates that the annual rate of deforestation is about 1.3 million square km per decade. About half of that is primary forest – forest that has not been cut previously (or at least recently).

Dry forests in general are easier to deforest and occupy than moist forests and so have been particularly targeted by human actions. An example of dry forest is the cork-oak forests of the western Mediterranean, which provide a habitat for many threatened species, including the Iberian lynx (*Lynx pardinus*) and the cinereous vulture (*Aegypius monachus*). In addition, the cork industry – with an annual production of 15 billion corks – is a sustainable industry that provides a source of income for more than 100,000 people. The use of screw tops and synthetic stoppers, however, is leading to the demise of the industry, with large areas of cork-oak forests at a heightened risk of desertification (see page 206) and forest fires. A continuation of the decline in the cork market could lead to the loss of three quarters of the forests within ten years.

Tropical deforestation

The greatest deforestation is occurring in the tropics, where a wide variety of forests exists. These range from rainforests that are hot and wet year-round, to forests that are merely humid and moist, to those in which trees in varying proportions lose their leaves in the dry season, and to dry open woodlands. Worldwide, humid forests once covered an area of about 18 million square km (7 million square miles). Of this, about 10 million square km (3.9 million square miles) remained in the early twenty-first century. At the current annual rates of deforestation, most of these forests will be cleared within the century. Indeed, in some places, such as West Africa and the coastal humid forests of Brazil, very little forest remains today.

The Amazon Rainforest is the largest remaining block of humid tropical forest, and about two thirds of it is in Brazil. (The rest lies along that country's borders to the west and to the north.) Detailed studies of Amazon deforestation from

1988 to 2005 show that the rate of forest clearing has varied from a low of about 11,000 square km (4,200 square miles) per year in 1991 to a high of about 30,000 square km (11,600 square miles) per year in 1995. The high figure immediately followed an El Niño event (see Chapter 5, page 265) that caused the Amazon basin to receive relatively little rain and so made its forests unusually susceptible to fires. Studies in the Amazon also reveal that 10,000–15,000 square km (3,900–5,800 square miles) are partially logged each year, a rate roughly equal to the low end of the forest-clearing estimates cited above. In addition, each year fires burn an area about half as large as the areas that are cleared. Even when the forest is not entirely cleared, what remains is often a patchwork of forests and fields or, in the event of more intensive deforestation, "islands" of forest surrounded by a "sea" of deforested areas.

The effects of forest clearing, selective logging, and fires interact. Selective logging increases the flammability of the forest because it converts a closed, wetter forest into a more open, drier one. This leaves the forest vulnerable to the accidental movement of fires from cleared adjacent agricultural lands and to the killing effects of natural droughts. As fires, logging, and droughts continue, the forest can become progressively more open until all the trees are lost.

The human activities that contribute to tropical deforestation include commercial logging and land clearing for cattle ranches and plantations of rubber trees, oil palms, and other economically valuable trees. Another major contributor is the practice of slash-and-burn, or "swidden" agriculture. This is a type of shifting agriculture – a system of cultivation whereby a plot of land is cleared and cultivated for a short period of time, then abandoned and allowed to revert to its natural vegetation while the cultivator moves on to another plot. The period of cultivation is usually terminated when the soil shows signs of exhaustion or, more commonly, when the field is overrun by weeds. The length of time that a field is cultivated is usually shorter than the period over which the land is allowed to regenerate by lying fallow. In slash-and-burn systems, small-scale farmers clear forests by burning them and then grow their crops in the soils fertilized by the ashes. Typically, the land produces for only a few years and then must be abandoned and new patches of forest burned.

Shifting agriculture has frequently been attacked in principle because it degrades the fertility of forestlands of tropical regions. Nevertheless, it is an adaptation to tropical soil conditions in regions where long-term, continued cultivation of the same field, without advanced techniques of soil con-servation and the use of fertilizers, would be extremely detri-mental to the fertility of the land. In such environments it may be preferable to cultivate a field for a short period and then abandon it before the soil is completely exhausted of nutrients. Tropical deforestation is further exacerbated by the recent drive to produce biofuels from palm and soybean oils to fuel cars and other vehicles from Europe and North America. The growing demand for biofuel – so-called "green energy" – has promoted the clearing of Southeast Asian rainforests to make way for palm plantations. In the Brazilian Amazon the plant-ing of soybeans had become a principal cause of rainforest loss.

Although forests may regrow after being cleared and then abandoned, this is not always the case. About 400,000 square km (154,000 square miles) of tropical deforested land exists in the form of steep mountain hillsides. The combination of steep slopes, high rainfall, and the lack of tree roots to bind the soil can lead to disastrous landslides that destroy fields, homes, and human lives. Steep slopes aside, only about a quarter of the humid forests that have been cleared are exploited as croplands. The rest are abandoned or used for grazing land that often can support only low densities of animals, because the soils underlying much of this land are extremely poor in nutrients. (To clear forests, the vegetation that contains most of the nutrients is often burned, and the nutrients literally "go up in smoke" or are washed away in the next rain.)

Global implications

Deforestation has important global consequences. Forests sequester carbon in the form of wood and other biomass as the trees grow, taking up carbon dioxide from the atmosphere. When forests are burned, their carbon is returned to the atmosphere as carbon dioxide, and the trees are no longer present to sequester more carbon. Approximately 55 billion tonnes of carbon are stored in terrestrial plants each year worldwide, most of it in forests. It is estimated that around 800 million hectares (20 billion acres) of forested land have been lost since the dawn of civilization; this translates to about 6 billion tonnes of carbon per year less net primary production than before land was cleared for agriculture and commerce. This estimated decrease in carbon storage can be compared with the 5–6 billion tonnes of carbon currently released per year by fossil-fuel burning.

The current estimate of net carbon loss from deforestation, reforestation, wood products decomposition, and abandonment of agricultural land is about 1.7 billion tonnes per year worldwide, or about one third the current loss from fossil-fuel burning. This figure could as much as double in the first half of the twenty-first century if the rate of deforestation is not controlled. Reforestation, on the other hand, could actually reduce the current carbon loss by up to 10 per cent without exorbitant demands on management practices.

Furthermore, most of the planet's valuable biodiversity (see Chapter 7) is within forests, particularly tropical ones. Moist tropical forests such as the Amazon Rainforest have the greatest concentrations of animal and plant species of any terrestrial ecosystem. Perhaps two thirds of Earth's species live only in these forests. As deforestation proceeds, it has the potential to cause the extinction of increasing numbers of these species.

Climate change and soil

As described earlier in this chapter, soils are strongly influenced by climate. The predicted temperature increases caused by global warming and the consequent change in rainfall patterns are expected to have a substantial impact on soils in the form of erosion. Many areas are receiving increased precipitation and in places the rainfall intensity has also increased. Some places, on the other hand, have become more arid, and there have been changes in the seasonal distribution of rainfall as well as in the severity and patterns of winds. All these factors are affecting patterns of vegetative growth and may lead to both wind- and water-induced erosion.

Once soil erosion has occurred it is usually irreversible. Most of the nutrients are found within the surface layer of the soil, and once this layer has been lost the soil is less able to support plant growth and is more susceptible to run-off. Soil erosion is therefore strongly connected to the phenomenon of desertification (see below). When vegetation has been lost from an area this in turn may influence the hydrologic cycle.

The other notable role of soils in global change phenomena is in the regulation of the carbon dioxide budget, which as we have seen is strongly influenced by the deforestation of land, especially in the humid tropics. Carbon in terrestrial biomass that is not used directly becomes carbon in litter (about 25

billion tonnes of carbon annually) and is eventually incorporated into soil humus. Soil respiration currently releases an
average of 68 billion tonnes of this carbon back into the
atmosphere. Loss of soil carbon exacerbated by erosion is
likely to increase the volume of carbon dioxide released into
the atmosphere by soils.

In addition to carbon dioxide, some computer models
predict that methane and nitrous oxide emissions will also
be very important in future global climate change. About 70
per cent of the methane and 90 per cent of the nitrous oxide in
the atmosphere are derived from soil processes. Soils can also
function as repositories for these gases, and it is important to
appreciate the complexity of the source–repository relationship. For example, the application of nitrogen-containing
fertilizers reduces the ability of the soil to process methane.
Even the amount of nitrogen introduced into soil from acid
rain on forests is sufficient to produce this effect. However, the
extent of net emissions of methane and nitrous oxide and the
microbial trade-off between the two gases are undetermined at
the global scale.

Desertificaton

Desertification is the spread or encroachment of a desert environment into arid or semi-arid regions. It is a complex
process, which may be caused by climatic changes, human
influence, or both. Climatic factors include periods of tempor-
ary but severe drought and long-term climatic changes toward
aridity. Human factors include the artificial alteration of the
local climate, such as degradation of the land by removing
vegetation (which can lead to unnaturally high erosion), excessive cultivation or over-grazing, and the exhaustion of
surface-water or groundwater supplies for irrigation, industry,

or domestic use. Intensive farming practices, dependent on chemical fertilizers rather than the organic content in soils, can lead soils to become increasingly dusty and susceptible to wind erosion. Desertification usually begins in areas made susceptible by drought or overuse by human populations and spreads into arid and semi-arid regions. The marginal ecosystems on desert fringes are particularly susceptible to increased population pressure and, as land use intensifies in these areas, the fragile ecological balance may be tipped.

Desertification drains an arid or semi-arid land of its life-supporting capabilities. The process is characterized by a declining groundwater table, salinization of topsoil and water, diminution of surface water, increasing erosion, and the disappearance of native vegetation. Areas undergoing desertification may show all of these symptoms, but the existence of only one usually provides sufficient evidence that the process is taking place. The phenomenon is not limited to non-desert regions and can occur in areas within deserts where the delicate ecological balance is disturbed. The Sonoran and Chihuahuan deserts of the American Southwest, for example, have become observably more barren as the wildlife and plant populations have diminished.

The United Nations designated the year 2006 as the International Year of Deserts and Desertification, and the effect of climate change on desert wildlife and biodiversity and the exacerbation of desertification received special attention. On June 5, World Environment Day, the UN Environment Programme released the report *Global Deserts Outlook*, which indicated that deserts might be among the ecosystems most affected by climate change. The report stated that from 1976 to 2000, nine out of twelve deserts studied showed increased temperatures. A temperature increase of 1–7°C (1.8–12.6°F) is predicted in all world deserts for the period 2071–2100.

Rainfall patterns are expected to vary: for example, the Gobi Desert in China is likely to receive more rain, while the Sahara and Great Basin deserts in the western United States could become drier. Deserts fed by melting snow or ice, such as those in Central Asia and the Andean foothills, will be particularly vulnerable to climate change.

Unpredictable climatic events are more important than average conditions in deserts, and even small changes in precipitation and temperature can therefore have a marked impact. The many species of the 3.7 million square km (1.4 million square miles) of land that make up the world's deserts would be adversely affected should the report's projected scenarios prove correct.

The climate in deserts is also strongly linked to environments in other parts of the world. For example, decreased rainfall in some deserts is likely to cause an increase in dust storms, some of which travel for thousands of miles. Most dust particles in the Earth's atmosphere originate from northern Africa and Asian deserts. Phosphorus and silicon carried by desert dust enhance growth in oceanic phytoplankton, and also contribute to the nutrient content of tropical soils.

Agriculture

The expected effects of climate change, in particular global warming, will have a marked impact on agriculture in various parts of the world. In temperate regions, agricultural productivity might increase modestly for some crops in response to a local warming of 1–3°C (1.8–5.4°F), but productivity will generally decrease with further warming. For tropical and subtropical regions, however, models predict decreases in crop productivity for even small increases in local warming. In some cases, adaptations such as altered planting practices will be required to ameliorate losses in productivity for modest amounts of warming.

An increased incidence of drought and flood events is likely to lead to further decreases in agricultural productivity and to decreases in livestock production, particularly among subsistence farmers in tropical regions. In regions such as the African Sahel, decreases in agricultural productivity have already been observed as a result of shortened growing seasons, which have occurred as a result of warmer and drier climatic conditions.

Other human land-use issues

Urbanization

For many decades there has been a nearly universal flow of populations from rural into urban areas. While definitions of urban areas differ from country to country and region to region, the most highly urbanized societies in the world are those of western and northern Europe, Australia, New Zealand, temperate South America, and North America: in all of these the fraction of the population living in urban areas now exceeds 75 per cent, and it has reached 85 per cent in West Germany. In industrialized nations the countryside has sometimes been virtually depopulated, to the point that, for example, in 1970 only 6.7 per cent of employed people in the United States were in the fields of agriculture, fisheries, and forestry. In much of tropical Latin America, 50–65 per cent of the population lives in cities. In many of the developing countries of Asia and Africa the urbanization process has only recently begun, with often less than one third of the population living in urban areas. However, in parts of the developing world the rate of growth of urban areas is now twice that of the population as a whole. Thus, in a population growing 3 per cent annually (doubling in about 23.1 years), it is likely that the urban growth rate is at least 6 per cent annually (doubling in about 11.6 years). The global acceleration of the process of urbanization has created vast slums in many urban centres.

Urbanization is associated with a number of environmental problems. Some of these relate to the urban environment itself: in cities in developing countries where economies are poor, water and air pollution, and inadequate sanitation and waste handling pose threats to human health. But there are also broader environmental impacts. Conversion of land to urban use and to support the urban infrastructure puts additional pressure on adjacent areas that may be more ecologically fragile, and may cause conflict over resources with neighbouring rural populations. Water systems may be polluted. In coastal areas in particular, urbanization often involves the destruction of sensitive ecosystems and can alter hydrology of the region. More wide-ranging effects result from the intensive extraction of natural resources – including firewood, building materials and water – to support the demands of the urban economy. Over time, these factors may have an extensive impact on the hydrological and carbon cycles.

Land pollution

Land pollution involves the degradation of land by human activities, particularly from industrial, mining and domestic waste disposal. This contamination of land is a direct consequence of urbanization and industrialization. A high concentration of non-biodegradeable material interferes with organic life in that area, and landfilling without proper controls can result in the leaching of chemicals into soils, which may in turn contaminate waterways. Land pollution may also result in the accumulation on land of substances in dispersed solid or liquid form that are injurious to life. This has been particularly noticeable with those chemicals (e.g. DDT) that are spread for the purpose of exterminating pests but then accumulate to the extent that they can do damage to many other forms of life.

Landfill often requires the availability of low-lying ground and frequently involves the destruction of marshland or swamps that have high biological value. Other methods of disposal include ocean dumping, which creates water pollution and destroys marine habitats; and burning, which increases air pollution. Obviously, none of these methods is entirely satisfactory, although using landfill to create artificial landscapes, which then are covered with soil and planted with various kinds of vegetation, is a possibility that remains to be fully developed. It is the great quantity of debris produced by urban communities, more than a shortage of raw materials, that forces the development of more effective means for recycling wastes.

5

THE CHANGING PLANET: HYDROSPHERE

In June 2005, a report commissioned by the UK's Royal Society showed that the oceans were becoming more acidic as they absorbed some of the excess carbon dioxide that was being released into the atmosphere through the burning of fossil fuels. The change could be catastrophic for marine ecosystems and for economies that rely on reef tourism or fishing. Seawater is naturally alkaline, with an average pH of 8.2. (On the pH scale, values above 7 are alkaline, values below 7 are acidic, and lower values correspond to greater acidity.) The study suggested that by the year 2100 expected increases in atmospheric carbon dioxide would lead to a fall of 0.5 in the average pH of ocean surface waters.

The increasing acidity might affect animals with high oxygen demands, such as squid, since dissolved oxygen would become more difficult to extract from water. The change might also have serious consequences for organisms with calcium-carbonate shells, including lobsters, crabs, shellfish, certain plankton species, and coral polyps, because increasing acidity would affect how readily calcium carbonate dissolves in seawater.

In addition to changes in pH, climate change is affecting the oceans and consequently weather systems in a number of ways, chiefly through the melting of ice caps and the resulting effects on ocean currents.

The oceans are the most prominent part of a larger region of the Earth-atmosphere system called the hydrosphere. This chapter explores the general nature of the hydrosphere, the role of both the oceans and sea ice, and the possible effects on the hydrosphere of increasing greenhouse gas emissions and global warming.

The hydrosphere

The hydrosphere is defined as the discontinuous layer of water at or near Earth's surface. It includes all liquid and frozen surface waters, groundwater held in soil and rock, and atmospheric water vapour. Water is the most abundant substance at the surface of Earth. About 1.4 billion cubic kilometres (326 million cubic miles) of water in liquid and frozen form make up the oceans, lakes, streams, glaciers, and groundwaters on our planet.

The Earth is unique in the solar system because of its distance from the sun and period of rotation. These combine to subject our planet to a solar radiation level that maintains it at a mean surface temperature of 16°C (61°F), which varies little over annual and diurnal cycles (that is those of a 24-hour period). This allows water to exist on the Earth in all three of its phases – solid, liquid, and gaseous. No other planet in the solar system has this feature. The liquid phase predominates on Earth. By volume, 97.957 per cent of the water on the planet exists as oceanic water and associated sea ice. The gaseous phase and droplet water in the atmosphere constitute 0.001 per cent. Fresh water in lakes and streams makes up 0.036 per cent, while groundwater is ten times more abundant at 0.365 per cent. Glaciers and ice caps constitute 1.641 per cent of the Earth's total water volume.

Each of the above is considered to be a reservoir of water. Water continuously circulates between these reservoirs in the hydrologic cycle, which is driven by energy from the sun. This cycle consists of water reservoirs, the processes by which water is transferred from one reservoir to another (or transformed from one state to another), and the rates of transfer associated with such processes. Evaporation, precipitation, movement of the atmosphere, and the downhill flow of river water, glaciers, and groundwater keep water in motion between the reservoirs and maintain the cycle. The transfer paths penetrate the entire hydrosphere, extending upward to about 15 km (9 miles) in Earth's atmosphere and downward to depths in the order of 5 km in its crust.

The large range of volumes in these reservoirs and the rates at which water cycles between them combine to create important conditions on Earth. If small changes occur in the rate at which water is cycled into or out of a reservoir, the volume of a reservoir changes. These volume changes may be relatively large and rapid in a small reservoir or small and slow in a large reservoir. A small percentage change in the volume of the oceans may produce a large proportional change in the land-ice reservoir, thereby promoting glacial and interglacial stages.

The hydrologic cycle

The present-day hydrologic cycle at Earth's surface is illustrated in Figure 5.1. Some 496,000 cubic km of water evaporate from the land and ocean surface annually, remaining for about ten days in the atmosphere before falling as rain or snow. The amount of solar radiation necessary to evaporate this water is half of the total solar radiation received at Earth's surface. About one-third of the precipitation falling on land runs off to the oceans, primarily in rivers, while direct groundwater discharge to the oceans accounts for only about 0.6 per cent of the total discharge. A small amount of precipitation is temporarily stored in the waters of rivers and lakes. The remaining precipitation over land – 0.073×10^6 cubic km per year – returns to the atmosphere by evaporation. Over the oceans, evaporation exceeds precipitation: the net difference represents transport of water vapour over land.

The various reservoirs in the hydrologic cycle have different water residence times. Residence time is defined as the amount of water in a reservoir divided by either the rate of addition of water to the reservoir or the rate of loss from it. The oceans have a water residence time of 37,000 years, which reflects the large amount of water in the oceans. In the atmosphere, as noted above, the residence time of water vapour relative to total evaporation is only ten days. Lakes, rivers, ice, and groundwaters have residence times lying between these two extremes and are highly variable.

There is considerable variation in the degree of evaporation and precipitation over the globe. In order to have precipitation, there must be sufficient atmospheric water vapour and enough rising air to carry the vapour to an altitude where it

The present-day surface hydrologic cycle.

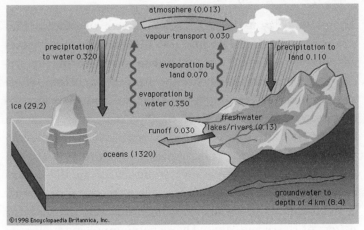

Figures are in millions of cubic kilometres

Figure 5.1 © Encyclopædia Britannica, Inc.

can condense and precipitate. The latitudinal variation of precipitation and evaporation is related to the global wind belts. The trade winds, for example, are initially cool, but they warm up as they blow toward the equator. These winds pick up moisture from the ocean, increasing ocean-surface salinity and causing seawater to sink. When the trade winds reach the equator they rise, and the water vapour in them condenses and forms clouds. Net precipitation is high near the equator and also in the belts of the prevailing westerlies where there is frequent storm activity. Evaporation exceeds precipitation in the subtropics where the air is stable, and near the poles where the air is both stable and has a low water-vapour content because of the cold. The Greenland and Antarctic ice sheets formed as a result of the very low evaporation rates at the poles, which means that precipitation exceeds evaporation in those regions.

The oceans

When viewed from space, the predominance of the oceans on Earth is evident. The oceans and their marginal seas cover nearly 71 per cent of the Earth's surface, with an average depth of 3,795 metres (12,450 feet). The exposed land that occupies the remaining 29 per cent of the planetary surface has a mean elevation of only 840 metres (2,756 feet).

Relative distribution of the oceans

Oceanic researchers generally recognize the existence of three major oceans, the Pacific, Atlantic, and Indian. (The Arctic Ocean is considered to be an extension of the Atlantic.) Arbitrary boundaries separate these three bodies of water in the southern hemisphere. One boundary extends southward to Antarctica from the Cape of Good Hope, while another stretches southward from Cape Horn. The last one passes through Malaysia and Indonesia to Australia, and then on to Antarctica. Many subdivisions can be made to distinguish the limits of seas and gulfs that have historical, political, and sometimes ecological significance. However, water properties, ocean currents, and biological populations do not necessarily recognize these boundaries. Indeed, neither do many researchers. For example, the oceanic area surrounding the Antarctic is considered by some to be the Southern Ocean.

If area–volume analyses of the oceans are to be made, then boundaries must be established to separate individual regions. In 1921 Erwin Kossina, a German geographer, published tables giving the distribution of oceanic water with depth for the oceans and adjacent seas. Kossina's work in bathymetry was updated in 1966 by H.W. Menard and S.M. Smith.

The latter only slightly changed the numbers derived by Kossina. This was remarkable, since the original effort relied entirely on the sparse depth measurements accumulated by individual wire soundings, while the more recent work had the benefit of acoustic depth soundings collected since the 1920s. This type of analysis, called hypsometry, allows quantification of the surface-area distribution of the oceans and their marginal seas with depth.

The distribution of oceanic surface area with five increments of latitude shows that the distribution of land and water on the Earth's surface is markedly different in the northern and southern hemispheres. The southern hemisphere may be called the water hemisphere, and the northern hemisphere the land hemisphere. This is especially the case in the temperate latitudes.

This asymmetry of land and water distribution between the northern and southern hemispheres makes the two hemispheres behave very differently in response to the annual variation in solar radiation received by the Earth. The southern hemisphere shows only a small change in surface temperature from summer to winter at temperate latitudes. This variation is controlled primarily by the ocean's response to seasonal changes in heating and cooling. The northern hemisphere has one change in surface temperature controlled by its oceanic area and another controlled by its land area. In the temperate latitudes of the northern hemisphere, the land area is much warmer than the oceanic area in summer and much colder in winter. This situation creates large-scale seasonal changes in atmospheric circulation and climate in the northern hemisphere that are not found in the southern hemisphere.

Origin and evolution of the oceans

The chemical history of the oceans has been divided into three stages. The first is an early stage in which the Earth's crust was cooling and reacting with volatile or highly reactive gases of an acidic, reducing nature (high hydrogen content) to produce the oceans and an initial sedimentary rock mass. This stage lasted until about 3.5 billion years ago. Fossils dated from the Precambrian age some 3.3 billion years ago show that bacteria and cyanobacteria (blue-green algae) existed, indicating the presence of water during this period. The second stage was a period of transition from the initial to essentially modern conditions, and is estimated to have ended 2 to 1.5 billion years ago. Since that time it is likely that there has been little change in seawater composition.

Early oceans

The initial formation of Earth by the agglomeration of solid particles occurred about 4.6 billion years ago. This initially cool, unsorted conglomerate was slowly warmed by radio-active and compressional heating, leading to the gradual separation and migration of materials to form the Earth's core, mantle, and crust. The formation of the Earth's core is thought to have taken about 500 million years. It is likely that core formation resulted in the escape of an original primitive atmosphere and its replacement by one derived from the loss of volatile substances from the Earth's interior. Whether most of this degassing took place during core for-mation or soon afterward, or whether there has been signifi-cant degassing of the Earth's interior throughout geologic time, is uncertain. It is thought that the Earth, after initial cold agglomeration, reached temperatures such that the whole

planet approached the molten state. As the initial crust of the
Earth solidified, volatile gases would be released to form an
atmosphere that would contain water (H_2O), later to become
the hydrosphere; carbon gases, such as carbon dioxide (CO_2),
methane (CH_4), and carbon monoxide (CO); sulphur gases,
mostly hydrogen sulphide (H_2S); and halogen compounds,
such as hydrochloric acid (HCl). Nitrogen may also have been
present, along with minor amounts of other gases. Gases of
low atomic number, such as hydrogen and helium, would
escape the Earth's gravitational field. Substances degassed
from the planetary interior have been called "excess volatiles"
because their masses cannot be accounted for simply by rock
weathering.

The early atmosphere would have been rich in carbon
dioxide and carbon monoxide. Photodissociation (separation
due to the energy of light) of water vapour into molecular
hydrogen (H_2) and molecular oxygen (O_2) in the upper
atmosphere allowed the hydrogen to escape and led to a
progressive increase of the partial pressure of oxygen at the
Earth's surface. The reaction of this oxygen with the materials of the surface gradually caused the vapour pressure of
water vapour to increase to a level at which liquid water
could form.

At an initial crustal temperature of about 600°C, however,
almost all these compounds, including water, would be in the
atmosphere. The sequence of events that occurred as the crust
cooled is difficult to construct. Below 100°C all the water
would have condensed, and the acid gases would have
reacted with the original igneous crustal minerals to form
sediments and an initial ocean. There are at least two possible
pathways by which these initial steps could have been accomplished.

One pathway assumes that the 600°C atmosphere contains, together with other compounds, water (as vapour), carbon dioxide, and hydrochloric acid in the ratio of 20:3:1 and would cool to the critical temperature of water. The water vapour therefore would have condensed into an early hot ocean. At this stage, the hydrochloric acid would be dissolved in the ocean (about 1 mole per litre), but most of the carbon dioxide would still be in the atmosphere, with about 0.5 mole per litre in the ocean water. This early acid ocean would react vigorously with crustal minerals, dissolving out silica and cations (atoms or molecules with a positive charge) and creating a residue that consisted principally of aluminous clay minerals that would form the sediments of the early ocean basins. This pathway of reaction assumes that reaction rates are slow relative to cooling. A second pathway of reaction, which assumes that cooling is slow, is also possible. In this scenario, at a temperature of about 400°C most of the water vapour would be removed from the atmosphere by hydration reactions with pyroxenes and olivines (types of silicates). Under these conditions, water vapour would not condense until some unknown temperature was reached, and the Earth might have had at an early stage in its history an atmosphere rich in carbon dioxide but no ocean: the surface would have been much like that of present-day Venus.

The pathways described are two of several possibilities for the early surface environment of the Earth. In either case, after the Earth's surface had cooled to 100°C, it would have taken only a short time geologically for the acid gases to be used up in reactions involving igneous rock minerals. The presence of bacteria and algae in the fossil record of rocks older than 3 billion years attests to the fact that the Earth's surface had cooled to temperatures lower than 100°C by this time and that the neutralization of the original acid gases had taken place. If most of the degassing of primary volatile substances from the Earth's interior occurred early, the chloride released by reaction of hydrochloric acid with rock minerals would be found in the oceans and seas or in evaporite deposits, and the oceans would have a salinity and volume comparable to those that they have today.

One component missing from the early terrestrial surface was free oxygen, because it would not have been a constituent released from the cooling crust. As described earlier, early production of oxygen was by photodissociation of water in the atmosphere as a result of absorption of ultraviolet light. Hydrogen produced by this process would escape into space, and the oxygen would react with the early reduced gases, producing sulphur dioxide and water. Oxygen production by photodissociation gave the early reduced atmosphere a start toward present-day conditions, but it was not until the appearance of photosynthetic organisms, approximately 3.3 billion years ago, that it was possible for the accumulation of oxygen in the atmosphere to proceed at a rate sufficient to lead to today's oxygenated environment.

Modern oceans

It is likely that the oceans achieved their modern chemical characteristics about 2 to 1.5 billion years ago. The chemical and mineralogical compositions and the relative proportions of sedimentary rocks of this age differ little from their counterparts of the Palaeozoic era (from 540 to 245 million years ago). Calcium sulphate deposits of the late Precambrian age (about 1.5 billion to 540 million years ago) attest to the fact that the acid sulphur gases had been neutralized to sulphate by this time. Chemically precipitated ferric oxides in late Precambrian sedimentary rocks indicate available free oxygen, whatever its percentage. The chemistry and mineralogy of middle and late Precambrian shales are similar to those of Palaeozoic shales. The carbon isotopic signature of carbonate rocks has been remarkably constant for more than 3 billion years, indicating exceptional stability in size and fluxes related to organic carbon. The sulphur isotopic signature of sulphur phases in rocks strongly suggests that the sulphur cycle, involving the reduction of sulphate by bacteria, was in operation 2.7 billion years ago. It therefore appears that continuous cycling of sediments similar to those of today has occurred for 1.5 to 2 billion years and that these sediments have controlled hydrospheric, and particularly oceanic, composition.

It was once thought that the saltiness of the modern oceans simply represents the storage of salts derived from rock weathering and transported to the oceans by fluvial processes. With increasing knowledge of the age of Earth, however, it was realized that, at the present rate of delivery of salts to the ocean or even at much reduced rates, the total salt content and the mass of individual salts in the oceans could be attained in geologically short time intervals compared with the planet's age. The total mass of salt in the oceans can be accounted for at today's rates of stream delivery in about 12 million years. The mass of dissolved silica in ocean water can be doubled in just 20,000 years by the addition of stream-derived silica; to double the sodium content would take 70 million years.

It became apparent, therefore, that the oceans were not merely an accumulator of salts. Rather, as water evaporated from the oceans, the salts introduced must be removed in the form of minerals deposited in sediments. Accordingly, the concept of the oceans as a chemical system changed from that of a simple accumulator to that of a steady-state system in which rates of inflow of materials equal rates of outflow. The steady-state concept permits influx to vary with time, but the inflow would be matched by nearly simultaneous and equal variation of efflux. Calculations of rates of addition of elements to the oceanic system and removal from it show that for at least 100 million years the oceanic system has been in a steady state with approximately fixed rates of major element inflow and outflow and, thus, fixed chemical composition.

In recent years, this steady-state conceptual view of the oceans has undergone some modification. In particular, it has been necessary to treat components of ocean water in terms of all their influxes and effluxes and to be more aware of the timescale of application of the steady-state concept. Indeed, the recent increase in the carbon dioxide concentration of the atmosphere due to the burning of fossil fuels may induce a change in the pH and dissolved inorganic carbon concentrations of surface ocean water on a timescale measured in hundreds of years. If fossil-fuel burning were to cease, a return to the original state of seawater composition could take thousands of years. Ocean water is not in steady state with respect to carbon on these timescales, but on a longer geological timescale it certainly could be. Even on this longer timescale, however, oceanic composition has varied because of natural changes in the carbon dioxide level of the atmosphere and because of other factors.

It appears that the best description of modern seawater composition is that of a chemical system in a dynamic quasi-steady state. Changes in composition may occur over time, but the system always seems to return to a time-averaged, steady-state composition. In other words, since 1.5 to 2 billion years ago, evolutionary chemical changes in the hydrosphere have been small when viewed against the magnitude of previous change.

It should be noted that rivers supply dissolved constituents to the oceans, whereas high- and low-temperature reactions between seawater and submarine basalts, and reactions in sediment pore waters, may add or remove constituents from ocean water. Biological processes involved in the formation of opaline silica skeletons of diatoms and radiolarians, and the carbonate skeletons of planktonic foraminiferans and coccolithophores (types of algae and protozoa) chiefly remove calcium and silica from seawater. Exchange reactions between river-borne clays entering seawater are particularly significant for sodium and calcium ions. Most of the carbon imbalance in ocean water represents carbon released to the ocean–atmosphere system during the precipitation of carbonate minerals.

In the case of iron, it has been documented that "dissolved" iron carried by rivers is rapidly precipitated as hydroxides in the mixing zone with seawater, and that the reduced dissolved iron released from anaerobic sediments is also rapidly precipitated under the prevailing oxic conditions (i.e. those with oxygen present). Iron is also precipitated as iron smectites, hydrated iron oxides, and nontronite (iron-rich montmorillonite) in the deep sea. It is thus likely that iron is removed by these processes.

Chemical and physical properties of seawater

The chemical composition of seawater

The chemical composition of seawater is influenced by a wide variety of chemical transport mechanisms. Rivers add dissolved and particulate chemicals to the oceanic margins. Wind-borne particulates are carried to mid-ocean regions thousands of kilometres from their continental source areas. Hydrothermal solutions that have circulated through crustal materials beneath the sea floor add both dissolved and particulate materials to the deep ocean. Organisms in the upper

ocean convert dissolved materials to solids, which eventually
settle to greater oceanic depths. Particulates in transit to the sea
floor, as well as materials both on and within the sea floor,
undergo chemical exchange with surrounding solutions. These
local and regional chemical input and removal mechanisms
lead to a variation in the spatial and temporal concentrations
of each element in the oceans. Conversely, physical mixing in
the oceans (thermohaline and wind-driven circulation; see
"Circulation of the ocean waters", page 238) tends to homo-
genize the chemical composition. The opposing influences of
physical mixing and of biogeochemical input and removal
mechanisms result in a substantial variety of chemical distributions in the oceans.

Dissolved inorganic substances

In contrast to the behaviour of most oceanic substances, the
concentrations of the principal inorganic constituents of the
oceans are remarkably constant. For 98 per cent of the
oceans' volume, the concentrations of their inorganic constituents vary by less than 3 per cent. Furthermore, with the
exception of inorganic carbon, the principal constituents
have very nearly fixed ion concentration ratios. Calculations
indicate that, for the main inorganic constituents of seawater,
the time required for thorough oceanic mixing is quite short
compared with the time that would be required for input or
removal processes to significantly change a constituent's
concentration. The concentrations of the principal inorganic
constituents of the oceans vary primarily in response to a
comparatively rapid exchange of water through precipitation
and evaporation.

Salinity is used by oceanographers as a measure of the total
salt content of seawater. Practical salinity (symbol S), is
determined through measurements of a ratio between the

electrical conductivity of seawater and the electrical conductivity of a standard solution. Practical salinity can be used to calculate precisely the density of seawater samples. Because of the constant relative proportions of the principal constituents, salinity can also be used directly to calculate the concentrations of the major ions in seawater. The measure of practical salinity was originally developed to provide an approximate measure of the total mass of salt in one kilogram of seawater. (For example, seawater with S equal to 35 contains approximately 35 grams of salt and 965 grams of water.)

Although only 11 constituents (principally chloride (Cl^-), sodium (Na^+), magnesium (Mg^{2+}) and sulphate [SO_4^{2-}]) account for more than 99.5 per cent of the dissolved solids in seawater, many other constituents are of great importance to the biogeochemistry of the oceans. Such chemicals as inorganic phosphorus ($HPO^{2-}/_4$ and $PO^{3-}/_4$) and inorganic nitrogen ($NO^-/_3$, $NO^-/_2$, and $NH^+/_4$) are essential to the growth of marine organisms. Nitrogen and phosphorus are incorporated into the tissues of marine organisms in approximately a 16:1 ratio and are eventually returned to solution in approximately the same proportion. As a consequence, in much of the oceanic waters dissolved inorganic phosphorus and nitrogen exhibit a close covariance (their concentrations are closely linked to each other). Dissolved inorganic phosphorus distributions in the Pacific Ocean strongly bear the imprint of phosphorus incorporation by organisms in the surface waters of the ocean and of the return of the phosphorus to solution via a rain of biological debris remineralized in the deep ocean. Inorganic phosphate concentrations in the western Pacific range from somewhat less than 0.1 micromole per kilogram (1×10^{-7} mole/kg) at the surface to approximately 3 micromoles per kilogram (3×10^{-6} mole/kg) at depth. Inorganic nitrogen ranges between somewhat less than 1 micromole/kg and 45 micromoles/kg along the same section of ocean.

A variety of elements essential to the growth of marine organisms, as well as some elements that have no known biological function, exhibit nutrient-like behaviour broadly similar to nitrate and phosphate. Silicate is incorporated into the hard structural parts of certain types of marine organisms (diatoms and radiolarians) that are abundant in the upper ocean. Dissolved silicate concentrations range between less than 1 micromole/kg (1×10^{-6} mole/kg) in surface waters to approximately 180 micromoles/kg (1.8×10^{-4} mole/kg) in the deep North Pacific. The concentration of zinc, a metal essential to a variety of biological functions, ranges between approximately 0.05 nanomole/kg (5×10^{-11} mole/kg) in the surface ocean to as much as 6 nanomoles/kg (6×10^{-9} mole/kg) in the deep Pacific. The distribution of zinc in the oceans is observed to generally parallel silicate distributions. Cadmium, though having no known biological function, generally exhibits distributions that are covariant with phosphate and concentrations that are even lower than those of zinc.

Many elements, including the essential trace metals iron, cobalt, and copper, show surface depletions but in general exhibit behaviour more complex than that of phosphate, nitrate, and silicate. Some of the complexities observed in elemental oceanic distributions are attributable to the adsorption (the accumulation as a thin film) of elements on the surface of sinking particles. Adsorptive processes, either exclusive of or in addition to biological uptake, serve to remove elements from the upper ocean and deliver them to greater depths. The distribution patterns of a number of trace elements are complicated by their participation in oxidation-reduction (electron-exchange) reactions. In general, electron-exchange reactions lead to profound changes in the solubility and reactivity of trace metals in seawater. Such reactions are important to the oceanic behaviour of a variety of elements, including iron, manganese, copper, cobalt, chromium, and cerium.

The processes that deliver dissolved, particulate, and gaseous materials to the oceans ensure that they contain, at some concentration, very nearly every element that is found in the Earth's crust and atmosphere. The principal components of the atmosphere, nitrogen (78.1 per cent), oxygen (21.0 per cent), argon (0.93 per cent), and carbon dioxide (0.035 per cent), occur in seawater in variable proportions, depending on their solubilities and oceanic chemical reactions. In equilibrium with the atmosphere, the concentrations of the unreactive gases, nitrogen and argon, in seawater (0°C, salinity 35) are 616 micromoles/kg and 17 micromoles/kg, respectively. For seawater at 35°C, these concentrations would decrease by approximately a factor of two.

The solubility behaviours of argon and oxygen are quite similar to each other. For seawater in equilibrium with the atmosphere, the ratio of oxygen and argon concentrations is approximately 20:45. Since oxygen is a reactive gas essential to life, oxygen concentrations in seawater that are not in direct equilibrium with the atmosphere are quite variable. Although oxygen is produced by photosynthetic organisms at shallow, sunlit ocean depths, oxygen concentrations in near-surface waters are established primarily by exchange with the atmosphere. Oxygen concentrations in the oceans generally exhibit minimum values at intermediate depths and relatively high values in deep waters. This distribution pattern results from a combination of biological oxygen utilization and physical mixing of the ocean waters. Estimates of the extent of oxygen utilization in the oceans can be obtained by comparing concentrations of oxygen with those of argon, since the latter are influenced only by physical processes. The physical processes that influence oxygen distributions include, in particular, the large-scale replenishment of oceanic bottom waters with cold, dense, oxygen-rich waters sinking toward the bottom from high latitudes. Due to the release of nutrients that accompanies the consumption of oxygen by biological debris, dissolved oxygen concentrations generally appear as a mirror image of dissolved nutrient concentrations.

While the atmosphere is a vast repository of oxygen compared with the oceans, the total carbon dioxide content of the oceans is very large compared with that of the atmosphere. Carbon dioxide reacts with water in seawater to form carbonic acid (H_2CO_3), bicarbonate ions (HCO^-_3), and carbonate ions (CO^{2-}_3). Approximately 90 per cent of the total organic carbon in seawater is present as bicarbonate ions. The formation of bicarbonate and carbonate ions from carbon dioxide is accompanied by the liberation of hydrogen ions (H^+). Reactions between hydrogen ions and the various forms of inorganic carbon buffer the acidity of seawater. The relatively high concentrations of both total inorganic carbon and boron – as $B(OH)_3$ and $B(OH)^-_4$ – in seawater are sufficient to maintain the pH of seawater between 7.4 and 8.3. This is quite important, because the extent and rate of many reactions in seawater are highly pH-dependent. Carbon dioxide produced by the combination of oxygen and organic carbon generally produces an acidity maximum (pH minimum) near the depth of the oxygen minimum in seawater. In addition to exchange with the atmosphere and, through respiration, with the biosphere, dissolved inorganic carbon concentrations in seawater are influenced by the formation and dissolution of the calcareous shells ($CaCO_3$) of organisms (foraminiferans, coccolithophores, and pteropods) abundant in the upper ocean.

Dissolved organic substances

Processes involving dissolved and particulate organic carbon are of central importance in shaping the chemical character of seawater. Marine organic carbon principally originates in the uppermost 100 metres of the oceans where dissolved inorganic carbon is photosynthetically converted to organic materials. The "rain" of organic-rich particulate materials, resulting directly and indirectly from photosynthetic production, is a principal factor behind the distributions of many organic and inorganic substances in the oceans. A large fraction of the vertical flux of materials in the uppermost waters is converted to dissolved substances within the upper 400 metres of the oceans. Dissolved organic carbon (DOC) accounts for at least 90 per cent of the total organic carbon in the oceans. Estimates of DOC appropriate to the surface of the open ocean are roughly 100–500 micromoles of carbon per kilogram of seawater. DOC concentrations in the deep ocean are five to ten times lower than surface values.

Dissolved organic carbon occurs in an extraordinary variety of forms; in general its composition is controversial and poorly understood. Conventional techniques have indicated that, in surface waters, about 15 per cent of DOC can be identified as carbohydrates and combined amino acids. At least 1–2 per cent of DOC in surface waters occurs as lipids (fat-soluble molecules) and 20–25 per cent as relatively unreactive humic substances (from soil humus). The relative abundances of reactive organic substances, such as amino acids and carbohydrates, are considerably reduced in deep ocean waters. Dissolved and particulate organic carbon in the surface ocean participates in diurnal cycles related to photosynthetic production and photochemical transformations.

The influence of dissolved organic matter on ocean chemistry is often out of proportion to its oceanic abundance.

Photochemical reactions involving DOC can influence the chemistry of vital trace nutrients such as iron, and, even at dissolved concentrations in the order of one nanomole/kg (1 × 10^{-9} mole/kg), dissolved organic substances in the upper ocean waters are capable of greatly altering the bioavailability of essential trace nutrients such as copper and zinc.

Effects of human activities

Although the oceans constitute an enormous reservoir, human activities have begun to influence their composition on both a local and a global scale. The addition of nutrients (through the discharge of untreated sewage or the seepage of soluble mineral fertilizers, for example) to coastal waters results in increased phytoplankton growth, high levels of dissolved and particulate organic materials, decreased penetration of light through seawater, and alteration of the community structure of bottom-dwelling organisms. Through industrial and auto-motive emissions, lead concentrations in the surface ocean have increased dramatically on a global scale compared with pre-industrial levels. Certain toxic organic compounds, such as polychlorinated biphenyls (PCBs), are found in seawater and marine organisms and are attributable solely to the activities of humankind. Although most radioactivity in seawater is nat-ural (approximately 90 per cent as potassium-40 and less than 1 per cent each as rubidium-87 and uranium-238), strontium-90 and certain other artificial radioisotopes have unique environmental pathways and potential for bioaccumulation.

Among the most dramatic influences of human activities on a global scale is the remarkable increase of carbon dioxide levels in the atmosphere. The atmospheric carbon dioxide level is expected to reach double its pre-industrial level by the middle of the twenty-first century, with potentially profound consequences for global climate and agricultural patterns (see

"The hydrosphere and climate change", page 277). It is thought that the oceans, as a great reservoir of carbon dioxide, will ameliorate this consequence of human activities to some degree. However, predictions show that if global warming continues past a certain point, the oceans may stop absorbing carbon dioxide and instead become a carbon source (see "Carbon cycle feedbacks", Chapter 1, page 37).

The physical properties of seawater

Water is a unique substance. Not only is it the most abundant substance at the Earth's surface, but it also has the most naturally occurring physical states of any Earth material or substance (solid, liquid, and gas). It is essential for sustaining life on Earth and affects the physical environment in myriad ways, as evidenced by the sculpting of landscape features by moving water, the maintaining of the Earth's radiation balance by atmospheric water vapour transfer, and the transporting of inorganic and organic materials about the planet's surface by the oceans. The addition of salt to water changes its behaviour only slightly.

Salinity distribution

A discussion of the salinity – salt content – of the oceans requires an understanding of two important concepts: (1) the present-day oceans are considered to be in steady state, receiving as much salt as they lose (see "Modern oceans", page 222), and (2) the oceans have been mixed over such a long time period that the composition of sea salt is the same everywhere in the open ocean.

The range of salinity observed in the open ocean is 33–37 grams of salt per kilogram of seawater or parts per thousand ($^0/_{00}$). For the most part, the observed departure from a mean value of approximately 35 $^0/_{00}$ is caused by processes at the Earth's surface that locally add or remove fresh water. Regions of high evaporation have elevated surface salinities, while regions of high precipitation have depressed surface salinities. In near-shore regions close to large freshwater sources, the salinity may be lowered by dilution. This is especially true in areas where the region of the ocean receiving the fresh water is isolated from the open ocean by the geography of the land. For example, areas of the Baltic Sea may have salinity values depressed to 10 $^0/_{00}$ or less.

Increased salinity by evaporation is accentuated where isolation of the water occurs. This effect is found in the Red Sea, where the surface salinity rises to 41 $^0/_{00}$. Coastal lagoon salinities in areas of high evaporation may be much higher. The removal of fresh water by evaporation or the addition of fresh water by precipitation does not affect the constancy of composition of the sea salt in the open sea. A river draining a particular soil type, however, may bring to the oceans only certain salts that will locally alter the salt composition. In areas of high evaporation where the salinity is driven to very high values, precipitation of particular salts may alter the composition too. At high latitudes where sea ice forms seasonally, the salinity of the seawater is elevated during ice formation and reduced when the ice melts.

At depth in the oceans, salinity may be altered as seawater percolates into fissures associated with deep-ocean ridges and crustal rifts involving volcanism. This water then returns to the ocean as superheated water carrying dissolved salts from the magmatic material within the crust. It may lose much of its dissolved load to precipitates on the sea floor and gradually blend in with the surrounding seawater, sharing its remaining dissolved substances. Salt concentrations as high as 256 $^0/_{00}$ have been found in hot but dense pools of brine trapped in depressions at the bottom of the Red Sea. The composition of the salts in these pools is not the same as the sea salt of the open oceans.

The salinities of the open oceans found at the greater depths are quite uniform in both time and space, with average values of 34.5–35 $^0/_{00}$. These salinities are determined by surface processes such as those described above, when the water, now at depth, was last in contact with the surface.

In addition, the intertropical convergence zone (ITCZ) (see "Zonal surface winds", Chapter 6, page 302), which has high precipitation centred about 5°N and supports the tropical rainforests of the world, leaves its imprint on the oceans as a latitudinal depression of surface salinity. At approximately 30–35°N and 30–35°S, the subtropical zones called the horse latitudes are belts of high evaporation that produce major deserts and grasslands on the continents and cause the surface salinity to rise. At 50–60°N and 50–60°S, precipitation again increases.

Temperature distribution

Mid-ocean surface temperatures vary with latitude in response to the balance between incoming solar radiation and outgoing long-wave radiation. There is an excess of incoming solar radiation at latitudes less than approximately 45° and an excess of radiation loss at latitudes higher than approximately 45°. Superimposed on this radiation balance are seasonal changes in the intensity of solar radiation and the duration of daylight hours, owing to the tilt of the Earth's axis to the plane of the ecliptic and the rotation of the planet about this axis. The combined effect of these variables is that average ocean surface temperatures are higher at low latitudes than at high latitudes. Because the sun, with respect to the Earth, migrates annually between the tropic of Cancer and the tropic of Capricorn, the yearly change in heating of the Earth's surface is small at low latitudes and large at mid- and higher latitudes.

Water has an extremely high heat capacity (it requires a lot of energy to increase its temperature by a given interval), and heat is mixed downward during summer surface-heating conditions and upward during winter surface cooling. This heat transfer reduces the actual change in ocean surface temperatures over the annual cycle. In the tropics the ocean surface is warm year-round, varying seasonally about 1–2°C. At mid-latitudes the mid-ocean temperatures vary about 8°C over the year. At the polar latitudes the surface temperature remains near the ice point of seawater – about -1.9°C.

Land temperatures have a large annual range at high latitudes because of the low heat capacity of the land surface. As a result, proximity to land, isolation of water from the open ocean, and processes that control stability of the surface water combine to increase the annual range of near-shore ocean surface temperatures.

In winter, prevailing winds carry cold air masses off the continents in temperate and subarctic latitudes, cooling the adjacent surface seawater below that of the mid-ocean level. In summer, the opposite effect occurs, as warm continental air masses move out over the adjacent sea. This creates a greater annual range in sea surface temperatures at mid-latitudes on the western sides of the oceans of the northern hemisphere but has only a small effect in the southern hemisphere as there is little land present. Instead, the oceans of the southern hemisphere act to control the air temperature, which in turn influences the land temperatures of the temperate zone and reduces the annual temperature range over the land.

Currents carry water with the characteristics of one latitudinal zone to another zone. The northward displacement of warm water to higher latitudes by the Gulf Stream of the North Atlantic and the Kuroshio (Japan Current) of the North Pacific creates sharp changes in temperature along the current boundaries or thermal fronts, where these northward-moving flows meet colder water flowing southward from higher latitudes. Cold water currents flowing from higher to lower latitudes also displace surface isotherms (cartographic lines showing areas of equal temperature) from near constant latitudinal positions. At low latitudes the trade winds act to move water away from the lee coasts of the landmasses to produce areas of coastal upwelling of water from depth and reduce surface temperatures.

Temperatures in the oceans decrease with increasing depth. There are no seasonal changes at the greater depths. The temperature range extends from 30°C at the sea surface to -1°C at the seabed. Like salinity, the temperature at depth is determined by the conditions that the water encountered when it was last at the surface. In the low latitudes the temperature change from top to bottom in the oceans is large. In high temperate and Arctic regions, the formation of dense water at the surface that sinks to depth produces nearly isothermal conditions with depth.

Areas of the oceans that experience an annual change in surface heating have a shallow wind-mixed layer of elevated temperature in the summer. Below this nearly isothermal layer 10–20 metres thick, the temperature decreases rapidly with depth, forming a shallow seasonal thermocline (a layer of sharp vertical temperature change). During winter cooling and increased wind mixing at the ocean surface, convective over-turning and mixing erase this shallow thermocline and deepen the isothermal layer. The seasonal thermocline re-forms when summer returns. At greater depths, a weaker nonseasonal thermocline is found separating water from temperate and subpolar sources.

Below this permanent thermocline, temperatures decrease slowly. In the very deep ocean basins, the temperature may be observed to increase slightly with depth. This occurs when the deepest parts of the oceans are filled by water with a single temperature from a common source. This water experiences an adiabatic temperature rise (occurring without the addition or subtraction of heat) as it sinks. Such a temperature rise does not make the water column unstable because the increased temperature is caused by compression, which increases the density of the water. For example, surface seawater of 2°C sinking to a depth of 10,000 metres increases its temperature by about 1.3°C. When measuring deep-sea temperatures, the adiabatic temperature rise, which is a function of salinity, initial temperature, and pressure change, is calculated and subtracted from the observed temperature to obtain the potential temperature. Potential temperatures are used to identify a common type of water and to trace this water back to its source.

Circulation of the ocean waters

General observations
The general circulation of the oceans defines the average movement of seawater, which, like the atmosphere, follows a specific pattern. Superimposed on this pattern are oscillations of tides and waves, which are not considered part of the general circulation. There are also meanders and eddies that represent temporal variations of the general circulation. The ocean circulation pattern exchanges water of varying characteristics, such as temperature and salinity, within the interconnected network of oceans and is an important part of the heat and freshwater fluxes of the global climate. Horizontal movements are called currents, which range in magnitude from a few centimetres per second to as much as 4 metres per second. A characteristic surface speed is about 5–50 cm per second.

Currents diminish in intensity with increasing depth. Vertical movements, often referred to as "upwelling" and "downwelling", exhibit much lower speeds, amounting to only a few metres per month. As seawater is nearly incompressible, vertical movements are associated with regions of convergence and divergence in the horizontal flow patterns.

Ocean circulation derives its energy at the sea surface from two sources that define two circulation types: (1) wind-driven circulation, forced by wind stress on the sea surface which induces a momentum exchange, and (2) thermohaline circulation, driven by the variations in water density imposed at the sea surface by the exchange of ocean heat and water with the atmosphere, inducing a buoyancy exchange. These two circulation types are not fully independent, since the sea-air buoyancy and momentum exchange are dependent on wind speed. The wind-driven circulation is the more vigorous of the two and is configured as large gyres that dominate an ocean region. It is strongest in the surface layer. The thermohaline circulation is more sluggish, with a typical speed of 1 cm per second, but this flow extends to the sea floor and forms circulation patterns that envelop the global ocean.

Distribution of ocean currents

Maps of the general circulation at the sea surface are constructed from a vast amount of data obtained from inspecting the residual drift of ships after course direction and speed are accounted for in a process called "dead reckoning". This information is amplified by satellite-tracked drifters at sea. The pattern is nearly entirely that of wind-driven circulation.

Deep-ocean circulation consists mainly of thermohaline circulation. The currents are inferred from the distribution of seawater properties, which trace the spreading of specific water masses. The distribution of the density of a volume of

Major surface currents of the world's oceans.

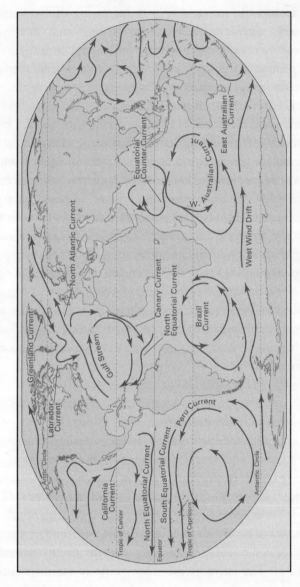

Figure 5.2 © Merriam-Webster, Inc.

water is used to estimate the presence and location of deep currents. Direct observations of sub-surface currents are made by deploying current meters from bottom-anchored moorings and by setting out neutral buoyant instruments whose drift at depth is tracked acoustically.

Causes of ocean currents

The general circulation of ocean currents is governed by the equation of motion, one of Sir Isaac Newton's fundamental laws of mechanics applied to a continuous volume of water. This equation states that the product of mass and current acceleration equals the vector sum of all forces that act on the mass. Besides gravity, the most important forces that cause and affect ocean currents are horizontal pressure-gradient forces (caused by differences in seawater density owing to variations of temperature and salinity), Coriolis forces (the apparent force that results from Earth's rotation) and frictional forces at boundary layers between the ocean, the land and the atmosphere.

Pressure gradients

The hydrostatic pressure, p, at any depth below the sea surface is given by the equation $p = g\rho z$, where g is the acceleration of gravity, ρ is the density of seawater, which increases with depth, and z is the depth below the sea surface. This is called the hydrostatic equation, which is a good approximation for the equation of motion for forces acting along the vertical. Horizontal differences in density (due to variations of temperature and salinity) measured along a specific depth cause the hydrostatic pressure to vary along a horizontal plane or geopotential surface – a surface perpendicular to the direction of the gravity acceleration. Horizontal gradients of pressure, though much smaller than vertical changes in pressure, give rise to ocean currents.

Coriolis force

The rotation of the Earth about its axis causes moving particles to behave in a way that can be understood only by adding a rotational dependent force. To an observer in space, a moving body would continue to move in a straight line unless the motion were acted upon by some other force. To an Earth-bound observer, however, this motion cannot be along a straight line because the reference frame is the rotating Earth. This is similar to the effect that would be experienced by an observer standing on a large turntable if an object moved over the turntable in a straight line relative to the "outside" world. An apparent deflection of the path of the moving object would be seen. If the turntable rotated counterclockwise, the apparent deflection would be to the right of the direction of the moving object, relative to the observer fixed on the turntable. This remarkable effect is evident in the behaviour of ocean currents. It is called the Coriolis force, named after Gustave-Gaspard Coriolis, a nineteenth-century French engineer and mathematician. For the Earth, horizontal deflections due to the rotationally induced Coriolis force act on particles moving in any horizontal direction. There also are apparent vertical forces, but these are of minor importance to ocean currents. Because the Earth rotates from west to east about its axis, an observer in the northern hemisphere would notice a deflection of a moving body toward the right. In the southern hemisphere, this deflection would be toward the left. At the equator there would be no apparent horizontal deflection.

Frictional forces

Movement of water through the oceans is slowed by friction, with surrounding fluid moving at a different velocity. A faster-moving fluid layer tends to drag along a slower-moving layer, and a slower-moving layer will tend to reduce the speed of a faster-moving layer. This momentum transfer between the layers is referred to as frictional force. The momentum transfer is a product of turbulence that moves kinetic energy to smaller scales until at the centimetre scale it is dissipated as heat. The wind blowing over the sea surface transfers momentum to the water. This frictional force at the sea surface (i.e. the wind stress) produces the wind-driven circulation. Currents moving along the ocean floor and the sides of the ocean are also subject to the influence of boundary-layer friction. The motionless ocean floor removes momentum from the circulation of the ocean waters.

Geostrophic currents

For most of the ocean volume away from the boundary layers, which have a characteristic thickness of 100 metres, frictional forces are of minor importance.

On a nonrotating Earth, water would be accelerated by a horizontal pressure gradient and would flow from high to low pressure. On the rotating Earth, however, the Coriolis force deflects the motion, and the acceleration ceases only when the speed of the current is just fast enough to produce a Coriolis force that can exactly balance the horizontal pressure-gradient force. This is called geostrophic balance. From this balance, it follows that the current direction must be perpendicular to the pressure gradient because the Coriolis force always acts perpendicular to the motion. In the northern hemisphere this direction is such that the high pressure is to the right when looking in current direction, while in the southern hemisphere it is to the left. This type of current is called a geostrophic current.

Ekman layer

The wind exerts stress on the ocean surface proportional to its speed and direction, setting the surface water in motion. This motion extends to a depth of about 100 metres in what is called the Ekman layer, after the Swedish oceanographer V. Walfrid Ekman. The average water particle within the Ekman layer moves at an angle of 90° to the wind; this movement is to the right of the wind direction in the northern hemisphere and to its left in the southern hemisphere. This phenomenon is called Ekman transport, and its effects are widely observed in the oceans.

Since the wind varies from place to place, so does the Ekman transport, forming convergence and divergence zones of surface water. A region of convergence forces surface water downward in a process called downwelling, while a region of divergence draws water from below into the surface Ekman layer in a process known as upwelling. Upwelling and downwelling also occur where the wind blows parallel to a coastline. The principal upwelling regions of the world are along the eastern boundary of the subtropical ocean waters, as, for example, the coastal region of Peru and north-western Africa. Upwelling in these regions cools the surface water and brings nutrient-rich subsurface water into the sunlit layer of the ocean, resulting in a biologically productive region. Upwelling and high productivity are also found along divergence zones at the equator and around Antarctica. The primary downwelling regions are in the subtropical ocean waters, e.g. the Sargasso Sea in the North Atlantic. Such areas are devoid of nutrients and are poor in marine life.

Wind-driven circulation
Wind stress induces a circulation pattern that is similar for each ocean. In each case, the wind-driven circulation is divided into large gyres that stretch across the entire ocean: subtropical gyres extend from the equatorial current system to the maximum westerlies in a wind field near 50° latitude, and subpolar

gyres extend poleward of the maximum westerlies (see below). The depth penetration of the wind-driven currents depends on the intensity of ocean stratification: for those regions of strong stratification, such as the tropics, the surface currents extend to a depth of less than 1,000 metres; within the low-stratification polar regions, the wind-driven circulation reaches all the way to the sea floor.

Equatorial currents

Near the equator the currents are for the most part directed toward the west – the North Equatorial Current in the northern hemisphere and the South Equatorial Current in the southern hemisphere. Near the thermal equator, where the warmest surface water is found, there occurs the eastward-flowing Equatorial Counter Current. This current is slightly north of the geographic equator, drawing the northern fringe of the South Equatorial Current to 5°N. The offset to the northern hemisphere matches a similar offset in the wind field. The east-to-west wind across the tropical ocean waters induces Ekman transport divergence at the equator, which cools the surface water there.

At the geographic equator a jet-like current is found just below the sea surface, flowing toward the east counter to the surface current. This is called the Equatorial Undercurrent. It attains speeds of more than 1 metre per second at a depth of nearly 100 metres. It is driven by higher sea level in the western margins of the tropical ocean, producing a pressure gradient, which in the absence of a horizontal Coriolis force drives a west-to-east current along the equator. The wind field reverses the flow within the surface layer, inducing the Equatorial Undercurrent.

Equatorial circulation undergoes variations following the irregular periods of roughly three to eight years of the

Southern Oscillation (fluctuations of atmospheric pressure over the tropical Indo-Pacific region). Weakening of the east-to-west wind during a phase of the Southern Oscillation allows warm water in the western margin to slip back to the east by increasing the flow of the Equatorial Counter Current. Surface-water temperatures and sea level decrease in the west and increase in the east. This event is called El Niño. The combined El Niño–Southern Oscillation (ENSO) effect (see page 264) has received much attention because it is associated with global-scale climatic variability. In the tropical Indian Ocean, the strong seasonal winds of the monsoons induce a similarly strong seasonal circulation pattern.

Subtropical gyres
These are anticyclonic (high-pressure) circulation features. The Ekman transport within these gyres forces surface water to sink, giving rise to the subtropical convergence near 20–30° latitude. The centre of the subtropical gyre is shifted to the west. This westward intensification of ocean currents was explained by the American meteorologist and oceanographer Henry M. Stommel (1948) as resulting from the fact that the horizontal Coriolis force increases with latitude. This causes the poleward-flowing western boundary current to be a jet-like current that attains speeds of two to four metres per second. This current transports the excess heat of the low latitudes to higher latitudes. The flow within the equatorward-flowing interior and eastern boundary of the subtropical gyres is quite different. It is more of a slow drift of cooler water that rarely exceeds 10 cm per second. Associated with these currents is coastal upwelling that results from offshore Ekman transport.

The strongest of the western boundary currents is the Gulf Stream in the North Atlantic Ocean. It carries about 30 million cubic metres of ocean water per second through the Straits of Florida and roughly 80 million cubic metres per second as it flows past Cape Hatteras off the coast of North Carolina. Responding to the large-scale wind field over the North Atlantic, the Gulf Stream separates from the continental margin at Cape Hatteras. After separation, it forms waves or meanders that eventually generate many eddies of warm and cold water. The warm eddies, composed of thermocline water normally found south of the Gulf Stream, are injected into the waters of the continental slope off the coast of the north-eastern United States. They drift to the south-east at rates of approximately 5–8 cm per second, and after a year they rejoin the Gulf Stream north of Cape Hatteras. Cold eddies of slope water are injected into the region south of the Gulf Stream and drift to the south-west. After two years they re-enter the Gulf Stream just north of the Antilles Islands. The path that they follow defines a clockwise-flowing recirculation gyre seaward of the Gulf Stream. (For additional details on the Gulf Stream, see also "Gulf Stream and Kuroshio systems", page 259.)

Among the other western boundary currents, the Kuroshio of the North Pacific is perhaps the most like the Gulf Stream, having a similar transport and array of eddies. The Brazil and East Australian currents are relatively weak. The Agulhas Current has a transport close to that of the Gulf Stream. It remains in contact with the margin of Africa around the southern rim of the continent, then separates from the margin and curls back to the Indian Ocean in what is called the Agulhas Retroflection. Not all the water carried by the Agulhas returns to the east; about 10–20 per cent is injected into the South Atlantic Ocean as large eddies that slowly migrate across it.

Subpolar gyres

The subpolar gyres are cyclonic (low-pressure) circulation features. The Ekman transport within these features forces upwelling and surface-water divergence. In the North Atlantic the subpolar gyre consists of the North Atlantic Current at its equatorward side and the Norwegian Current that carries relatively warm water northward along the coast of Norway. The heat released from the Norwegian Current into the atmosphere maintains a moderate climate in northern Europe. Along the east coast of Greenland is the southward-flowing cold East Greenland Current. It loops around the southern tip of Greenland and continues flowing into the Labrador Sea. The southward flow that continues off the coast of Canada is called the Labrador Current. This current separates for the most part from the coast near Newfoundland to complete the subpolar gyre of the North Atlantic. Some of the cold water of the Labrador Current, however, extends further south.

In the North Pacific the subpolar gyre is composed of the northward-flowing Alaska Current, the Aleutian Current (also known as the Subarctic Current), and the southward-flowing cold Oyashio Current. The North Pacific Current forms the separation between the subpolar and subtropical gyres of the North Pacific.

In the southern hemisphere, the subpolar gyres are less defined. Large cyclonic flowing gyres lie poleward of the Antarctic Circumpolar Current and can be considered counterparts to the Northern Hemispheric subpolar gyres. The best-formed is the Weddell Gyre of the South Atlantic sector of the Southern Ocean. The Antarctic coastal current flows toward the west. The northward-flowing current off the east coast of the Antarctic Peninsula carries cold Antarctic coastal water into the circumpolar belt. Another cyclonic gyre occurs north of the Ross Sea.

Antarctic Circumpolar Current

The Southern Ocean links the major oceans by a deep circumpolar belt in the 50–60°S range. In this belt flows the Antarctic Circumpolar Current from west to east, encircling the globe at high latitudes. It transports 125 million cubic metres of seawater per second over a path of about 24,000 km and is the most important factor in diminishing the differences between oceans. The Antarctic Circumpolar Current is not a well-defined single-axis current but rather consists of a series of individual filaments separated by frontal zones. It reaches the sea floor and is guided along its course by the irregular bottom topography. Large meanders and eddies develop in the current as it flows. These features induce poleward transfer of heat, which may be significant in balancing the oceanic heat loss to the atmosphere above the Antarctic region further south.

Seasonal and interannual ocean–atmosphere interactions

The notion of a connection between the temperature of the surface layers of the oceans and the circulation of the lowest layer of the atmosphere, the troposphere, is one that has been touched on previously. The surface mixed layer of the ocean is a huge reservoir of heat when compared with the overlying atmosphere. The heat capacity of an atmospheric column of unit area cross-section extending from the ocean surface to the outermost layers of the atmosphere is equivalent to the heat capacity of a column of seawater of 2.6-metre depth. The surface layer of the oceans is continuously being stirred by the overlying winds and waves, and thus a surface mixed layer is formed that has vertically uniform properties in temperature and salinity. This mixed layer, which is in direct contact with the atmosphere, has a minimum depth of 20 metres in summer and a maximum depth exceeding 100 metres in late winter in the mid-latitudes. In lower latitudes the seasonal variation in the mixed layer is less marked than at higher latitudes, except in regions such as the Arabian Sea where the onset of the south-western Indian Ocean monsoon may produce large changes in the depth of the mixed layer. Temperature anomalies (i.e. deviations from the normal seasonal temperature) in the surface mixed layer have a long residence time compared with those of the overlying turbulent atmosphere, and may persist for a number of consecutive seasons and even for years.

Observational studies to investigate the relationship between anomalies in ocean surface temperature and the tropospheric circulation have been undertaken primarily in the Pacific and Atlantic. They have identified large-scale ocean surface temperature anomalies that have similar spatial scales to monthly and seasonal anomalies in atmospheric circulation. The longevity of the ocean surface temperature anomalies, as compared with the shorter dynamical and thermodynamical "memory" of the atmosphere, has suggested that they may be an important predictor for seasonal and interannual climate anomalies.

Ocean surface temperature and climate anomalies

It is useful to consider some examples of the association between anomalies in ocean surface temperature and irregular changes in climate. The Sahel, a region that borders the southern fringe of the Sahara in Africa, experienced a number of devastating droughts during the 1970s and 1980s, which can be compared with a much wetter period during the 1950s. Data were obtained that showed the difference in ocean surface temperature during the period from July to September between the driest and wettest rainfall seasons in the Sahel after 1950. Of particular note were the higher-than-normal surface temperatures in the tropical South Atlantic, Indian, and south-east Pacific oceans and the lower-than-normal temperatures in the North Atlantic and Pacific oceans. This example illustrates that climate anomalies in one region of the world may be linked to ocean surface temperature changes on a global scale. Global atmospheric modelling studies under-taken during the mid-1980s have indicated that the positions of the main rainfall zones in the tropics are sensitive to anomalies in ocean surface temperature.

Shorter-lived climate anomalies, on timescales of months to one or two years, also have been related to ocean surface temperature anomalies. The equatorial oceans have the largest influence on these climate anomalies because of the evapora-tion of water. A relatively small change in ocean surface temperature, say, of 1°C, may result in a large change in the evaporation of water into the atmosphere. The increased water vapour in the lower atmosphere is condensed in regions of upward motion known as convergence zones. This process liberates latent heat of condensation, which in turn provides a major fraction of the energy to drive tropical circulation and is one of the mechanisms responsible for the El Niño–Southern Oscillation phenomenon (see page 264).

Given the sensitivity of the tropical atmosphere to variations in tropical sea surface temperature, there also has been con-siderable interest in their influence on extratropical circulation. The sensitivity of the tropospheric circulation to surface tem-perature in both the tropical Pacific and Atlantic oceans has been shown in theoretical and observational studies alike. Figures were prepared to demonstrate the correlation between the equatorial ocean surface temperature in the east Pacific (the location of El Niño) and the atmospheric circulation in the middle troposphere during winter. The atmospheric pattern was a characteristic circulation type known as the Pacific-North American (PNA) mode. Such patterns are intrinsic modes of the atmosphere, which may be forced by thermal anomalies in the tropical atmosphere and which in their turn are forced by tropical ocean surface temperature anomalies. As noted earlier, enhanced tropical sea surface temperatures increase evaporation into the atmosphere. In the 1982–83 El Niño event a pattern of circulation anomalies occurred throughout the northern hemisphere during winter. These modes of the atmosphere, however, account for much less than 50 per cent of the variability of the circulation in mid-latitudes, though in certain regions (northern Japan, southern Canada, and the southern United States), they may have sufficient amplitude for them to be used for predicting seasonal surface temperature perhaps up to two seasons in advance.

The response of the atmosphere to mid-latitude ocean surface anomalies has been difficult to detect unambiguously because of the complexity of the turbulent westerly flow between 20° and 60° latitude in both hemispheres. This flow has many properties of nonlinear chaotic systems and thus exhibits behaviour that is difficult to predict beyond a couple of weeks. The atmosphere alone can exhibit large fluctuations on seasonal and longer timescales without any change in external forcing conditions, such as ocean surface temperature. Notwithstanding this inherent problem, some effects of ocean surface temperature anomalies on the atmosphere have been observed and modelled.

The influence of the oceans on the atmosphere in the mid-latitudes is greatest during autumn and early winter when the ocean mixed layer releases to the atmosphere the large quantities of heat that it has stored up over the previous summer. Anomalies in ocean surface temperature are indicative of either a surplus or a deficiency of heat available to the atmosphere. The response of the atmosphere to ocean surface temperature, however, is not geographically random. The circulation over the North Atlantic and northern Europe during early winter has been found to be sensitive to large ocean surface temperature anomalies south of Newfoundland. When a warm positive anomaly exists in this region, an anomalous surface anticyclone occurs in the central Atlantic at a similar latitude to the temperature anomaly, and an anomalous cyclonic circulation is located over the North Sea, Scandinavia, and central Europe. With colder-than-normal water south of Newfoundland, the circulation patterns are reversed, producing cyclonic circulation over the central Atlantic and anticyclonic circulation over Europe. The sensitivity of the atmosphere to ocean surface temperature anomalies in this particular region is thought to be related to the position of the overlying storm tracks and jet stream. The region is the most active in the northern hemisphere for the growth of storms associated with very large heat fluxes from the surface layer of the ocean.

Another example of a similar type of air-sea interaction event has been documented over the North Pacific Ocean. A statistical seasonal relationship exists between the summer ocean temperature anomaly in the Gulf of Alaska and the atmospheric circulation over the Pacific and North America during the following autumn and winter. The presence of warmer-than-normal ocean surface temperature in the Gulf of Alaska results in increased cyclone development during the subsequent autumn and winter. The relationship has been established by means of monthly sea-surface-temperature and atmospheric-pressure data collected over 30 years in the North Pacific Ocean.

The air–sea interaction events in both the North Pacific and North Atlantic oceans discussed above raise questions as to how the anomalies in ocean surface temperature in these areas are initiated, how they are maintained, and whether they yield useful information for atmospheric prediction beyond the normal timescales of weather forecasting (namely one to two weeks). Statistical analyses of previous case studies have shown that ocean surface temperature anomalies initially develop in response to anomalous atmospheric forcing. Once developed, however, the temperature anomaly of the ocean surface tends to reinforce and thereby maintain the anomalous atmospheric circulation. The mechanisms thought to be responsible for this behaviour in the ocean are the surface wind drift, wind mixing, and the interchange of heat between the ocean and atmosphere. The question of prediction is therefore difficult to answer, as these events depend on a synchronous and interconnected behaviour between the atmosphere and the surface layer of the ocean, which allows for positive feedback between the two systems.

Tropical cyclones represent still another example of air–sea interactions. These storm systems are known as hurricanes in the North Atlantic and eastern North Pacific and as typhoons in the western North Pacific. The winds of such systems revolve around a centre of low pressure in an anticlockwise direction in the northern hemisphere and in a clockwise direction in the southern hemisphere. The winds attain velocities in excess of 115 km per hour, or 65 knots, in most cases. Tropical cyclones may last from a few hours to as long as two weeks, their average lifetime being six days.

The oceans provide the source of energy for tropical cyclones both by direct heat transfer from their surface (known as sensible heat) and by the evaporation of water. This water is subsequently condensed within a storm system, thereby releasing latent heat energy. When a tropical cyclone moves over land, this energy is severely depleted and the circulation of the winds is consequently weakened.

Such storms are truly phenomena of the tropical oceans. They originate in two distinct latitude zones, between 4°S and

22°S and between 4°N and 35°N. They are absent in the equatorial zone between 4°S and 4°N. Most tropical cyclones are spawned on the poleward side of the region known as the intertropical convergence zone (ITCZ).

More than two thirds of observed tropical cyclones originate in the northern hemisphere. The North Pacific has more than one third of all such storms, while the south-east Pacific and South Atlantic are normally devoid of them. Most northern hemispheric tropical cyclones occur between May and November, with peak periods in August and September. The majority of southern hemispheric cyclones occur between December and April, with peaks in January and February.

Conditions associated with cyclone formation

The formation of tropical cyclones is strongly influenced by the temperature of the underlying ocean or, more specifically, by the thermal energy available in the upper 60 metres of ocean waters. Typically, the underlying ocean should have a temperature in excess of 26°C in this layer. This temperature requirement, however, is only one of five that need to be met for a tropical cyclone to form and develop. The other preconditions relate to the state of the tropical atmosphere between the sea surface and a height of 16 km, the boundary of the tropical troposphere. They can be summarized as follows.

A deep convergence of air must occur in the troposphere between the surface and a height of 7 km that produces a cyclonic circulation in the lower troposphere overlain by an anticyclonic circulation in the upper troposphere. The stronger the inflow, or convergence, of the air, the more favourable are the conditions for tropical cyclone formation.

The vertical shear (difference in wind speed and direction) of the horizontal wind velocity between the lower troposphere and the upper troposphere should be at minimum. Under this condition the heat and moisture are retained rather than being exchanged and diluted with the surrounding air. Monsoonal and trade wind flows are characterized by a large vertical shear of the horizontal wind and so are not generally conducive to tropical cyclone development.

A strong vertical coupling of the flow patterns between the upper and lower troposphere is required. This is achieved by large-scale deep convection associated with cumulonimbus clouds.

A high humidity level in the middle troposphere from 3 to 6 kilometres in height is more conducive to the production of deep cumulonimbus convection and therefore to stronger vertical coupling in the troposphere.

All these conditions may be met but still not lead to cyclone formation. It is thought that the most important factor is the presence of a large-scale cyclonic circulation in the lower troposphere. The above conditions occur for a period of 5 to 15 days and are followed by less favourable conditions for a duration of 10 to 20 days.

Once a tropical cyclone has formed, it usually follows certain distinct stages during its lifetime. In its formative stage the winds are below hurricane force and the central pressure is about 1,000 millibars. The formative period is extremely variable in length, ranging from 12 hours to a few days. This stage is followed by a period of intensification, when the central pressure drops rapidly below 1,000 millibars. The winds increase rapidly, and they may achieve hurricane force within a radius of 30–50 km of the storm centre. At this stage the cloud and rainfall patterns become well organized into narrow bands that spiral inward toward the centre. In the mature phase the central pressure stops falling and, as a consequence, the winds no longer increase. The region of hurricane force winds, however, expands to occupy a radius of 300 km or more. This expansion is not symmetrical around the storm centre: the strongest winds occur toward the right-hand side of the centre in the direction of the cyclone's path. The period of maturity may last for one to three days. The terminal stage of a tropical cyclone is usually reached when the storm strikes land, followed by a resultant increase in energy dissipation by surface friction and a reduction in its energy supply of moisture. A reduction in moisture input into the storm system may also take place when it moves over a colder segment of the ocean. A tropical cyclone may regenerate in higher latitudes as an extratropical depression, but it loses its identity as a tropical storm in the process.

The paths of tropical cyclones show a wide variation. In both the North Atlantic and the North Pacific, the paths tend to be initially north-westward and then recurve toward the north-east at higher latitudes. It is now known that the tracks of tropical cyclones are largely determined by the large-scale tropospheric flow. This fact opens up the possibility that, with the aid of high-resolution numerical models, accurate predictions of their tracks may become feasible. The development of polar-orbiting and geostationary satellites (see Chapter 3, page 173) has made it possible to accurately track cyclones over the remotest areas of the tropical oceans.

Effects of tropical cyclones on ocean waters

A tropical cyclone can affect the thermal structure and currents in the surface layer of the ocean waters in its path. Cooling of the surface layer occurs in the wake of such a storm. Maximum cooling occurs on the right of a hurricane's path in the northern hemisphere. In the wake of Hurricane Hilda's passage through the Gulf of Mexico in 1964 at a translational speed of only 5 knots, the surface waters were cooled by as much as 6°C. Tropical cyclones that have higher translational velocities cause less cooling of the surface. The surface cooling is caused primarily by wind-induced upwelling of cooler water from below the surface layer. The warm surface water is simultaneously transported toward the periphery of the cyclone, where it downwells into the deeper ocean layers. Heat loss across the air–sea interface and the wind-induced mixing of the surface water with those of the cooler subsurface layers make a significant but smaller contribution to surface cooling.

In addition to surface cooling, tropical cyclones may induce large horizontal surge currents and vertical displacements of the thermocline. The surge currents have their largest amplitude at the surface, where they may reach velocities approaching one metre per second. The horizontal currents and the vertical displacement of the thermocline observed in the wake of a tropical cyclone oscillate close to the inertial period. These oscillations remain for a few days after the passage of the storm and spread outward from the rear of the system as an internal wake on the thermocline. The vertical motion may transport nutrients from the deeper layers into the sunlit surface waters, which in turn promotes phytoplankton blooms. The ocean surface temperature normally recovers to its pre-cyclone value within ten days of a storm's passage.

Influence on atmospheric circulation and rainfall

Tropical cyclones play an important role in the general circulation of the atmosphere, accounting for 2 per cent of the global annual rainfall and 4–5 per cent of the global rainfall in August and September at the height of the northern hemispheric cyclone season. For a local area, the occurrence of a single tropical cyclone can have a major impact on the region's annual rainfall. Furthermore, tropical cyclones contribute approximately 2 per cent of the kinetic energy of the general circulation of the atmosphere, some of which is exported from the tropics to higher latitudes.

Gulf Stream and Kuroshio systems

Gulf Stream

This major current system is a western boundary current that flows poleward along a boundary separating the warm and more saline waters of the Sargasso Sea to the east from the colder, slightly fresher continental slope waters to the north and west. The warm, saline Sargasso Sea, composed of a water mass known as North Atlantic Central Water, has a temperature that ranges from 8°C to 19°C and a salinity of 35.10 and 36.70 parts per thousand. This is one of the two dominant water masses of the North Atlantic Ocean, the other being the North Atlantic Deep Water, which has a temperature of 2.2°C to 3.5°C and a salinity of 34.90–34.97 parts per thousand, and which occupies the deepest layers of the ocean (generally below 1,000 metres). The North Atlantic Central Water occupies the upper layer of the North Atlantic Ocean between roughly 20°N and 40°N. The stream or "lens" of this water is at its lowest depth of 1,000 metres in the north-west Atlantic and becomes progressively shallower to the east and south. To the north it shallows abruptly and outcrops at the surface in winter, and it is at this point that the Gulf Stream is most intense.

The Gulf Stream flows along the rim of the warm North Atlantic Central Water northward from the Florida Straits along the continental slope of North America to Cape Hatteras. There, it leaves the continental slope and turns north-eastward as an intense meandering current that extends toward the Grand Banks of Newfoundland. Its maximum velocity is typically between one and two metres per second. At this stage, a part of the current loops back on itself, flowing south and east.

Another part flows eastward toward Spain and Portugal, while the remaining water flows north-eastward as the North Atlantic Drift (also called the North Atlantic Current) into the northernmost regions of the North Atlantic Ocean between Scotland and Iceland.

The southward-flowing currents are generally weaker than the Gulf Stream and occur in the eastern lens of the North Atlantic Central Water or the subtropical gyre (see "Subtropical gyres", page 246). The circulation to the south on the southern rim of the subtropical gyre is completed by the westward-flowing North Equatorial Current, part of which flows into the Gulf of Mexico; the remaining part flows northward as the Antilles Current. This subtropical gyre of warm North Atlantic Central Water is the hub of the energy that drives the North Atlantic circulation. It is principally forced by the overlying atmospheric circulation, which at these latitudes is dominated by the clockwise circulation of a subtropical anticyclone. This circulation is not steady and fluctuates in particular on its poleward side, where extratropical cyclones in the westerlies periodically make incursions into the region. On the western side, hurricanes (during the period from May to November) occasionally disturb the atmospheric circulation. However, because of the energy of the subtropical gyre and its associated currents, these short-term fluctuations have little influence on it.

The gyre obtains most of its energy from the climatological wind distribution over periods of one or two decades. This wind distribution drives a system of surface currents in the uppermost 100 metres of the ocean. Nonetheless, these currents are not simply a reflection of the surface wind circulation as they are influenced by the Coriolis force (see page 242). The wind-driven current decays with depth, becoming negligible below 100 metres. The water in this surface layer is transported to the right and perpendicular to the surface wind stress because of the Coriolis force. Hence an eastward-directed wind on the poleward side of the subtropical anticyclone would transport the surface layer of the ocean to the south. On the equatorward side of the anticyclone, the trade winds would cause a contrary drift of the surface layer to the north and west. Thus surface waters under the subtropical anticyclone are driven toward the mid-latitudes at about 30°N. These surface waters, which are warmed by solar heating and have a high salinity by virtue of the predominance of evaporation over precipitation at these latitudes, then converge and are forced downward into the deeper ocean.

Over many decades this process forms a deep lens of warm, saline North Atlantic Central Water. The shape of the lens of water is distorted by other dynamical effects, the principal one being the change in the vertical component of the Coriolis force with latitude, known as the beta effect. This effect involves the displacement of the warm water lens toward the west, so that the deepest part of the lens is situated to the north of the island of Bermuda rather than in the central Atlantic Ocean. This warm lens of water plays an important role, establishing as it does a horizontal pressure-gradient force in and below the wind-drift current. The sea level over the deepest part of the lens is about one metre higher than outside the lens. The Coriolis force in balance with this horizontal pressure gradient force gives rise to a dynamically induced geostrophic current, which occurs throughout the upper layer of warm water. The strength of this geostrophic current is determined by the horizontal pressure gradient through the slope in sea level. The slope in sea level across the Gulf Stream has been measured by satellite radar altimeter to be one metre over a horizontal distance of 100 kilometres, which is sufficient to cause a surface geostrophic current of one metre per second at 43°N.

The large-scale circulation of the Gulf Stream system is, however, only one aspect of a far more complex and richer structure of circulation. Embedded within the mean flow is a variety of eddy structures that not only put kinetic energy into circulation but also carry heat and other important properties, such as nutrients for biological systems. The best known of these eddies are the Gulf Stream rings, which develop in meanders of the current east of Cape Hatteras. Though the eddies were mentioned as early as 1793 by Jonathan Williams, a grandnephew of Benjamin Franklin, they were not systematically studied until the early 1930s, by the oceanographer Phil E. Church. Intensive research programmes were finally undertaken during the 1970s.

Gulf Stream rings have either warm or cold cores. The warm rings are typically 100 to 300 km in diameter and have a clockwise rotation. They consist of waters from the Gulf Stream and Sargasso Sea and form when the meanders in the Gulf Stream pinch off on its continental slope side. They move generally westward, flowing at the speed of the slope waters, and are reabsorbed into the Gulf Stream at Cape Hatteras after a typical lifetime of about six months. The cold core rings, composed of a mixture of Gulf Stream and continental slope waters, are formed when the meanders pinch off to the south of the Gulf Stream. They are a little larger than their warm-core counterparts, characteristically having diameters of 200–300 km and an anticlockwise rotation. They move generally south-westward into the Sargasso Sea and have lifetimes of one to two years. The cold-core rings are usually more numerous than warm-core rings, typically numbering ten each year as compared with five warm-core rings annually.

Kuroshio

This western boundary current is similar to the Gulf Stream in that it produces both warm and cold rings. The warm rings are generally 150 km in diameter and have a lifetime similar to their Gulf Stream counterparts. The cold rings form at preferential sites and in most cases drift south-westward into the western Pacific Ocean. Occasionally a cold ring has been observed to move north-westward and eventually be reabsorbed into the Kuroshio.

Poleward transfer of heat

A significant characteristic of the large-scale North Atlantic circulation is the poleward transport of heat. Heat is transferred in a northward direction throughout the North Atlantic. This heat is absorbed by the tropical waters of the Pacific and Indian oceans, as well as of the Atlantic, and is then transferred to the high latitudes, where it is finally given up to the atmosphere.

The mechanism for the heat transfer is principally by thermohaline circulation rather than by wind-driven circulation (see "Circulation of the ocean waters", page 238). Circulation of the thermohaline type involves a large-scale overturning of the ocean, with warm and saline water in the upper 1,000 metres moving northward and being cooled in the Labrador, Greenland, and Norwegian seas. The density of the water in contact with the atmosphere is increased by surface cooling, and the water subsequently sinks below the surface layer to the lowest depths of the ocean. This water is mixed with the surrounding water masses by a variety of processes to form North Atlantic Deep Water. The water moves slowly southward as the lower limb of the thermohaline circulation, and it is this overturning circulation, rather than the horizontal wind-driven circulation discussed above, that is responsible for the warm winter climate of north-western Europe (notably the British Isles and Norway).

The North Atlantic Drift, which is an extension of the Gulf Stream system to the south, provides this northward flow of warm and saline waters into the polar seas. This feature makes the circulation of the North Atlantic Ocean uniquely different from that of the Pacific Ocean, which has a less effective thermohaline circulation. Although there is a northward transfer of heat in the North Pacific, the subtropical wind-driven gyre in the upper ocean is mainly responsible for it. Thus the Kuroshio on the western boundary of the North Pacific gyre is principally driven by the surface wind circulation of the North Pacific.

Studies of sediment cores obtained from the ocean floor have indicated that the ocean surface temperature was as much as 10°C cooler than it is today in the northernmost region of the North Atlantic Ocean during the last glacial maximum some 18,000 years ago. This difference in surface temperature would indicate that the warm North Atlantic Drift was much reduced then compared with the present, and hence the thermohaline circulation was considerably weaker. In contrast, the Gulf Stream was probably more intense than it is today and exhibited a large shift from its present path to an eastward flow at 40°N.

El Niño–Southern Oscillation and climatic change

As explained earlier, the oceans can moderate the climate of certain regions. Not only do they affect such geographic variations, but they also influence temporal changes in climate. The timescales of climate variability range from a few years to millions of years, and include the so-called ice age cycles that repeat every 20,000 to 40,000 years. These are interrupted by interglacial periods of "optimum" climate, such as the one we are experiencing at present (see Chapter 2, page 140). The climatic modulations that occur at shorter scales include such periods as the Little Ice Age from the early sixteenth to the mid-nineteenth centuries, when the global average temperature was approximately 1°C (1.8°F) lower than it is today. Several climate fluctuations on the scale of decades have occurred in the twentieth century, such as warming from 1910 to 1940, cooling from 1940 to 1970, and the warming trend since 1970.

Although many of the mechanisms of climate change are understood, it is usually difficult to pinpoint the specific causes. Scientists acknowledge that climate can be affected by factors external to the land–ocean–atmosphere climate system, such as variations in solar brightness, the shading effect of aerosols injected into the atmosphere by volcanic activity, or the increased atmospheric concentration of green-house gases produced by human activities. However, none of these factors explains the periodic variations observed during the twentieth century, which may simply be manifestations of the natural variability of climate. The existence of natural variability at many timescales makes the identification of causative factors such as human-induced warming more difficult. Whether change is natural or caused, the oceans play a key role in, and have a moderating effect on, influencing factors.

El Niño

The shortest, or interannual, timescale of climatic change relates to natural variations that are perceived as years of unusual weather (e.g. excessive heat, drought, or storminess). Such changes are so common in many regions that any given year is just as likely to be considered as exceptional as typical. The best example of the influence of the oceans on interannual climate anomalies is the occurrence of El Niño conditions in the eastern Pacific Ocean at irregular intervals of about three to ten years. The stronger El Niño episodes of enhanced ocean temperatures (2–8°C [about 3.6–14°F] above normal) are typically accompanied by altered weather patterns around the globe, such as droughts in Australia, north-eastern Brazil, and the highlands of southern Peru; excessive summer rainfall along the coast of Ecuador and northern Peru; severe winter storminess along the coast of central Chile; and unusual winter weather along the west coast of North America.

The effects of El Niño have been documented in Peru since the Spanish conquest in 1525. The Spanish term "la corriente de El Niño" was introduced by fishermen of the Peruvian port of Paita in the nineteenth century; it refers to a warm, south-ward ocean current that temporarily displaces the normally cool, northward-flowing Humboldt, or Peru, Current. (The name "El Niño" is a pious reference to the Christ child, chosen because of the typical appearance of the countercurrent during the Christmas season.) By the end of the nineteenth century Peruvian geographers recognized that every few years this countercurrent is more intense than normal, extends further south, and is associated with torrential rainfall over the otherwise dry northern desert. The abnormal countercurrent also was observed to bring tropical debris, as well as such flora and fauna as bananas and aquatic reptiles, from the coastal region of Ecuador further north. Increasingly during the twentieth century, El Niño has come to connote an exceptional year rather than the original annual event.

As Peruvians began to exploit the guano of marine birds for fertilizer in the early twentieth century, they noticed El Niño-related deteriorations in the normally high marine produc-tivity of the coast of Peru, manifested by large reductions in the bird populations that depend on anchovies and sardines for sustenance. The preoccupation with El Niño increased after the mid-twentieth century, as the Peruvian fishing industry rapidly expanded to exploit the anchovies directly. (Fish meal produced from the anchovies was exported to industrialized nations as a feed supplement for livestock.) By 1971 the Peruvian fishing fleet had become the largest in its history; it had extracted very nearly 13 million tonnes of anchovies in that year alone. Peru was catapulted into first place among fishing nations, and scientists expressed serious concern that fish stocks were being depleted beyond self-sustaining levels, even for the extremely productive marine ecosystem of Peru. The strong El Niño of 1972–73 captured world attention because of the drastic reduction in anchovy catches to a small fraction of prior levels. The anchovy catch did not return to previous levels, and the effects of plummeting fish meal exports reverberated throughout the world commodity markets.

El Niño was only a curiosity to the scientific community in the first half of the twentieth century, as it was thought to be geographically limited to the west coast of South America. There was little data, mainly gathered coincidentally from foreign oceanographic cruises, and it was generally believed that El Niño occurred when the normally northward coastal winds off Peru, which cause the upwelling of cool, nutrient-rich water along the coast, decreased, ceased, or reversed in direction. When systematic and extensive oceanographic measurements were made in the Pacific in 1957–58 as part of the International Geophysical Year, it was found that El Niño had occurred during the same period and was also associated with extensive warming over most of the Pacific equatorial zone. Eventually, tide-gauge and other measurements made throughout the tropical Pacific showed that the coastal El Niño was but one manifestation of basinwide ocean circulation changes that occur in response to a massive weakening of the westward-blowing trade winds in the western and central equatorial Pacific, and not to localized wind anomalies along the Peru coast.

One of the most intense El Niño events of the twentieth century began in mid-1982 and ended in mid-1983. Sea-surface temperatures in the eastern tropical Pacific and much of the equatorial zone further west were 5–10°C (9–18°F) above normal. Australia was hit by severe drought, typhoons occurred as far east as Tahiti, and central Chile suffered from record rainfall and flooding. The west coast of North America was unusually stormy during the winter of 1982–83, and fish catches were dramatically altered from Mexico to Alaska.

Southern Oscillation

Wind anomalies are a manifestation of an atmospheric counterpart to the oceanic El Niño. At the turn of the twentieth century, the British climatologist Gilbert Walker set out to determine the connections between the Asian monsoon and other climatic fluctuations around the globe, in an effort to predict unusual monsoon years that bring drought and famine to the Asian sector. Unaware of any connection to El Niño, he discovered a coherent interannual fluctuation of atmospheric pressure over the tropical Indo-Pacific region, which he termed the Southern Oscillation (SO). During years of reduced rainfall over northern Australia and Indonesia, the pressure in that region (at what are now Darwin and Jakarta) was anomalously high and wind patterns were altered. Simultaneously, in the eastern South Pacific pressures were unusually low, negatively correlated with those at Darwin and Jakarta. A Southern Oscillation Index (SOI), based on pressure differences between the two regions (east minus west), showed low, negative values at such times, which were termed the "low phase" of the SO. During more normal "high-phase" years, the pressures were low over Indonesia and high in the eastern Pacific, with high, positive values of the SOI. In papers published during the 1920s and 1930s, Walker gave statistical evidence for widespread climatic anomalies around the globe being associated with the SO pressure "seesaw".

In the 1950s, years after Walker's investigations, it was noted that the low-phase years of the SOI corresponded with periods of high ocean temperatures along the Peruvian coast. However, no physical connection between the SO and El Niño was recognized until Jacob Bjerknes, in the early 1960s, tried to understand the large geographic scale of the anomalies observed during the 1957–58 El Niño event. Bjerknes, a meteorologist, formulated the first conceptual model of the large-scale ocean–atmosphere interactions that occur during El Niño episodes. His model has been refined through intensive research since the early 1970s.

During a year or two prior to an El Niño event (high-phase years of the SO), the westward trade winds typically blow more intensely along the equator in the equatorial Pacific, causing warm upper-ocean water to accumulate in a thickened surface layer in the western Pacific where sea level rises. Meanwhile, the stronger, upwelling-favourable winds in the eastern Pacific induce colder surface water and lowered sea levels off South America. Towards the end of the year preceding an El Niño, the area of intense tropical storm activity over Indonesia migrates eastward toward the equatorial Pacific west of the International Date Line (which corresponds in general to the 180th meridian of longitude), bringing episodes of eastward wind reversals to that region of the ocean. These wind bursts excite extremely long ocean waves, known as Kelvin waves (imperceptible to an observer), that propagate eastward toward the coast of South America, where they cause the upper ocean layer of relatively warm water to thicken and the sea level to rise.

The tropical storms of the western Pacific also occur in other years, though less frequently, and produce similar Kelvin waves, but an El Niño event does not result and the waves continue poleward along the coast toward Chile and California, detectable only in tide-gauge measurements. Something else occurs prior to an El Niño that is not fully understood: as the Kelvin waves travel eastward along the equator, an anomalous eastward current carries warm western Pacific water further east, and the warm surface layer deepens in the central equatorial Pacific (east of the International Date Line). Additional surface warming takes place as the upwelling-favourable winds bring warmer subsurface water to the surface. (The subsurface water is warmer now, rather than cooler, because the overlying layer of warmer water is now significantly deeper than before.) The anomalous warming creates conditions favourable for the further migration of the tropical storm centre toward the east, giving renewed vigour to eastward winds, more Kelvin waves, and additional warming. Each increment of anomalies in one medium (e.g. the ocean) induces further anomalies in the other (the atmosphere) and vice versa, giving rise to an unstable growth of anomalies through a process of positive feedbacks. During this time, the SO is found in its low phase.

After several months of these unstable ocean–atmosphere interactions, the entire equatorial zone becomes considerably warmer (2–5°C [3.6–9°F]) than normal, and a sizeable volume of warm upper-ocean water is transported from the western to the eastern Pacific. As a result, sea levels fall by 10–20 cm in the west and rise by larger amounts off the coast of South America, where sea surface temperature anomalies may vary from 2°C to 8°C (3.6–14°F) above normal. Anomalous conditions typically persist for 10 to 14 months before returning to normal. This warming off South America occurs even though the upwelling-favourable winds there continue una-bated. The upwelled water is warmer during this period, and its associated nutrients are less plentiful. As a result, local marine ecosystems are less productive than usual. The current focus of oceanographic research is on understanding the circumstances leading to the demise of the El Niño event and the onset of another such event several years later. The most widely held hypothesis is that a second class of long equatorial ocean waves – Rossby waves, with a shallow sur-face layer (see Chapter 6, page 321) – is generated by the El Niño and that they propagate westward to the landmasses of Asia. There, the Rossby waves reflect off the Asian coast eastward along the equator in the form of upwelling Kelvin waves, resulting in a thinning of the upper-ocean warm layer and a cooling of the ocean as the winds bring deeper, cooler water to the surface. This process is thought to initiate one to two years of colder-than-average conditions until Rossby waves of a contrary sense (i.e. with a thickened surface layer) are again generated, functioning as a switching mechanism, this time to start another El Niño sequence.

Another goal of scientists is to understand climate change on the scale of centuries or longer, and to make projections about the changes that will occur within the next few generations. Yet determinations of current climatic trends from recent data are made difficult by natural variability at shorter timescales, such as the El Niño phenomenon. Many scientists are attempting to understand the mechanisms of change during an El Niño event from improved global measurements so as to determine how the ocean–atmosphere engine operates at longer timescales. Others are studying prehistoric records preserved in trees, sediments, and fossil corals in an effort to reconstruct past variations, including those like the El Niño. Their aim is to remove such short-term variations in order to be able to make more accurate estimates of long-term trends.

Sea ice

Sea ice (frozen seawater) is present within the Arctic Ocean and its adjacent seas as far south as China and Japan, and in the seas surrounding Antarctica. Most sea ice occurs as pack ice, which is very mobile, drifting across the ocean surface under the influence of the wind and ocean currents and moving vertically under the influence of tides, waves, and swells. There is also landfast ice, or fast ice, which is immobile, since it is either attached directly to the coast or sea floor or locked in place between grounded icebergs. Fast ice grows in place by freezing of seawater or by pack ice becoming attached to the shore, sea floor, or icebergs. Fast ice moves up and down in response to tides, waves, and swells, and pieces may break off and become part of the pack ice. A third type of sea ice, known as marine ice, forms far below the ocean surface at the bottom of ice shelves in Antarctica. Occasionally seen in icebergs that calve from the ice shelves, marine ice can appear green owing to organic matter within it.

Sea ice that is not more than one winter old is known as first-year ice. Sea ice that survives one or more summers is known as multiyear ice. Most Antarctic sea ice is first-year pack ice. Multiyear ice is common in the Arctic, where most of it occurs as pack ice in the Arctic Ocean.

Sea ice undergoes large seasonal changes in extent as the ocean freezes and the ice cover expands in the autumn and winter, followed by a period of melting and retreat in the spring and summer. Northern-hemisphere sea ice extent typically ranges from approximately 8 million square km (3.1

million square miles) in September to approximately 15 million square km (6 million square miles) in March. Southern-hemisphere sea ice extent ranges from approximately 4 million square km (1.5 million square miles) in February to approximately 20 million square km (7.7 million square miles) in September. In September 2007 the sea ice extent in the northern hemisphere declined to roughly 4.1 million square km, a figure some 50 per cent below mean sea ice coverage for that time of year. Globally, the minimum and maximum sea ice extents are about 20 million square km and 30 million square km (12 million square miles), respectively. Measured routinely using data obtained from orbiting satellite instruments, the minimum and maximum sea ice extent figures vary annually and by decade. These figures are important factors for understanding polar and global climatic variation and change.

Sea ice formation and features

Ice salinity, temperature, and interactions

As seawater freezes and ice forms, liquid brine and air are trapped within a matrix of pure ice crystals. Solid salt crystals subsequently precipitate in pockets of brine within the ice. The brine volume and chemical composition of the solid salts are temperature dependent.

Liquid ocean water has an average salinity of 35 parts per thousand. New ice such as nilas (a type of new ice that is up to 10 cm [4 inches] thick and looks dark grey) has the highest average salinity (12–15 parts per thousand); as ice grows thicker during the course of the winter, the average salinity of the entire ice thickness decreases as brine is lost from it. Brine loss occurs by temperature-dependent brine-pocket migration, brine expulsion, and, most importantly, by gravity drainage via a network of cells and channels. At the end of winter, Arctic first-year ice has an average salinity of 4–6 parts per thousand. Antarctic first-year ice is more saline, perhaps because ice growth rates are more rapid than in the Arctic, and granular ice traps more brine.

In summer, gravity drainage of brine increases as the ice temperature and permeability increase. In the Arctic, summer gravity drainage is enhanced by flushing, as snow and ice meltwater percolate into the ice. Consequently, after a few summers the ice at the surface is completely desalinated and the average salinity of Arctic multiyear ice drops to 3–4 parts per thousand. Antarctic multiyear ice is more saline because the snow rarely melts completely at the ice surface, and brine flushing is uncommon. Instead of percolating into the ice, snow meltwater refreezes onto the ice surface, forming a layer of hard, glassy ice. In contrast, even though it forms from platelets in seawater, marine ice contains little or no salt. The reasons for this remain unclear, but possible explanations include the densification of the ice crystals or their desalination by convection within the "mushy" crystal layer.

Because sea ice is porous and permeable and the brine held within it contains nutrients, sea ice often harbours rich and complex ecosystems. Viruses, bacteria, algae, fungi, and protozoans inhabit sea ice, taking advantage of the differences in salinity, temperature, and light levels. Algae are perhaps the most obvious manifestation of the sea ice ecosystem because they are pigmented and darken the ice. Algae are found at the top, bottom, and interior of Antarctic sea ice, but they are found primarily at the bottom of Arctic sea ice, where they can occur as strands many metres in length. Sea ice algae are important as a concentrated food source for krill and other zooplankton. Melting sea ice rich in algae may also be important for seeding phytoplankton blooms in the previously ice-covered ocean.

Pack ice drift and thickness

The large-scale drift of sea ice in the Arctic Ocean is dominated by the Beaufort Gyre (a roughly circular current flowing clockwise within the surface waters of the Beaufort Sea in the western or North American Arctic) and the Transpolar Drift (the major current flowing into the Atlantic Ocean from the eastern or Eurasian Arctic). The clockwise rotation of the Beaufort Gyre and the movement of the Transpolar Drift, the result of large-scale atmospheric circulation, are dominated by a high-pressure centre over the western Arctic Ocean. The pattern is not constant but varies in both strength and position about every decade or so, as the high-pressure centre weakens and moves closer to Alaska and the Canadian Arctic. This decadal shift in the high-pressure centre is known as the Arctic Oscillation.

The Transpolar Drift exports large volumes of ice from the Arctic Ocean south through the Fram Strait and along the east coast of Greenland into the North Atlantic Ocean. Ice drift speeds, determined from buoys placed on the ice, average 10–15 km (about 6–9 miles) per day in the Fram Strait. Ice can drift in the Beaufort Gyre for as much as seven years at rates that vary between zero at the centre to an average of 4–5 km (about 2.5–3 miles) per day at the edge. Together, the Beaufort Gyre and Transpolar Drift strongly influence the Arctic Ocean ice thickness distribution, which has been determined largely from submarine sonar measurements of the ice draft. Ice draft is a measurement of the ice thickness below the water line and often serves as a close proxy for total ice thickness. The average draft increases from about 1 m (about 3 feet) near the Eurasian coast to 6–8 m (about 20–26 feet) along the coasts of north Greenland and the Canadian Arctic islands, where the ice is heavily ridged.

In Antarctica the large-scale sea circulation is dominated by westward motion along the coast and eastward motion further offshore in the West Wind Drift (also known as the Antarctic Circumpolar Current). The average drift speed is 20 km (about 12 miles) per day in the westward flow and 15 km (about 9 miles) per day in the eastward flow. Where katabatic winds (see "Local wind systems", Chapter 6, page 299) force the ice away from the coast and create polynyas (semipermanent open areas in sea ice), local sea ice motion is roughly perpendicular to the shore. There are gyres in the Ross Sea and Weddell Sea where the westward-moving ice is deflected to the north and meets the eastward-moving ice further offshore. Unlike the Beaufort Gyre in the Arctic Ocean, these gyres do not appear to recirculate ice. Ice-thickness data from drilling on floes, visual estimates by observers on ships, and a few moored sonars indicate that Antarctic sea ice is thinner than Arctic sea ice. Typically, Antarctic first-year ice is less than 1 m (about 3 feet) thick, while multiyear ice is less than 2 m (about 6.5 feet) thick.

Interactions with the oceans, atmosphere, and climate

The growth and decay of sea ice influence local, regional, and global climate through interactions with the atmosphere and ocean. Whereas snow-covered sea ice is an effective insulator that restricts heat loss from the relatively warm ocean to the colder atmosphere, there is significant turbulent heat and mass transfer from polynyas and leads (typically linear features in pack ice, a few metres to hundreds of metres wide and extending for hundreds of kilometres) to the ocean and atmosphere during the winter months. These losses are manifested as frost smoke from evaporation and condensation at the water surface, and they affect atmospheric processes hundreds of metres above and hundreds of kilometres downstream from leads and polynyas. Brine rejected from ice growing within leads and polynyas drives the deep mixing of the ocean. Rejected brine also affects global ocean circulation and ventilation processes by increasing the salt concentration of the water into which it is released. The conversion of both new and young ice into pressure ridges creates rough top and bottom surfaces that enhance the transfer of momentum from the atmosphere to the ocean. Ridges at the ice surface act as sails and catch the wind. The subsequent movement of the ice floes transfers energy to the underlying water via the keels on the underside of the ice.

Snow and ice reduce the amount of solar radiation available for organisms residing in the ice and water. This decrease in the amount of available energy affects and often reduces the productivity of plants, animals, and micro-organisms. Snow has a high albedo (it reflects a significant proportion of solar radiation back to the atmosphere), and thus the temperature at the surface remains cool. In the Arctic the surface albedo decreases in summer as the snow melts completely, ponds of meltwater form on the ice surface that absorb a greater share of incoming solar radiation, and the overall ice concentration (the ratio of ice area to open-water area) decreases. The increase in radiation absorption by meltwater ponds and the open ocean accelerates the melting process and further reduces surface albedo. This ice-albedo positive feedback plays a key role in the interaction of sea ice with climate.

The hydrosphere and climate change

Globally, humans affect the hydrosphere by emitting greenhouse gases into the atmosphere. Of the greenhouse gases released by anthropogenic activities, carbon dioxide has received much attention (see Chapter 1, page 18). It has been shown from the measurements of carbon dioxide in air bubbles trapped in ice and from the continuous measurement of carbon dioxide concentrations in air samples that the present atmospheric concentration of 384 parts per million (ppm) is 37 per cent higher than its pre-industrial value. The component of the hydrosphere most greatly affected by this emission of carbon dioxide is the ocean.

Before human activities had substantially affected the carbon dioxide cycle, there was a net flux of carbon dioxide from the oceans through the atmosphere to the land, where the gas was used in the net production of organic matter and the chemical weathering of minerals in continental rocks. Because of fossil-fuel burning and land-use practices, the net transfer from the ocean to the land has been reversed, and the ocean has now become an important sink of carbon dioxide. However, it is predicted that if global warming continues to a certain point, the oceans may cease to be a carbon sink and would become a net source (see "Carbon cycle feedbacks", Chapter 1, page 37). The oceans are currently gaining 2,340 million tonnes of carbon per year, and in 2005 a report by the UK's Royal Society showed that they were becoming more acidic as a result. This increasing acidity could have disastrous consequences for marine species, especially those with calcium-carbonate shells such as lobsters, crabs, and shellfish, because the calcium carbonate will dissolve more readily in more-acidic water.

Based on greenhouse climate models and other considerations, it is possible that atmospheric carbon dioxide concentrations may reach double their pre-industrial value by the end of the twenty-first century and, along with those of other greenhouse gases (e.g. methane and nitrous oxide), give rise to a global mean surface-temperature increase of 1.8–4°C. This projected temperature increase would be two to three times greater at the poles than at the equator and greater in the Arctic than in the Antarctic. The effect of the potential rise in surface temperature would be to speed up the hydrologic cycle and probably the rate of chemical weathering of continental rocks. Increases of 4–7 per cent in the global mean evaporation and precipitation rates might occur with a doubling of the carbon dioxide level and a few degrees' rise in global mean temperature. The effect on the water balance would be regional in nature, with some places becoming wetter and others drier. The IPPC's 2007 Fourth Assessment Report stated that most ocean–atmosphere circulation models for a warmer climate predict increased summer dryness and winter wetness in most parts of the northern middle and high latitudes. Because a warmer atmosphere has a greater water-holding capacity, there is more chance of intense precipitation and flooding. In a warmer climate, precipitation tends to be concentrated into more intense events, with longer dry periods in between.

Melting ice caps and rising sea levels

Global warming could further affect the hydrologic cycle by the melting of ice and snow in the Greenland and Antarctic ice caps and in mountain glaciers, resulting in the transfer of water to the oceans. This process, together with thermal expansion of the oceans because of global warming, could lead to a slow rise in sea level of about 0.21–0.48 metres (0.7–1.6 feet) by 2100. If the West Antarctic ice sheet were to disintegrate, a much larger and more rapid rise in sea level of 5–6 metres (about 16–20 feet) could occur over the next several hundred years. The melting of all glacial ice would raise the sea level about 56 metres (184 feet). It is also possible that an increase in global temperature could result in a reduction in the overall extent and thickness of sea ice in the Arctic and circum-Antarctic regions. Complete melting of the Arctic sea ice might occur, causing a northward shift in storm tracks and a reduction in northern-hemispheric precipitation during the spring and autumn.

While the current generation of models predicts that such global sea-level changes might take several centuries to occur, it is possible that the rate could accelerate as a result of processes that tend to hasten the collapse of ice sheets. One such process is the development of moulins, or large vertical shafts in the ice, that allow surface meltwater to penetrate to the base of the ice sheet. A second process involves the vast ice shelves off Antarctica that buttress the grounded continental ice sheet of Antarctica's interior. If these ice shelves collapse, the continental ice sheet could become unstable, slide rapidly toward the ocean, and melt, thereby further increasing mean sea level. Thus far, neither process has been incorporated into the theoretical models used to predict sea-level rise.

The potential rise in sea levels could be catastrophic for many ecosystems and human populations. A global rise in sea level of one metre, for example, would almost completely inundate the coastal areas of Bangladesh. Island nations and continental beaches and cities would be endangered. Agricultural lands could be displaced, just as patterns of arid, semi-arid, and wet lands might become modified. It is essential that society plan for such potential changes so that, if they do occur, appropriate adjustments can be made to accommodate them.

Impacts of recent changes to sea ice

Submarine sonar data obtained since 1958 have revealed that the average ice draft in the Arctic Ocean in the 1990s decreased by over 1 metre (about 3 feet) and that ice volume was 40 per cent lower than during the period 1958–76. The greatest ice draft reduction occurred in the central and eastern Arctic. Remote sensing also revealed a reduction of 3 per cent per decade in Arctic sea ice extent from 1978, with particularly rapid losses occurring from the late 1980s. This included the eastern Arctic, where both the ice concentration and the duration of the ice-covered season also decreased. Computer simulations suggest that sea ice changes in this region were due to changes in atmospheric circulation, and thus ice dynamics, rather than higher air temperatures. Yet it is not clear whether these changes are a result of natural variability – i.e. the Arctic Oscillation – or whether they represent a regime shift that will persist and perhaps become even more severe in the future.

Since computer models of climate change predict that the consequences of global warming will occur earlier and be most pronounced in the polar regions, particularly the Arctic, monitoring and understanding the behaviour of sea ice are important. Continued reductions of Arctic sea ice extent could have potentially severe ecological impacts. One such event may have arisen in western Hudson Bay, Canada, where a significant decline in the physical condition and reproductive success of polar bears occurred as the duration and extent of sea ice cover decreased during the 1980s and 1990s. On the other hand, a reduction in sea ice could be advantageous for oil and mineral exploration, production, and transport and for navigation through the Northern Sea Route (Northeast Passage), a water route connecting the Atlantic and Pacific Oceans along the northern coast of Europe and Russia; and the Northwest Passage, a similar route along the northern coast of North America.

Whaling records suggest that Antarctic sea ice extent decreased by approximately 25 per cent between the mid-1950s and early 1970s, whereas ice core samples suggest a 20 per cent decrease in sea ice extent since 1950. Since then, remote sensing data have indicated an increase in Antarctic sea ice extent parallel to the decrease in Arctic sea ice extent through the 1980s and 1990s. Yet the increase in Antarctic sea ice extent has not been uniformly distributed. A reduction in sea ice extent west of the Antarctic Peninsula has been correlated with slight declines in Adélie penguin numbers and a significant rise in the Chinstrap penguin population. There is speculation that if ice extent continues to decrease in this region, krill numbers will diminish significantly as they lose their under-ice habitat and face growing competition from salps (free-floating, thumb-sized filter feeders).

Climatic impacts of icebergs

An iceberg is a floating mass of freshwater ice that has broken from the seaward end of either a glacier or an ice shelf. Icebergs are found in the oceans surrounding Antarctica, in the seas of the Arctic and sub-Arctic, in Arctic fjords, and in lakes fed by glaciers.

Impacts on ice sheets and sea level

Apart from local weather effects, such as fog production, icebergs have two main impacts on climate. Iceberg production affects the mass balance, or change to the overall extent, of the parent ice sheets, and melting icebergs influence both ocean structure and global sea level.

The Antarctic Ice Sheet has a volume of 28 million cubic km (about 6.7 million cubic miles), which represents 70 per cent of the total fresh water (including groundwater) in the world. The mass of the ice sheet is kept in balance by a process of gain and loss – gain from snowfall over the whole ice sheet and ice loss from the melting of ice at the bottom of the ice shelf and from the calving of icebergs from the edges of the ice shelf. The effect of summer run-off and from sublimation off the ice surface is negligible.

Annual snowfall estimates for the Antarctic continent start at 1,000 cubic km (240 cubic miles). If the Antarctic Ice Sheet is in neutral mass balance, the annual rate of loss from melting and iceberg calving must be close to this value; indeed, estimates of iceberg flux do start at this value, though some run much higher. Such apparently large fluxes are still less than the mean flow rate of the Amazon River, which is 5,700 cubic km (about 1,370 cubic miles) per year. In Antarctica the annual loss amounts to only one ten-thousandth of its mass, so the ice sheet is an enormous passive reservoir. However, if losses from iceberg calving and ice-shelf melting are greater than gains from snowfall, global sea levels will rise.

At present, the size of the contribution from Antarctica, and even whether it is positive or negative, is uncertain. Consequently, Antarctic ice flux has not been included as a term in the sea-level predictions of the Fourth Assessment Report of the IPCC. What is more certain is that the retreat of glaciers in the Arctic and mountain regions has contributed about 50 per cent to current rates of sea-level rise. (The rest is due to thermal expansion of water as the ocean warms.) An increasing contribution is coming from a retreat of the Greenland Ice Sheet, and part of this contribution is occurring as an iceberg flux.

Impact on ocean structure

In considering the effect of iceberg melt upon ocean structure, it is found that the total Antarctic melt is equivalent to the addition of 0.1 metre (0.3 foot) of fresh water per year at the surface. This is like adding 0.1 metre of extra annual rainfall.

The dilution that occurs, if averaged over a mixed layer 100–200 metres (330–660 feet) deep, amounts to a decrease of 0.015–0.03 part per thousand (ppt) of salt. Melting icebergs thus make a small but measurable contribution to maintaining the Southern Ocean pycnocline (the density boundary separating low-salinity surface water from higher-salinity deeper water) and to keeping surface salinity in the Southern Ocean to its observed low value of 34 ppt or below.

It is interesting to note that the annual production of Antarctic iceberg ice is about one-tenth of the annual production of Antarctic sea ice. Sea ice has a neutral effect on overall ocean salinity, because it returns to liquid during the summer months. Nevertheless, when sea ice forms, it has an important differential effect in that it increases ocean salinity where it forms. This is often near the Antarctic coast. Increased salinity encourages the development of convection currents and the formation of bottom water (masses of cold and dense water). Icebergs, on the other hand, always exert a stabilizing influence on the salinity of the water column. This stabilizing influence manifests itself only when the icebergs melt, and this occurs at lower latitudes.

6

THE CHANGING PLANET: ATMOSPHERE

On October 5, 2006, the World Health Organization (WHO) unveiled its new air-quality guidelines, last issued in 1997, and at the same time it challenged governments to improve urban air quality in order to protect public health. WHO estimated that air pollution caused about two million premature deaths each year, with more than half of the deaths in less-developed countries. The new guidelines' value for sulphur-dioxide exposure over 24 hours was reduced from 125 to 20 $\mu g/m^3$ (micrograms per cubic metre), while the value for ozone exposure over 8 hours was reduced from 120 to 100 $\mu g/m^3$. The guidelines also recommended that the annual mean for PM10 emissions (small particulates produced mainly from the burning of fossil fuels) be less than 20 $\mu g/m^3$.

The European Environment Agency reported that summer smog in 2006 reached its second-worst level in a decade. The EU alert threshold for ozone of 240 $\mu g/m^3$ was exceeded 190 times, compared with 127 times in 2005 and 99 in 2004. During Europe's 2003 heat wave, the alert threshold was exceeded 720 times. The target value of 120 $\mu g/m^3$ was exceeded at most stations. The highest ozone level – 370 µg/m^3 – was recorded in Italy.

For four days in August 2007, 1.3 million cars were removed from Beijing traffic and some 800 extra buses put into service for the 2008 Summer Olympic Games. The measure reduced air pollution by 15–20 per cent. (It also reduced congestion, which allowed traffic to move much faster. Buses, for example, were able to average 20 km/hour [12 miles/hour] rather than the customary 14 km/hour [9 miles/hour]. The experimental scheme, part of the city's preparations for the 2008 Olympics, banned cars that had licence plates with odd numbers on Saturday and Monday and cars with even-numbered plates on Friday and Sunday.

Air pollution is one of the more noticeable of the human-induced changes in the Earth's atmosphere. Air pollution, and related concepts such as ozone depletion and global warming, affect various facets of Earth's atmosphere. This chapter describes the structure of the atmosphere and explains the many and complex patterns of air current that drive the climate and weather conditions we experience at the surface of the Earth. It also discusses the changes that have occurred as a result of human activities.

The atmosphere

Evolution and structure

The atmosphere is the envelope of gas and aerosols (microscopic suspended particles of liquid or solids, e.g. dust, soot, smoke, or chemicals) that extends from the ocean, land, and ice-covered surface of Earth outward into space. The density of the atmosphere decreases outward, because the gravitational attraction of the planet, which pulls the gases and aerosols inward, is greatest close to the surface. Earth's atmosphere has been able to contain water in each of its three phases (solid, liquid, and gas), which has been essential for the development of life on the planet.

The evolution of Earth's current atmosphere is not completely understood. It is thought that it resulted from a gradual release of gases both from the planet's interior and from the metabolic activities of life forms – as opposed to the primordial atmosphere, which developed by outgassing during the original formation of the planet. Current volcanic gaseous emissions include water vapour (H_2O), carbon dioxide (CO_2), sulphur dioxide (SO_2), hydrogen sulphide (H_2S), carbon monoxide (CO), chlorine (Cl), fluorine (F), and diatomic nitrogen (N_2), as well as traces of other substances. Approximately 85 per cent of volcanic emissions are in the form of water vapour. In contrast, carbon dioxide is about 10 per cent of the effluent.

During the early evolution of the atmosphere on Earth, water must have been able to exist as a liquid, since the oceans have been present for at least 3 billion years (see "Origin and evolution of the oceans", Chapter 5, page 219). Given that solar output 4 billion years ago was only about 60 per cent of what it is today, enhanced levels of carbon dioxide and perhaps ammonia (NH_3) must have been present in order to retard the loss of infrared radiation into space. The initial life forms that evolved in this environment will have been anaerobic (i.e. surviving in the absence of oxygen). In addition, they must have been able to resist the biologically destructive ultraviolet radiation in sunlight, which was not absorbed by a layer of ozone as it is now.

Once organisms developed the capability for photosynthesis, oxygen was produced in large quantities. This capability arose in primitive forms of plants between 2 and 3 billion years ago. Prior to the evolution of photosynthetic organisms, oxygen was produced in limited quantities as a by-product of the decomposition of water vapour by ultraviolet radiation. The build-up of oxygen in the atmosphere also permitted the development of the ozone layer in the upper atmosphere, as O_2 molecules were dissociated into monatomic oxygen (O, consisting of single oxygen atoms) and recombined with other O_2 molecules to form triatomic ozone molecules (O_3). The ozone layer extends about 10–50 km (6–30 miles) in altitude.

The current molecular composition of Earth's atmosphere is 78.08 per cent diatomic nitrogen (N_2), 20.95 per cent diatomic oxygen (O_2), 0.93 per cent argon (A), about 0–4 per cent water (H_2O), and 0.038 per cent carbon dioxide (CO_2). Inert gases such as neon (Ne), helium (He), and krypton (Kr), and other constituents such as nitrogen oxides, compounds of sulphur, and compounds of ozone, are found in lesser amounts.

Earth's atmosphere is bounded at the bottom by water and land. Heating of this surface is accomplished by three physical processes – radiation, conduction, and convection (vertical mixing) – and the temperature at the interface of the atmosphere and surface is a result of this heating. The relative contributions of each process depend on the wind, temperature, and moisture structure in the atmosphere immediately above the surface, the intensity of solar insolation (radiation received at the surface), and the physical characteristics of the surface.

The convection process in the atmosphere is also referred to as turbulence. It is a mechanism of heat flux that occurs in the atmosphere in two forms. When the surface is substantially warmer than the overlying air, mixing will spontaneously occur in order to redistribute the heat. This process, referred to as free convection, occurs when the environmental lapse rate (the rate of change of an atmospheric variable, such as temperature or density, with increasing altitude) of temperature decreases at a rate greater than 1°C per 100 metres (approximately 1°F per 150 feet). This rate is called the adiabatic lapse rate (the rate of temperature change occurring within a rising or descending air parcel).

Wind

Relationship of wind to pressure and governing forces

Earth's changing wind patterns are governed by Newton's second law of motion, which states that the sum of the forces acting on a body equals the product of the mass of that body and the acceleration caused by those forces. There is a balance between the force created by horizontal differences in pressure (the horizontal pressure-gradient force) and an apparent force that results from Earth's rotation (the Coriolis force, see also Chapter 5, page 242). The observer on the ground experiences the Coriolis force as a deflection of the relative motion to the right in the northern hemisphere and to the left in the southern hemisphere. Of particular significance in this simple model of wind-pressure relationships is the fact that the geostrophic wind (a result of the combined pressure-gradient and Coriolis forces) blows in a direction parallel to the isobars, with the low pressure on the observer's left as he or she looks downwind in the northern hemisphere, and on his or her right in the southern hemisphere. Wind speed increases as the distance between isobars decreases (or pressure gradient increases).

Large-scale, observed winds tend to behave much as the geostrophic- or gradient-flow models predict in most of the atmosphere. The most notable exceptions occur in low latitudes, where the Coriolis parameter becomes very small, and in the lowest kilometre of the atmosphere, where friction becomes important. The friction induced by airflow over the underlying surface reduces the wind speed and alters the simple balance of forces.

Cyclones and anticyclones

Cyclones and anticyclones are regions of relatively low and high pressure, respectively. They occur over most of the Earth's surface in a variety of sizes ranging from very large, semipermanent ones to smaller, highly mobile systems. Common to both cyclones and anticyclones are the characteristic circulation patterns. The geostrophic-wind and gradient-wind models dictate that, in the northern hemisphere, flow around a cyclone (cyclonic circulation) is counterclockwise, and flow around an anticyclone (anticyclonic circulation) is clockwise. Circulation directions are reversed in the southern hemisphere. In the presence of friction, the additional component of motion toward lower pressure produces a "spiralling" effect toward the low-pressure centre and away from the high-pressure centre.

Cyclones

The cyclones that form outside the equatorial belt, known as extratropical cyclones (see below), may be regarded as large eddies in the broad air currents that flow in the general direction from west to east around the middle and higher latitudes of both hemispheres. They are an essential part of the mechanism by which the excess heat received from the sun in Earth's equatorial belt is conveyed toward higher latitudes. Higher latitudes radiate more heat to space than they receive from the sun, and heat must reach them by winds from the lower latitudes if their temperature is to be continually cool rather than cold. If there were no cyclones and anticyclones, the north–south movements of the air would be much more limited, and there would be little opportunity for heat to be carried poleward by winds of subtropical origin. Under such circumstances the temperature of the lower latitudes would increase and the polar regions would cool – the temperature gradient between them would intensify.

Strong horizontal gradients of temperature are particularly favourable for the formation and development of cyclones. The temperature difference between polar regions and the equator builds up until it becomes sufficiently intense to generate new cyclones. As their associated cold fronts sweep equatorward and their warm fronts (the advancing masses of cooler and warmer air, respectively) move poleward, the new cyclones reduce the temperature difference.

Thus, the wind circulation on Earth represents a balance between the heating effects of solar radiation occurring at the equator and in the polar regions. Wind circulation, through the effect of cyclones, anticyclones, and other wind systems, also periodically destroys this temperature contrast.

Cyclones of a somewhat different character occur closer to the equator, generally forming in latitudes of 10–30°N and 10–30°S over the oceans. They are generally known as tropical cyclones when their winds equal or exceed 119 km (74 miles) per hour. They are also known as hurricanes if they occur in the Atlantic Ocean and the Caribbean Sea, as typhoons in the western Pacific Ocean and the China Sea, and as cyclones off the coasts of Australia. These storms are of smaller diameter than the extratropical cyclones, ranging from 100 km to 500 km (60–300 miles) in diameter, and are accompanied by winds of sometimes extreme violence.

Extratropical cyclones
Of the two types of large-scale cyclones, extratropical cyclones are the most abundant and exert influence on the broadest scale; they affect the largest percentage of Earth's surface.

Furthermore, this class of cyclones is the principal cause of day-to-day weather changes experienced in middle and high latitudes and thus is the focus of much of modern weather forecasting.

The seeds for many current ideas concerning extratropical cyclones were sown between 1912 and 1930 by a group of Scandinavian meteorologists working in Bergen, Norway. This so-called Bergen school, founded by the Norwegian meteorologist and physicist Vilhelm Bjerknes, formulated a model for a cyclone that forms as a disturbance along a zone of strong temperature contrast (a front), which in turn constitutes a boundary between two contrasting air masses. In this model the masses of colder polar air and warmer tropical air around the globe are separated by the polar front. This transition region possesses a strong temperature gradient and thus is a reservoir of potential energy ('stored' energy) that can be readily tapped and converted into the kinetic energy (energy expressed in motion) associated with extratropical cyclones.

For this reservoir to be tapped, a cyclone (called a wave, or frontal, cyclone) must develop much in the way shown in Figure 6.1. The feature that is of primary importance prior to cyclone development (cyclogenesis) is a front, represented in the initial stage (A) as a heavy black line with alternating triangles or semicircles attached to it. This stationary or very slow-moving front forms a boundary between cold and warm air and thus is a zone of strong horizontal temperature gradient (sometimes referred to as a baroclinic zone).

Cyclone development is initiated as a disturbance along the front, which distorts the front into the wavelike configuration (B: wave appearance). As the pressure within the disturbance continues to decrease, the disturbance assumes the appearance of a cyclone and forces poleward and equatorward movements of warm and cold air, respectively, which are represented by

Evolution of a wave (frontal) cyclone.

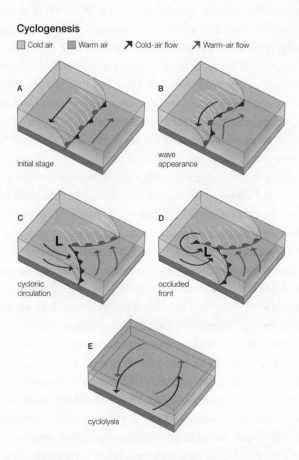

Figure 6.1 © Encyclopædia Britannica, Inc.

mobile frontal boundaries. As depicted in the cyclonic circulation stage (C), the front that signals the advancing cold air
(cold front) is indicated by the triangles, while the front
corresponding to the advancing warm air (warm front) is
indicated by the semicircles. As the cyclone continues to
intensify, the cold dense air streams rapidly equatorward,
yielding a cold front with a typical slope (which describes
the "shape" of the cold front) of 1 to 50 (meaning for every 50
units of distance, the cold air is 1 unit high in altitude) and a
propagation speed (how fast it moves relative to the Earth's
surface) that is often 8 to 15 metres per second (about 18 to 34
miles per hour) or more.

At the same time, the warm, less-dense air moving in a
equatorward direction flows up over the cold air ahead of the
cyclone to produce a warm front with a typical slope of 1 to
200 and a typically much slower propagation speed of about
2.5 to 8 metres per second (6 to 18 miles per hour). This
difference in propagation speeds between the two fronts allows
the cold front to overtake the warm front and produce yet
another, more complicated, frontal structure, known as an
occluded front. An occluded front (D) is represented by a line
with alternating triangles and semicircles on the same side.
This occlusion process may be followed by further storm
intensification. The separation of the cyclone from the warm
air toward the equator, however, eventually leads to the
storm's decay and dissipation (E) in a process called cyclolysis.

The life cycle of such an event is typically several days,
during which the cyclone may travel from several hundred to a
few thousand kilometres. In its path and wake occur dramatic
weather changes. Warm frontal weather is most frequently
characterized by stratiform clouds (long, grey clouds with a
uniform base) which ascend as the front approaches and
potentially yield rain or snow. The passing of a warm front

brings a rise in air temperature and clearing skies. The warmer air, however, may also harbour the ingredients for rain shower or thunderstorm formation, a condition that is enhanced as the cold front approaches. The passage of the cold front is marked by the influx of colder air, the formation of stratocumulus clouds (sheets of grey, humped clouds) with some lingering rain or snow showers, and then eventual clearing.

While this is an oft-repeated scenario, it is important to recognize that many other weather sequences can also occur. For example, the stratiform clouds of a warm front may have embedded cumulus formations and thunderstorms; the warm sector might be quite dry and yield few or no clouds; the pre-cold-front weather may closely resemble that found ahead of the warm front; or the post-cold-front air may be completely cloud free. Cloud patterns oriented along fronts and spiralling around the cyclone vortex are consistently revealed in satellite pictures of Earth. The actual formation of any area of low pressure requires that mass in the column of air lying above Earth's surface be reduced. This loss of mass then reduces the surface pressure.

Anticyclones

While cyclones are typically regions of inclement weather, anticyclones are usually meteorologically quiet regions. Generally larger than cyclones, anticyclones exhibit persistent downward motions and yield dry, stable air that may extend horizontally many hundreds of kilometres. In most cases, an actively developing anticyclone forms over a ground location in the region of cold air behind a cyclone as it moves away. This anticyclone forms before the next cyclone advances into the area, and is known as a cold anticyclone.

A result of the downward air motion in an anticyclone, however, is compression of the descending air. As a consequence of this compression, the air is warmed. Thus, after a few days, the air composing the anticyclone at levels of 2–5 km (1–3 miles) above the ground tends to increase in temperature, and the anticyclone is transformed into a warm anticyclone. Warm anticyclones move slowly, and cyclones are diverted around their periphery.

During their transformation from cold to warm status, anticyclones usually move out of the main belt followed by cyclones in middle latitudes and often amalgamate with the quasi-permanent bands of relatively high pressure found in both hemispheres around latitude 20–30° – the so-called subtropical anticyclones. On some occasions the warm anticyclones remain in the belt normally occupied by the mid-latitude westerly winds. The normal cyclone tracks are then considerably modified; atmospheric depressions (areas of low pressure) are either blocked in their eastward progress or diverted to the north or south of the anticyclone. Anticyclones that interrupt the normal circulation of the westerly wind in this way are called blocking anticyclones, or blocking highs. They frequently persist for a week or more, and the occurrence of a few such blocking anticyclones may dominate the character of a season. Blocking anticyclones are particularly common over Europe, the eastern Atlantic, and the Alaskan area.

The descent and warming of the air in an anticyclone might be expected to lead to the dissolution of clouds and the absence of rain. Near the centre of the anticyclone, the winds are light and the air can become stagnant. Air pollution can build up as a result. The US city of Los Angeles, for example, often has poor air quality because it is frequently under a stationary anticyclone. In winter the ground cools, and the lower layers of the atmosphere also become cold. Fog may be formed as the air is cooled to its dew point in the stagnant air.

Under other circumstances, the air trapped in the first kilometre above Earth's surface may pick up moisture from the sea or other moist surfaces, and layers of cloud may form in areas near the ground up to a height of about 1 km (0.6 mile). Such layers of cloud can be persistent in anticyclones (except over the continents in summer), but they rarely grow thick enough to produce rain. If precipitation occurs, it is usually drizzle or light snow.

Anticyclones are often regions of clear skies and sunny weather in summer; at other times of the year, cloudy and foggy weather – especially over wet ground, snow cover, and the ocean – may be more typical. Winter anticyclones produce colder than average temperatures at the surface, particularly if the skies remain clear. Anticyclones are responsible for periods of little or no rain, and such periods may be prolonged in association with blocking highs.

Cyclone and anticyclone climatology

Migrating cyclones and anticyclones tend to be distributed around certain preferred regions, known as tracks, that emanate from preferred cyclogenetic and anticyclogenetic regions. Favoured cyclogenetic regions in the northern hemisphere are found on the lee side of mountains and off the east coasts of continents. Cyclones then track east or south-east before eventually turning toward the north-east and decaying. The tracks are displaced further northward in July, reflecting the more northward position of the polar front in summer.

Continental cyclones usually intensify at a rate of 0.5 mb (0.05 kPa) per hour or less, although more dramatic examples can be found. Marine cyclones, on the other hand, often experience explosive development in excess of 1 mb (0.1 kPa) per hour, particularly in winter.

Anticyclones tend to migrate equatorward out of the cold airmass regions and then eastward before decaying or merging with a warm anticyclone. Like cyclones, warm anticyclones also slowly migrate poleward with the warm season.

In the southern hemisphere, where most of Earth's surface is covered by oceans, the cyclones are distributed fairly uniformly through the various longitudes. Typically, cyclones form initially in latitudes 30–40°S and move in a generally south-eastward direction, reaching maturity in latitudes near 60°S. Thus, the Antarctic continent is usually ringed by a number of mature or decaying cyclones.

The belt of ocean from 40°S to 60°S is a region of persistent, strong westerly winds that form part of the circulation to the north of the main cyclone centres. These are the "roaring forties", where the westerly winds are interrupted only at intervals by the passage south-eastward of developing cyclones.

Scale classes

Wind systems occur in spatial dimensions ranging from tens of metres to thousands of kilometres and possess residence times that vary from seconds to weeks. Since the atmosphere exhibits such a large variety of both spatial and temporal scales, efforts have been made to group various phenomena into scale classes, according to their typical size and lifespan.

The class describing the largest and longest-lived of these phenomena is known as the planetary scale. Such phenomena are typically a few thousand kilometres in size and have lifespans ranging from several days to several weeks. Examples of planetary-scale phenomena include semipermanent pressure centres (see "Atmospheric pressure", Chapter 2, page 56) and certain globe-encircling upper-air waves (see "Upper-level winds", page 320).

A second class is known as the synoptic scale. Spanning smaller distances (a few to several hundred kilometres) and possessing shorter lifespans (a few to several days), this class contains the migrating cyclones and anticyclones that control day-to-day weather changes. Sometimes the planetary and synoptic scales are combined into a single classification termed the large-scale, or macroscale. Large-scale wind systems are distinguished by the predominance of horizontal motions over vertical motions and by the pre-eminent importance of the Coriolis force in influencing wind characteristics. Examples of large-scale wind systems include the trade winds and the westerlies.

There is a third class of phenomena of even smaller size and shorter lifespan. In this class, vertical motions may be as significant as horizontal movement, and the Coriolis force often plays a less important role. Known as the mesoscale, this class is characterized by spatial dimensions of ten to a few hundred kilometres and lifespans of a day or less. Because of the shorter time scale and because the other forces may be much larger, the effect of the Coriolis force in mesoscale phenomena is sometimes neglected. Two of the best-known examples of mesoscale phenomena are the thunderstorm and its devastating by-product, the tornado. The next section focuses on less intense, though nevertheless commonly observed, wind systems that are found in rather specific geographic locations and thus are often referred to as local wind systems.

Local wind systems

The so-called "sea and land breeze circulation" is a local wind system typically encountered along coastlines adjacent to large bodies of water, and is induced by differences that occur between the water surface and the adjacent land surface due to heating or cooling. Water has a higher heat capacity than do the substances in the land surface (i.e. more units of heat are required to produce a given temperature change in a volume of water than the same volume of land). Daytime solar radiation penetrates to several metres into the water, the water vertically mixes, and the volume is slowly heated. In contrast, daytime solar radiation heats the land surface more quickly because it does not penetrate more than a few centimetres below the land surface. The land surface, now at a higher temperature relative to the air above it, transfers more heat to its overlying air mass and creates an area of low pressure. Thus, a circulation cell is induced. Because the surface airflow is from the water toward the land, it is called a sea breeze.

Since the landmass possesses a lower heat capacity than water, it cools more rapidly at night than does the water. Consequently, at night the cooler landmass yields a cooler overlying air mass and creates a zone of relatively higher pressure. This produces a circulation cell with air motions opposite to those found during the day. This flow from land to water is known as a land breeze. The land breeze is typically shallower than the sea breeze since the cooling of the atmosphere over land is confined to a shallower layer at night than the heating of the air during the day.

Sea and land breezes occur along the coastal regions of oceans or large lakes in the absence of a strong large-scale wind system during periods of strong daytime heating or night-time cooling. Those who live within 10–20 km (6–12 miles) of a coastline often experience the cooler, brisker winds of the sea breeze on a sunny afternoon, only to find it turn into a sultry land breeze late at night.

One of the features of the sea and land breeze is a region of low-level air convergence in the termination region of the surface flow. Such convergence often induces local upward motions and cloud formations. Thus, in sea and land breeze regions, it is not uncommon to see clouds lying off the coast at night; these clouds are then dissipated by the daytime sea breeze, which forms new clouds, perhaps with showers occurring over land in the afternoon.

Another group of local winds is induced by the presence of mountain and valley features on the Earth's surface. One subset of such winds, known as mountain winds or breezes, occurs as a result of differential heating or cooling along mountain slopes. During the day, solar heating of the sunlit slopes causes the overlying air to move upslope. These winds are referred to as anabatic flow. At night, as the slopes cool, the direction of airflow is reversed, and cool, downslope drainage motion occurs, known as katabatic flow.

Such winds may be relatively gentle or may occur in strong gusts, depending on the topographic configuration. These winds are one type of katabatic flow. In an enclosed valley, the cool air that drains into the valley may give rise to thick fog. Fog persists until daytime heating reverses the circulation and creates clouds associated with the upslope motion at the mountain top.

Another subset of katabatic flow, called foehn winds (also known as chinook winds east of the Rocky Mountains and as Santa Ana winds in southern California), is induced by adiabatic temperature changes (those that take place without the addition or subtraction of heat) occurring as air flows over a mountain. When air is lifted, it enters a region of lower pressure and expands, inducing a reduction of temperature (adiabatic cooling). When air subsides, it contracts and experiences adiabatic warming.

As air ascends on the windward side of the mountain, its cooling rate may be moderated by heat that is released during the formation of precipitation. However, having lost much of its moisture, the descending air on the leeward side of the mountain adiabatically warms faster than it was cooled on the windward ascent. Thus, the effect of this wind, if it reaches the surface, is to produce warm, dry conditions. Usually, such winds are gentle and produce a slow warming. On occasion, however, foehn winds may exceed 185 kilometres (115 miles) per hour and produce air-temperature increases of tens of degrees – sometimes more than 20°C (36°F) – within only a few hours.

Other types of katabatic wind can occur when the underlying geography is characterized by a cold plateau adjacent to a relatively warm region of lower elevation. Such conditions are found in areas in which major ice sheets or cold elevated land surfaces border warmer large bodies of water. Air over the cold plateau cools and forms a large dome of cold dense air. Unless held back by background wind conditions, this cold air will spill over into the lower elevations with speeds that vary from gentle (a few kilometres per hour) to intense (93–185 km [58–115 miles] per hour), depending on the incline of the slope.

Zonal surface winds

As described in Chapter 2 (see "Atmospheric pressure", page 56), semipermanent atmospheric pressure centres indicate the source regions for major, relatively uniform bodies of air known as air masses. On average, certain geographic locations can expect to experience winds that emanate from one prevailing direction, largely dictated by the presence of major semipermanent pressure systems. Such prevailing winds have long been known in marine environments because of their influence on the great sailing ships. Tropical and subtropical regions are characterized by a general band of low pressure lying near the equator. This band is bounded by centres of high pressure that may extend poleward into the middle latitudes. Between these low- and high-pressure regions is the region of the tropical winds. Of these, the most extensive are the trade winds – so named because of their favourable influence on trade ships travelling across the subtropical North Atlantic. Trade winds flow westward and somewhat in the direction of the equator on the equatorward side of the subtropical high-pressure centres.

The "root of the trades", occurring on the eastern side of a subtropical high-pressure centre, is characterized by subsiding air. This produces the very warm, dry conditions above a shallow layer of oceanic stratus clouds found in the eastern extremes of the subtropical Atlantic and Pacific ocean basins. As the trade winds progress westward, however, subsidence abates, the air mass becomes more humid, and scattered showers appear. These showers occur particularly on islands with elevated terrain features that interrupt the flow of the warm moist air. The equatorward flow of the trade winds of the northern and southern hemispheres often results in a convergence of the two air streams in a region known as the intertropical convergence zone (ITCZ). Deep convective clouds, showers, and thunderstorms occur along the ITCZ. When the air reaches the western extreme of the high-pressure centre, it turns poleward and then eventually returns eastward in the middle latitudes.

The poleward-moving air is now warm and laden with moist maritime tropical air; it gives rise to the warm, humid, showery climate characteristic of the Caribbean region, eastern South America, and the western Pacific island chains.

The westerlies are associated with the changeable weather common to the middle latitudes. Migrating extratropical cyclones and anticyclones associated with contrasting warm moist air moving poleward from the tropics, and cold dry air moving equatorward from polar latitudes, yield periods of rain (sometimes with violent thunderstorms), snow, sleet, or freezing rain interrupted by periods of dry, sunny, and sometimes bitterly cold conditions.

These patterns are seasonally dependent, with more intense cyclones and colder air prevailing in winter but with a higher incidence of thunderstorms common in spring and summer. Furthermore, these migrations and the associated climate are complicated by the presence of landmasses and major mountain features, particularly in the northern hemisphere. The westerlies lie on the equatorward side of the semipermanent subpolar centres of low pressure. Poleward of these centres, the surface winds turn westward again over significant portions of the subpolar latitudes. As in the middle latitudes, the presence of major landmasses, notably in the northern hemisphere, results in significant variations in these polar easterlies.

Monsoons

As shown in the January and July maps of sea-level atmospheric pressure, particularly strong seasonal pressure variations occur over continents. Such seasonal fluctuations, commonly called monsoons, are more pronounced over land surfaces because these surfaces are subject to more significant seasonal temperature variations than water bodies. Since land surfaces warm and cool faster than water bodies, they often quickly modify the temperature and density of air parcels passing over them. Monsoons blow for approximately six months from the north-east and six months from the southwest, principally in South Asia and parts of Africa. Summer monsoons have a dominant westerly component and a strong tendency to converge, rise, and produce rain. Winter monsoons have a dominant easterly component and a strong tendency to diverge, subside, and cause drought. Both are the result of differences in annual temperature trends over land and sea.

The Indian Ocean monsoon

At the equator the area near India is unique in that dominant or frequent westerly winds occur at the surface almost constantly throughout the year; the surface easterlies reach only to latitudes near 20°N in February, and even then they have a very strong northerly component. They soon retreat northward, and drastic changes take place in the upper-air circulation. This is a time of transition between the end of one monsoon and the beginning of the next.

Late in March, the high-sun season reaches the equator and moves further north. With it go atmospheric instability, convectional (that is, rising and turbulent) clouds, and rain. The westerly subtropical jet stream still controls the flow of air across northern India, and the surface winds are northeasterlies.

As the high-sun season moves northward during April, India becomes particularly prone to rapid heating because the highlands to the north protect it from any incursions of cold air. There are three distinct areas of relative upper tropospheric warmth: (1) above the southern Bay of Bengal, (2) above the Plateau of Tibet, and (3) across the trunks of the various peninsulas that are relatively dry during this time. These three areas combine to form a vast heat-source region.

The relatively warm area above the southern Bay of Bengal occurs mostly at the 500–100-millibar level. (This atmospheric pressure region typically occurs at elevations between 5,500 and 16,200 metres [18,000 and 53,000 feet] but may vary according to changes in heating and cooling.)

In contrast, a heat sink appears over the southern Indian Ocean as the relatively cloud-free air cools by emitting

long-wavelength radiation. Monsoon winds at the surface
blow from heat sink to heat source. As a result, by May the
south-west monsoon is well established over Sri Lanka, off
the south-eastern tip of the Indian peninsula.

Also in May, the dry surface of Tibet (above 4,000 metres
[13,100 feet]) absorbs and radiates heat that is readily trans-
mitted to the air immediately above. At about 6,000 metres
(19,700 feet) an anticyclonic cell arises, causing a strong
easterly flow in the upper troposphere above northern India.
The subtropical jet stream suddenly changes its course to the
north of the anticyclonic ridge and the highlands, though it
may occasionally reappear southward of them for very brief
periods. This change of the upper tropospheric circulation
above northern India from westerly jet to easterly flow
coincides with a reversal of the vertical temperature and
pressure gradients between 600 and 300 millibars.

Because of India's inverted triangular shape, the land is
heated progressively as the sun moves northward. This accelerated spread of heating, combined with the general direction
of heat being transported by winds, results in a greater initial
monsoonal activity over the Arabian Sea (at late springtime),
where a real frontal situation often occurs, than over the Bay
of Bengal. The relative humidity of coastal districts in the
Indian region rises above 70 per cent, and some rain occurs.
Above the heated land, the air below 1,500 metres (5,000 feet)
becomes unstable, but it is held down by the overriding easterly flow. This does not prevent frequent thunderstorms from
occurring in late May.

During June the easterly jet becomes firmly established at
150–100 millibars, an atmospheric pressure region typically
occurring at elevations between 13,700 and 16,200 metres
(45,000 and 53,000 feet). It reaches its greatest speed at its
normal position to the south of the anticyclonic ridge, at about

15°N from China through India. These upper-air features that arise so far away from the equator are associated with the surface monsoon and are absent when there is no monsoonal flow.

Various factors, especially topography, combine to make up a complex regional pattern. Oceanic air flowing toward India below 6,000 metres (19,700 feet) is deflected in accordance with the Coriolis effect. The converging, moist oncoming stream becomes unstable over the hot land and is subject to rapid convection. Towering cumulonimbus clouds rise thousands of metres, producing violent thunderstorms and releasing latent heat in the surrounding air.

Later, in June and July, the monsoon is strong and well established to a height of 6,000 metres (19,900 feet), though less in the far north, with occasional thickening to 9,000 metres (29,500 feet). Weather conditions are cloudy, warm, and moist all over India. Rainfall varies between 400 mm and 500 mm (16–20 inches), but topography introduces some extraordinary differences. On the southern slopes of the Khasi Hills at only 1,300 metres (4,300 feet), where the moist air-streams are lifted and overturned, the village of Cherrapunji in Meghalaya state receives an average rainfall of 2,730 mm (107 inches) in July, with record totals of 897 mm (35 inches) in 24 hours in July 1915, more than 9,000 mm (354 inches) in July 1861, and 16,305 mm (642 inches) in the monsoon season of 1899.

It is mainly in July and August that waves of low pressure appear in the body of monsoonal air. Fully developed depressions appear once or twice per month. They travel from east to west more or less concurrently with high-level easterly waves and bursts of speed from the easterly jet, causing a local strengthening of the low-level monsoonal flow. The rainfall consequently increases and is much more evenly distributed than it is in June. Some of the deeper depressions become tropical cyclones before they reach the land, and these bring torrential rains and disastrous floods.

A totally different development arises when the easterly jet moves further north than usual. The monsoonal wind rising over the southern slopes of the Himalayas brings heavy rains and local floods. The weather over the central and southern districts, however, becomes suddenly drier and remains so for as long as the abnormal shift lasts. The opposite shift is also possible, with mid-latitude upper air flowing along the south face of the Himalayas and bringing drought to the northern districts. Such dry spells are known as "breaks" of the monsoon.

By August the intensity and duration of sunshine have decreased, temperatures begin to fall, and the surge of south-westerly air diminishes spasmodically almost to a standstill in the north-west. Cherrapunji still receives over 2,000 mm (79 inches) of rainfall at this time, however. In September, dry, cool, northerly air begins to circle the west side of the high-lands and spread over north-western India. The easterly jet weakens, and the upper tropospheric easterlies move much further south. The rainfall becomes extremely variable over most of the region, but showers are still frequent in the south-eastern areas and over the Bay of Bengal.

By early October, variable winds are very frequent every-where. At the end of the month, the entire Indian region is covered by northerly air and the winter monsoon takes shape. The surface flow is deflected by the Coriolis force and becomes a northeasterly flow. This causes an October–December rainy season for the extreme south-east of the Deccan (including the Madras coast) and eastern Sri Lanka, which cannot be ex-plained by topography alone because it extends well out over the sea. Tropical depressions and cyclones are important contributing factors. Most of India thus begins a sunny, dry, and dusty season.

Conversely, the western slopes of the Karakoram Range and Himalayas are then reached by the mid-latitude frontal depressions that come from the Atlantic and the Mediterranean. The winter rains they receive, moderate as they are, place them clearly outside the monsoonal realm.

Because crops and water supplies depend entirely on monsoonal rains, it became imperative that quantitative, long-range weather forecasts be available. Embedded in the weather patterns of other parts of the world are clues to the summer conditions in South Asia. These clues often appear in the months leading up to the onset of the monsoon. For a forecast to be released at the beginning of June, South American pressure data and Indian upper-wind data for the month of April are examined. Although widely separated from one another, these data are positively correlated and may be used as predictors of June conditions.

Forecasts may be further refined in May by comparing rainfall patterns in both Zimbabwe and Java to the easterly winds above the city of Kolkata (Calcutta) in West Bengal state. In this situation the correlation between rainfall and easterly winds is negative.

Air pollution and the winter monsoon in Asia

During the winter months in Asia, the Siberian anticyclone, which is typically centred near Lake Baikal in Russia, extends its influence over most of central and northern China. In these locations, subsiding air and strong temperature inversions prevail in the lowest kilometre of the atmosphere. Temperature inversions create regions of stagnant air that have the effect of trapping air pollution.

In China, Southeast Asia, and South Asia, the widespread practice of burning firewood and coal for cooking, heating, and industry produces significant air pollution in the lower troposphere. The prevailing northeasterly winds of the winter monsoon transport the bulk of this polluted air southward, where it contributes to what has become known as the "Asian Brown Cloud" over the Indian Ocean.

A strong reversal of winds, where northeasterly winds near Earth's surface shift to more westerly directions above 1 km (0.6 mile), is a characteristic feature of the northeast monsoon. Parcels of polluted air rising above the 1-km level are carried eastward, over the Pacific Ocean, where they disperse. When these parcels are caught in the strong winds of the subtropical jet stream, they have been known to traverse half the globe in a matter of 10 to 15 days. As a result, pollution related to the Asian winter monsoon has become widely recognized as a major scientific issue with global environmental implications.

The Malaysian–Australian monsoon

Southeast Asia and northern Australia are combined in one monsoonal system that differs from others because of the peculiar and somewhat symmetrical distribution of landmasses on both sides of the equator. In this respect, the northwest monsoon of Australia is unique. The substantial masses of water between Asia and Australia have a moderating effect on tropospheric temperatures, weakening the summer monsoon. The many islands (e.g. the Philippines and Indonesia) provide an infinite variety of topographic effects.

Typhoons that develop within the monsoonal air bring additional complications. It would be possible to exclude North China, Korea, and Japan from the monsoonal domain because their seasonal rhythm follows the normal mid-latitude pattern – a predominant outflow of cold continental air in winter and frontal depressions and rain alternating with fine, dry anticyclonic weather in the warm season. On the other hand, the seasonal reversal of wind direction in this area is almost as persistent as that in India. The winter winds of north-eastern Asia are much stronger because of the relative proximity of the Siberian anticyclone. The tropical ridge of high pressure is the natural boundary between these non-monsoonal areas and the monsoonal lands further south. The northern limit of the typical monsoon may be set at about 25°N latitude. Further north, the summer monsoon is not strong enough to overcome the effect of the travelling anti-cyclones normally typical of the subtropics.

As a result, monsoonal rains occur in June and also in late August and September, separated by a mild anticyclonic drought in July. In South China and the Philippines the trade winds prevail in the October–April (winter) period, strength-ened by the regional, often gusty, outflow of air from the stationary Siberian anticyclone. Their disappearance and replacement by opposite (southwesterly) winds in the May–September (summer) period is the essence of the monsoon.

These monsoonal streams are quite shallow, about 1,500 metres (4,900 feet) in winter and 2,000 metres (about 6,600 feet) in summer. They bring rain only when subject to con-siderable cooling, such as anywhere along the steep, windward slopes of the Philippines and Taiwan. On the larger islands there are contrasting effects: the slopes facing west receive most of their rainfall from May to October and experience drought from December to April, whereas the slopes facing east receive orographic rains (those produced when moist air is forced to rise by topography) from September to April and mainly convectional rains from May to October.

In Vietnam and Thailand the summer monsoon is more strongly developed because of the wider expanses of over-heated land. The southwesterly stream flows from May to October, reaching a thickness of 4–5 km (about 2.5–3 miles); it brings plentiful, but not extraordinary, rainfall. The period from November to February is the cool, dry season and the period from March to April is the hot, dry one; in the far south the coolness is but relative. Along the east coast and on the eastward slopes, more rain is brought by the winter monsoon. In the summer, somewhere between Thailand and Cambodia in the interior, there may be a faint line of convergence between the southwesterly Indian–Myanmar monsoon and the southeasterly Malaysian monsoon.

Monsoonal winds are weak over Indonesia because of the expanses of water and the low latitude, but their seasonal reversal is definite. From April to October the Australian air flows southeasterly, whereas north of the equator the flow becomes a southwesterly. The Malaysian–Australian monsoon generally maintains its dryness over the islands closer to Australia, but further north it carries increasing amounts of moisture. The northeasterly flow from Asia, which becomes northwesterly south of the equator, is laden with moisture when it reaches Indonesia, bringing cloudy and rainy weather between November and May. The wettest months are December in most of Sumatra and January elsewhere, but rainfall patterns are highly localized. In Java, for instance, at sea level alone there are two major regions: an "equatorial" west with no dry season, and a "monsoonal" east with extreme drought in August and September.

Because of its relatively small size and compact shape, Australia exhibits relatively simple monsoonal patterns. The north shore is subject to a clear-cut wind reversal between summer (November–April, northwesterly flow) and winter (May–October, southeasterly flow), but with two definite limitations: first, the northwesterly, rain-bearing monsoonal wind is often held offshore and is most likely to override the land to any depth during January and February; second, even in summer there often are prolonged spells of southeasterly trade winds issuing from travelling anticyclones, separating the brief monsoonal incursions.

The West African monsoon

The main characteristics of the West African seasons have been known to the scientific community for more than two centuries. The southwest winter monsoon flows as a shallow (less than 2,000 metres [about 6,600 feet]) humid layer of surface air overlain by the primary northeast trade wind, which blows from the Sahara and the Sahel as a deep stream of dry, often dusty air. As a surface northeasterly, it is generally known as the harmattan – gusty and dry in the extreme, cool at night and scorchingly hot by day. As in a thorough monsoonal development, upper tropospheric anticyclones occur at about 20°N, while the easterly jet stream may occur at about 10°N, much closer to the equator than they do in the Indian region.

The West African monsoon is the alternation of the southwesterly wind and the harmattan at the surface. Such alternation is normally found between latitudes 9°N and 20°N. Northeasterlies occur constantly further north, but only southwesterlies occur further south. Except for erratic rains in the high-sun season, the whole year is more or less dry at 20°N. The drought becomes shorter and less complete further south. At 12°N it lasts for about half the year, and at 8°N it disappears completely.

Further south a different, lighter drought begins to appear in the high-sun months, when the monsoonal southwesterly is strongest. This drought is due to the arrival of dry surface air issuing from anticyclones formed beyond the equator in the southern hemisphere and is thus similar to the monsoonal drought in Java. Like the "break" of the monsoon in southern India, however, it occurs beyond the equator. The moist southwesterly stream, particularly frequent between 5°N and 10°N, can reach much further north, bringing warm, humid nights and moderately hot, but still humid, days. The harmattan brings cooler nights, but the extreme daily heating causes a thermal range of 10–12°C (18–22°F). Even in the daytime, the harmattan may give a sensation of coolness to the skin as it evaporates moisture from the skin's surface.

The harmattan comes in spells that mostly last from a few days to more than a week. The advancing fringe of the southwest monsoon is too shallow (under 1,000 metres [3,300 feet]) for many thunderstorms and other disturbances to occur: they usually occur 200–300 km (about 125–185 miles) behind the fringe, where the moist air is deeper but the ground is still hot enough to make it very unstable. The tops of cumulonimbus clouds may reach 12,000 metres (about 39,000 feet), well above freezing level. Disturbances usually occur along a given longitude line that is slightly curved and may in fact form one long line squall. They also reach 12,000 metres or more, travelling steadily westward at 37–56 km (23–35 miles) per hour. This suggests that they originate in the primary trade wind aloft and, as in India, are probably related to the tropical easterly jet stream.

Humidity is very high, and the daily range of temperature remains around 4°C (7°F). If it were not for the change in wind direction when the southeast trades have crossed the equator, the monsoon system of West Africa could not be distinguished from the weather system, caused by the seasonal shift in the latitude of the intertropical convergence zone, experienced over most of Central Africa.

There is a rainy season (in this case, the monsoon season), which lasts two to three months at latitude 16°N on the west coast, three to four months at 14°N, six to seven months at 10°N, and eight months at 8°N. On the south coast, which is at latitude 4°N to 6°N, the southwest monsoon (as the inter-tropical convergence) may occur at any time, but the results are quite atypical for various reasons.

In the low-sun season (December–February), the south-westerly is rare and ineffective, and the weather is cloudy but dry. From April to June, the midday sun is at its highest, and insolation (radiation received at Earth's surface) is most intense. Because the southwest wind occurs most frequently, the consequent building up of clouds leads to the main rainy season. During July and August (the short drought), cloudy conditions prevail, but the air issues direct from anticyclones further south and is dry, in spite of the fact that its direction of flow does not change. Although cloudiness decreases after the second high-sun season in September and October, there is a period of occasional rains just sufficient to constitute a sec-ondary maximum.

Toward the north, conditions are more distinctly monsoonal: by latitude 8°N the two wet seasons have merged into one long "wet" with two subdued peaks, which last approximately seven to eight months (March–October). The "dry", which is con-trolled by northeast winds, lasts from November to early March. There is one rainfall maximum (in August or Septem-ber) a short distance further north, although the wet season is only a few weeks shorter.

Monsoonal tendencies in Europe and North America

In central Europe, where the average wind direction in summer differs some 30°–40° from that of the Atlantic, there are monsoonal tendencies that occur not as a contin-uous flow but rather intermittently within frontal depres-sions, bringing cool, cloudy weather, rain, and thunderstorms. Some see in this climatic pattern a true monsoon, but it is really only an "embryo monsoon" that results in weather singularities. The latitude is too high for a true monsoon to arise.

In North America the relatively low latitude and the orientation of the land–sea boundary on the Gulf of Mexico are quite favourable to monsoonal developments. During the summer, low atmospheric pressure is frequent over the heated land; the northeasterly trade winds are consequently deflected to become easterly, southeasterly, or even southerly winds. In general, Texas and the Gulf Coast of the United States may be completely overrun by a shallow sheet of oceanic air, which may continue for a long distance inland.

The rainfall regime does not reveal any marked monsoonal pattern. There are mostly two, three, or even four minor peaks in the sequence of monthly rainfall totals. In the winter there often occur "northers", which are offshore winds caused by the general anticyclonic flow of air from the cold land. Neither the summer onshore wind nor the winter offshore wind is persistent enough to constitute a monsoonal sequence, even though monsoonal tendencies are quite evident.

In Central America, a true monsoonal cycle occurs over a small area facing the Pacific Ocean between 5°N and 12°N. Not only is there a complete seasonal reversal of the wind, but the rainfall regime is typically monsoonal. The winter period (from November to January and from March to April according to latitude and other factors) is very dry. The rainy season begins earlier (May) in the south and progressively later further north, coming at the end of June in southern Mexico. It concludes at the end of September in the north and as late as early November in the south.

The result is a rainy season that increases in duration with decreasing latitude; it lasts three months in southern Mexico and from six to seven months in Costa Rica. Latitude for latitude, this is a subdued replica of the monsoon of India.

Monsoon variability
Diurnal variability

Landmasses in regions affected by monsoons warm up very rapidly in the afternoon hours, especially on days with cloud-free conditions; surface air temperatures of 35–40°C (95–104°F) are not uncommon. Under such conditions, warm air is slowly and continually steeped in the moist and cloudy environment of the monsoon. Consequently, over the course of a 24-hour period, energy from this pronounced diurnal (daily) change in terrestrial heating is transferred to the cloud, rain, and diurnal circulation systems.

The scale of this diurnal change extends from that of coastal sea breezes to continent-sized processes – the latter confirmed by satellite observations. For example, air from surrounding areas is drawn into the lower troposphere over warmer land areas of South Asia during summer afternoon hours. This build-up of afternoon heat is accompanied by the production of clouds and rain. In contrast, a reverse circulation, characterized by suppressed clouds and rain, is noted in the early morning hours.

Intra-annual variability

Monsoon rainfall and dry spells alternate on several timescales. One well-known timescale is found around periods of 40–50 or 30–60 days. This is called the Madden–Julian Oscillation (MJO), named after the American atmospheric scientists Roland Madden and Paul Julian in 1971. The phenomenon comes in the form of alternating cyclonic and anticyclonic regions that enhance and suppress rainfall, respectively, and flow eastward along the equator in the Indian and Pacific oceans.

The MJO has the ability to influence monsoonal circulation and rainfall by adding moisture during its cyclonic (wet) phase and reducing convection during its anticyclonic (dry) phase. At the surface in monsoon regions, both dry and wet spells result. These periods may alternate locally in the order of two or more weeks per phase.

Interannual variability

The variability of monsoon-driven rainfall in the Indian Ocean and Australia appears to parallel El Niño episodes (see Chapter 5, page 310). During El Niño events, which occur about every two to seven years, ocean temperatures rise over the central equatorial Pacific Ocean by about 3°C (5.4°F). Atypical conditions characterized by increased rising air motion, convection, and rain are created in the western equatorial Pacific. At the same time, a compensating lobe of descending air, producing below-normal rainfall, appears in the vicinity of eastern Australia, Malaysia, and India.

Figure 6.2 illustrates a well-known El Niño–monsoon rainfall relationship. Here, precipitation figures from above- and below-normal monsoon rainfall periods over India are expressed as a function of years. Years characterized by El Niño events are marked by darkened histogram barbs. The graph shows that many of the years with below-normal monsoon rainfall coincide with El Niño years.

This illustration provides only limited guidance to seasonal forecasters, since monsoon rainfall is close to normal during many El Niño and La Niña years. (La Niña, also called the Pacific Cold Episode, describes the appearance of cooler ocean temperatures in the central and eastern Pacific. These conditions, which recur every three to five years, are the opposite of El Niño.) Many other factors, aside from equatorial Pacific Ocean surface temperatures, contribute to the interannual variability of monsoon rainfall. Excessive spring snow and ice cover on the Plateau of Tibet is related to the deficient monsoon rainfall that occurs during the following summer season in India.

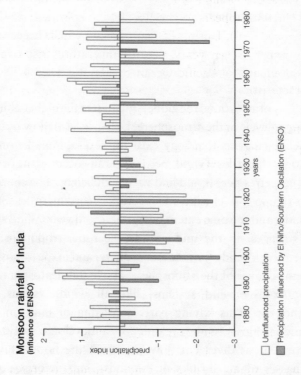

The influence of El Niño–Southern Oscillation (ENSO) on rainfall produced by the Indian summer monsoon.

Monsoon rainfall of India
(influence of ENSO)

precipitation index

years

□ Uninfluenced precipitation

■ Precipitation influenced by El Niño/Southern Oscillation (ENSO)

Figure 6.2 © Encyclopædia Britannica, Inc.

Furthermore, strong evidence exists that relates excessive snow and ice cover in western Siberia to deficient Asian summer rainfall. Warmer than normal sea-surface temperatures over the Indian Ocean may also contribute somewhat to above normal rainfall in South Asia. The interplay among these many factors makes forecasting monsoon strength a challenging problem for researchers.

Upper-level winds

Characteristics
The flow of air around the globe is greatest in the higher altitudes, or upper levels of the atmosphere. Upper-level airflow occurs in wavelike currents that may exist for several days before dissipating. Upper-level wind speeds generally occur in the order of tens of metres per second and vary with height. The characteristics of upper-level wind systems vary according to season and latitude, and to some extent hemisphere and year. Wind speeds are strongest in the mid-latitudes near the tropopause (the uppermost boundary of the troposphere) and in the mesosphere (the third layer of the atmosphere, above the stratosphere).

Upper-level wind systems, like all wind systems, may be thought of as having parts consisting of uniform flow, rotational flow (with cyclonic or anticyclonic curvature), convergent or divergent flow (in which the horizontal area of masses of air shrinks or expands), and deformation (in which the horizontal area of air masses remains constant while experiencing a change in shape). Upper-level wind systems in the mid-latitudes tend to have a strong component of uniform westerly flow, though this flow may change both direction and uniformity during the summer.

A series of cyclonic and anticyclonic vortices superimposed on the uniform westerly flow makes up a "wavetrain" – a succession of waves occurring at periodic intervals. The waves are called Rossby waves, after the Swedish American meteorologist C.G. Rossby, who first explained fundamental aspects of their behaviour in the 1930s. Waves whose wavelengths are around 6,000 km (3,700 miles) or less are called short waves, while those with longer wavelengths are called long waves. In addition, short waves progress in the same direction as the mean airflow, which is from west to east in the mid-latitudes; long waves retrogress (that is, move in the opposite direction of the mean flow). Although the undulating current of air is composed of a number of waves of varying wavelength, the dominant wavelength is usually around several thousand kilometres.

Near and underneath the tropopause, regions of divergence are found over regions of gently rising air at the surface, while regions of convergence aloft are found over regions of sinking air below. These regions are usually much more difficult to detect than those of rotational and uniform flow. While the horizontal wind speed is typically in the range of 10–50 metres per second (about 22–110 miles per hour), the vertical wind speed associated with the waves is only in the order of centimetres per second.

The winds at upper levels, where surface friction does not occur, tend to be approximately geostrophic. In other words, there is a near balance between the pressure gradient force, which directs air from areas of relatively high pressure to areas of relatively low pressure, and the Coriolis force, which deflects air from its straight-line path to the right in the northern hemisphere and to the left in the southern hemisphere. An important consequence of this geostrophic balance (see also Chapter 5, page 243) is that the winds blow parallel to isobars; lower pressures will be found to the left of the direction of the wind in the northern hemisphere and to the right of the wind in the southern hemisphere. Furthermore, wind speed increases as the spacing between isobars decreases.

In a wavetrain of westerly flow, the regions of cyclonic flow are associated with troughs of low pressure, whereas anticyclonic flows are characterized by ridges of high pressure. Rising motions tend to be found downstream from the troughs and upstream from the ridges, while sinking motions tend to be found downstream from the ridges and upstream from the troughs. The areas of rising motion tend to be associated with clouds and precipitation, whereas the areas of sinking motion tend to be associated with clear skies.

The vertical variation of the structure of the waves depends upon the temperature pattern. In general, owing to the net difference in incoming shorter-wavelength solar radiation and outgoing longer-wavelength infrared radiation between the polar and the equatorial regions, there is a horizontal temperature gradient in the troposphere. At both the surface and upper levels, the troposphere is warmest at low latitudes and coldest at high latitudes. The atmosphere is mainly in hydrostatic balance – equilibrium between the upward-directed pressure gradient force and the downward-directed force of gravity.

In addition to the general pole-to-equator temperature gradient found in the troposphere, there are zonally oriented temperature variations that are wavelike. Roughly speaking, the isotherms (cartographic lines indicating points of equal temperature) are nearly parallel to the isobars in the upper levels of the troposphere. Most frequently, relatively cold air lies just upstream from upper-level troughs and just downstream from upper-level ridges, while relatively warm air lies just upstream from upper-level ridges and just downstream from upper-level troughs.

When cold air is close to the upper-level cyclones and warm air is colocated with the upper-level anticyclones, both circulation patterns increase in intensity with height and are called cold-core and warm-core systems, respectively. Tropical cyclones, on the other hand, are warm-core systems that are most intense at the surface and that decrease in intensity with height. The vertical structure of upper-level waves has an important effect on smaller-scale features that may be embedded within them.

Jet streams

The upper-level wind flow is frequently concentrated into relatively narrow bands called jet streams, or jets. The jets, whose wind speeds are usually in excess of 30 metres per second (about 70 miles per hour) but can be as high as 107 metres per second (about 240 miles per hour), act to steer upper-level waves. Jet streams are of great importance to air travel because they affect the ground speed – the velocity relative to the ground – of aircraft. Since strong upper-level flow is usually associated with strong vertical wind shear (a sudden change in wind direction with increasing altitude in the atmosphere), jet streams in mid-latitudes are accompanied by strong horizontal temperature gradients.

Some regions of high vertical wind shear are marked by clear-air turbulence (CAT). Jet streams whose extents are relatively isolated are called jet streaks. Well-defined circulation patterns of rising and sinking air are usually found just upstream and downstream, respectively, from jet streaks (those that are not too curved). Rising motion is found to the left and right just downstream and upstream, respectively, and sinking motion is found to the right and left just downstream and upstream, respectively. Jets tend to be strongest near the tropopause, where the horizontal temperature gradient reverses.

The polar front jet moves in a generally westerly direction in mid-latitudes, and its vertical wind shear, which extends below its core, is associated with horizontal temperature gradients that extend to the surface. As a consequence, this jet manifests itself as a front that marks the division between colder air over a deep layer and warmer air over a deep layer. The polar front jet can be baroclinically (where the density of the atmosphere depends on temperature and pressure) unstable and break up into waves.

The subtropical jet is found at lower latitudes and at slightly higher elevation, owing to the increase in height of the tropopause at lower latitudes. The associated horizontal temperature gradients of the subtropical jet do not extend to the surface, so a surface front is not evident. In the tropics an easterly jet is sometimes found at upper levels, especially when a landmass is located poleward of an ocean, so the temperature increases with latitude. The polar front jet and the subtropical jet play a role in maintaining Earth's general atmospheric circulation. They are slightly different in each hemisphere because of differences in the distribution of landmasses and oceans.

Winds in the stratosphere and mesosphere

The winds in the stratosphere and mesosphere are usually estimated from temperature data collected by satellites. At these high levels the winds are assumed to be geostrophic. Overall, in the mid-latitudes, they have a westerly component in the winter and an easterly component in the summer. The highest zonal winds are around 60–70 metres per second (135–155 miles per hour) at 65–70 km (37–43 miles) above Earth's surface.

The west-wind component is stronger during the winter in the southern hemisphere. The axes of the strongest easterly and westerly wind components in the southern hemisphere tilt toward the south, with increased altitude during the northern hemisphere winter and the southern hemisphere summer. The zonal component of thermal wind shear is in accord with the zonal distribution of temperature.

During the winter there is, on average, an intense cyclonic vortex about the poles in the lower stratosphere. Over the North Pole this vortex has an embedded mean trough over north-eastern North America and over north-eastern Asia, whereas over the Pacific there is a weak anticyclonic vortex.

The winter cyclonic vortex over the South Pole is much more symmetrical than the one over the North Pole. During the summer there is an anticyclone above each pole that is much weaker than the wintertime cyclone. In the stratosphere, deviations from the mean behaviour of the winds occur during events called sudden warmings, when the meridional temperature gradient reverses on timescales as short as several days. This also has the effect of reversing the zonal wind direction.

Sudden warmings tend to occur during the early and middle parts of the winter and the transition period from winter to spring. The latter marks the changeover from the cold winter polar cyclone to the warm summer polar anticyclone. It is notable that long waves from the troposphere can propagate into the stratosphere during the winter when westerlies and sudden warmings occur, but this is not the case during the summer when easterly winds prevail. The zonal component of the winds in the stratosphere above equatorial and tropical regions is, in general, relatively weak.

This is not necessarily the case at any given time, because the winds reverse direction on average every 13–14 months. This phenomenon, which is known as the Quasi-biennial Oscillation, is caused by the interaction of vertically propagating waves with the mean flow. Its effect is greatest around 27 km

(17 miles) above Earth's surface in the equatorial region. The strongest easterlies are stronger than the strongest westerlies.

Ozone layer

Ozone (O_3) is always present in trace quantities in Earth's atmosphere, but its largest concentrations are in the ozonosphere. There it is formed primarily as a result of short-wave solar ultraviolet radiation (wavelengths shorter than 242 nanometres), which dissociates normal molecular oxygen (O_2) into two oxygen atoms. These oxygen atoms then combine with nondissociated molecular oxygen to yield ozone. Ozone, once it has been formed, can also be easily destroyed by solar ultraviolet radiation of wavelengths less than 320 nanometres.

Because of the strong absorption of solar ultraviolet radiation by molecular oxygen and ozone, solar radiation capable of producing ozone cannot reach the lower levels of the atmosphere, and the photochemical production of ozone is not significant below about 20 km (12 miles). This absorption of solar energy is very important in producing a temperature maximum at about 50 km, called the stratopause or mesopeak. Also, the presence of the ozone layer in the upper atmosphere, with its accompanying absorption, effectively blocks almost all solar radiation of wavelengths less than 290 nanometres from reaching Earth's surface, where it would injure or kill most living things.

Even though the ozone layer is about 40 km (25 miles) thick, the total amount of ozone, compared with more abundant atmospheric gases, is quite small. If all of the ozone in a vertical column reaching up through the atmosphere were compressed to sea-level pressure, it would form a layer only a few millimetres thick.

Changes to the atmosphere

Air pollution

Air pollution involves the release into the atmosphere of gases, finely divided solids, or finely dispersed liquid aerosols at rates that exceed the capacity of the atmosphere to dissipate them or to dispose of them through incorporation into solid or liquid layers of the biosphere. Air pollution results from a variety of causes, not all of which are within human control. Dust storms in desert areas and smoke from forest and grass fires contribute to chemical and particulate pollution of the air. Forest fires that swept the state of Victoria, Australia, in 1939 caused observable air pollution in Queensland, more than 3,000 kilometres (2,000 miles) away. Dust blown from the Sahara has been detected in West Indian islands. The discovery of pesticides in Antarctica, where they have never been used, suggests the extent to which aerial transport can carry pollutants from one place to another. Probably the most important natural source of air pollution is volcanic activity, which at times pours great amounts of ash and toxic fumes into the atmosphere. The eruptions of such volcanoes as Krakatoa, in the East Indies; Mount St Helens, in Washington; and Katmai, in Alaska, have been related to measurable climatic changes.

Air pollution may affect humans directly, causing a smarting of the eyes or coughing. More indirectly, its effects are experienced at considerable distances from the source – as, for example, the fallout of tetraethyl lead from urban car exhausts, which has been observed in the oceans and on the Greenland Ice Sheet.

Urban air pollution
It is the immediate effect of air pollution on urban atmospheres that is most noticeable and causes the strongest public

reaction. The city of Los Angeles is a case in point – noted for both the extent of its air pollution and the actions undertaken for control. Los Angeles lies in a coastal plain, surrounded by mountains that restrict the inward sweep of air and that separate a desert from the coastal climate. Fog moving in from the ocean is normal in the city. Temperature inversions characterized by the establishment of a layer of warm air on top of a layer of cooler air prevent the air near the ground from rising and thus effectively trap pollutants that have accumulated in the lower layer of air. In the 1940s, the air in Los Angeles became noticeably polluted, interfering with visibility and causing considerable discomfort. Attempts to control pollution, initiated during the 1950s, resulted in the successful elimination of such pollution sources as industrial effluents and the outdoor burning of waste. Nevertheless, pollution continued to increase as a result of the increased number of motor vehicles. Exhaust fumes from the engines of cars contain a number of polluting substances, including carbon monoxide and a variety of complex hydrocarbons, nitrogen oxides, and other compounds. When acted upon by sunlight, these substances undergo a change in composition producing the brown, photochemical smog for which Los Angeles is well known. Efforts to reduce pollution from car engines and to develop pollution-free engines may eventually eliminate the more serious air pollution problems. In the meantime, however, air pollution has driven many forms of agriculture from the Los Angeles basin, has had a serious effect upon the pine forests in nearby mountains, and has caused respiratory distress, particularly in children, elderly people, and those suffering from respiratory diseases.

Los Angeles is neither a unique nor the worst example of polluted air. Tokyo has such a serious air-pollution problem that oxygen is supplied to policemen who direct traffic at busy intersections. Milan, Ankara, Mexico City, and Buenos Aires face similar problems. Although New York City produces greater quantities of pollutants than Los Angeles, it has been spared from an air-pollution disaster only because of favourable climatic circumstances.

The task of cleaning up air pollution, though difficult, is not believed to be insurmountable. Methods of control include the use of fuels that are low in pollutants, such as low-sulphur forms of petroleum; more complete burning of fossil fuels, at best to carbon dioxide and water; the scrubbing of industrial smokestacks or precipitation of pollutants from them, often in combination with a recycling of the pollutants; and the shift to less-polluting forms of power generation, such as solar energy in place of fossil fuels. The example of London, as well as of other cities, has shown that major improvements in air quality can be achieved in ten years or less.

Climatic effects of polluted air

Less obvious than local concentrations of pollution are the climatic effects of air pollutants. As described in Chapter 1, as a result of the growing worldwide consumption of fossil fuels, atmospheric carbon dioxide levels have increased steadily since 1900, and the rate of increase is accelerating. The output of carbon dioxide is believed by some to have reached a point such that it may exceed both the capacity of plant life to remove it from the atmosphere and the rate at which it goes into solution in the oceans. The greenhouse effect is causing an increase in the temperature of the lower atmosphere, with consequent melting of the polar ice caps, raising of the sea level, and flooding of the coastal areas of the world.

Counterbalancing the effect of carbon dioxide is the increase of particulate matter in the air – a result of the output of smoke, dust, and other solids associated with human activity.

Such an increase may have the effect of increasing the albedo, or reflectance, of the atmosphere, causing a higher percentage of solar radiation to be reflected back into space. At present, however, the greater danger lies in the steady increase in carbon dioxide, with its associated atmospheric warming.

Acid rain

The emission of sulphur dioxide (SO_2) and nitrogen oxides to the atmosphere by human activities has led to the acidification of rain and freshwater aquatic systems. Such precipitation has become an increasingly serious environmental problem in many areas of North America, Europe, and Asia. Although this form of pollution is most severe in and around large urban and industrial areas, substantial amounts of acid precipitation may be transported great distances.

Sulphur dioxide and nitrogen oxide are released by cars, certain industrial operations (e.g. smelting and refining), and – primarily – the burning of fossil fuels such as coal and oil. The gases combine with water vapour in clouds to form sulphuric and nitric acids (H_2SO_4 and HNO_3). When precipitation falls from the clouds, it is highly acidic, with a pH value of about 5.6 or lower. (The pH scale ranges from 0 to 14; a pH of 7 is neutral, pH above 7 is alkaline, and pH below 7 is acidic.) At several locations in the eastern United States and western Europe, pH values between 2 and 3 have been recorded. In areas such as Los Angeles, San Francisco, and Whiteface Mountain in New York, fog is often ten or more times as acidic as the local precipitation.

Precipitation and fog of high acidity contaminate lakes and streams; they are particularly harmful to fish and other aquatic life in regions with thin soil and granitic rock, which provide little buffering to acidic inputs. Acid rain is blamed for the disappearance of fish from many lakes in the Adirondacks, for the widespread death of forests in European mountains, and for damaging tree growth in the United States and Canada. It has also been discovered that aluminium is leached from the soil in regions subjected to such acid precipitation, and that dissolved aluminium seems to be extremely toxic to aquatic organisms. All forms of acid precipitation have been found to damage various kinds of vegetation, including agricultural crops and trees – chiefly by inhibiting nitrogen fixation and leaching nutrients from foliage. In addition, these pollutants can corrode the external surfaces of buildings and other structures; marble structures and statues are especially vulnerable to their damaging effects.

Because the contaminants are carried long distances, the sources of acid rain are difficult to pinpoint and hence difficult to control. The drifting of pollutants causing acid rain across international boundaries has created disagreements between Canada and the United States and among European countries over the causes and solutions of the precipitation. The international scope of the problem has led to the signing of international agreements on the limitation of sulphur and nitrogen oxide emissions.

Damage to the ozone layer

Certain air pollutants, particularly chlorofluorocarbons (CFCs) and halons (chlorofluorobromine compounds), can diffuse into the ozonosphere and destroy ozone. Chlorofluorocarbons are any of several organic compounds composed of carbon, fluorine, and chlorine. They are well suited for use as aerosol-spray propellants, solvents, and foam-blowing agents because they are nontoxic and nonflammable and can be readily converted from a liquid to a gas and vice versa.

However, CFCs, once released into the atmosphere, accumulate in the stratosphere.

Ozone is destroyed as the result of chemical reactions on the surfaces of particles in polar stratospheric clouds (PSCs). These clouds are isolated within an atmospheric circulation pattern known as the "polar vortex", which develops during the long, cold Antarctic winter. The chemical reactions take place with the arrival of sunlight in spring and are facilitated by the presence of halogens (chlorine and fluorine). Ultraviolet radiation in the stratosphere causes the CFC molecules to dissociate, producing chlorine atoms and radicals (atoms or molecules that contain unpaired electrons). The chlorine atoms then react with ozone, initiating a process whereby a single chlorine atom can cause the conversion of thousands of ozone molecules to oxygen. This process of ozone destruction also occurs to a lesser extent in the Arctic.

In the mid-1980s scientists discovered that a "hole" developed periodically in the ozonosphere above Antarctica. It was found that the ozone layer there was thinned by as much as 40–50 per cent from its normal concentrations. This severe regional ozone depletion was explained as a natural phenomenon, but one that was probably exacerbated by the effects of CFCs and halons.

Stratospheric ozone shields life on Earth from the harmful effects of the sun's ultraviolet B radiation – a type of radiation shown to impair photosynthesis in plants, cause an increase in skin cancer in humans, and damage DNA molecules in living things. Even a relatively small decrease in the stratospheric ozone concentration can result in an increased incidence of skin cancer in humans and genetic damage in many organisms. Concern over increasing global ozone depletion led to international restrictions on the use of CFCs and halons and to scheduled reductions in their manufacture.

Potential effects of global warming

As discussed in Chapter 1 ("Causes of global warming", page 11), the build-up of greenhouse gases, mainly carbon dioxide (CO_2), methane (CH_4), and nitrous oxide (N_2O), is leading to a warming of Earth's surface and troposphere, as these gases absorb some of the infrared radiation produced by Earth's surface and prevent it escaping back into space. The average global temperature increased by $0.6°C$ ($1.1°F$) in the twentieth century, and the Intergovernmental Panel on Climate Change (IPCC) has warned that significant ecological and economic damage would result if temperatures rose by more than $2°C$ ($3.6°F$) by the end of this century. Emissions of carbon dioxide, the most significant greenhouse gas, result mainly from the burning of fossil fuels, as well as from large-scale deforestation and land clearing.

The change in global temperature is expected to impact on the complex atmospheric systems that influence the climate and weather conditions across the world. In its Fourth Assessment Report in 2007, the IPCC said that future tropical cyclones are likely to become more intense, with higher peak wind speeds and heavier precipitation, associated with continuing increases in tropical sea surface temperatures. There are signs that such changes are already taking place: the average number of category 4 and 5 hurricanes per year has increased over the past 30 years.

Models have also predicted that storms outside the tropics will become fewer but more intense, with more extreme wind events and higher ocean waves in several regions associated with deepened cyclones. Predictions indicate that storm tracks in both hemispheres will shift poleward by several degrees of latitude, leading in turn to changes in wind, precipitation and temperature patterns.

Regional predictions of future climate change remain limited by uncertainties in how the precise patterns of atmospheric winds and ocean currents will vary with increased surface warming. For example, some uncertainty remains in how the frequency and magnitude of El Niño–Southern Oscillation (ENSO) events (see Chapter 5, page 264) will adjust to climate change. Since ENSO is one of the most prominent sources of interannual variations in regional patterns of precipitation and temperature, any uncertainty in how it will change means a corresponding uncertainty in certain regional patterns of climate change. For example, increased El Niño activity would be likely to lead to more winter precipitation in some regions, such as the desert south-west of the United States. This might offset the drought predicted for those regions, but at the same time it might lead to less precipitation in other regions. Rising winter precipitation in the desert south-west of the United States might exacerbate drought conditions in locations as far away as South Africa.

7

THE DECLINE IN BIODIVERSITY

In May 2008, a report by the Worldwide Fund for Nature (WWF), the *Living Planet Index* (LPI), stated that global biodiversity had declined by more than a quarter in the last 35 years. The LPI tracks nearly 4,000 populations of wildlife species. It found that some marine species showed a particularly sharp decline, and that seabird populations had fallen by around 30 per cent in less than 20 years.

The World Conservation Union (IUCN) publishes the Red List of endangered species every year. In 2007 it showed that one quarter of all mammals, one third of amphibians, and one in eight birds were in danger of extinction. The total number of extinct species has risen to 785.

The Convention on Biological Diversity (CBD), an international treaty signed by 150 world leaders at the 1992 Rio Earth Summit and subsequently endorsed by the United Nations, aims to achieve a "significant reduction" in biodiversity loss by 2010. However, according to the Millennium Ecosystem Assessment, whose findings were reported in 2005, "unprecedented" additional efforts would be needed if the 2010 target is to be achieved.

It is recognized that the shocking decline in biodiversity on Earth in recent history is caused, whether directly or indirectly, by human activities. There are various ways in which humanity has impacted on the survival of other species, but the main cause is widely agreed to be habitat loss – through urbanization, agricultural intensification, deforestation, and pollution (see Chapter 6). However, in the last decade or so there has also been growing awareness of the effect of climate change on species' populations. This chapter explains the concept of biodiversity and describes the ways in which current and past levels of species extinction are assessed. It discusses the main causes of the decline in biodiversity, including climate change, and considers how the problem may be remedied.

What is biodiversity?

Biodiversity, also called biological diversity, is the variety of life found in a place on Earth. Often it also refers to the total variety of life on Earth. A common measure of this variety, called species richness, is the number of species in an area. Colombia and Kenya, for example, each have more than 1,000 breeding species of birds, whereas the forests of Great Britain and of eastern North America are home to fewer than 200. A coral reef off northern Australia may have 500 species of fish, while the rocky shoreline of Japan may be home to only 100 species. Such numbers capture some of the differences between places – the tropics, for example, have more biodiversity than temperate regions.

Although examining counts of species is perhaps the most common method used to compare the biodiversity of various places, a measure of biodiversity generally takes into account different weightings for different species. This is because some species are deemed more valuable or more interesting than others. One way this "value" or "interest" is assessed is by examining the diversity that exists above the species level, in the genera, families, orders, classes, and phyla to which species belong. For example, the number of animal species that live on land is much higher than the number of those that live in the oceans, because there are huge numbers of terrestrial insect species; insects comprise many orders and families, and they constitute the largest class of arthropods, which themselves constitute the largest animal phylum. In contrast, there are fewer animal phyla in terrestrial environments than in the oceans. No animal phylum is restricted to the land, but brachiopods (lamp shells), pogonophorans (beardworms), and other animal phyla occur exclusively or predominantly in marine habitats.

Some species have no close relatives and exist alone in their genus, whereas others occur in genera made up of hundreds of species. Given this, one can ask whether it is a species belonging to the former or latter category that is more important. On one hand, a taxonomically distinct species – the only one in its genus or family, for example – may be more likely to be distinct biochemically and so be a valuable source for medicines, for example, simply because there is nothing else quite like it. On the other hand, although the only species in a genus carries more genetic novelty, a species belonging to a large genus might possess something of the evolutionary vitality that has led its genus to be so diverse.

A second way to weight species biodiversity is to recognize the unique biodiversity of those environments that contain few species but unusual ones. Dramatic examples come from extreme environments such as the summits of active Antarctic volcanoes (e.g. Mount Erebus and Mount Melbourne in the Ross Sea region), hot springs (e.g. Yellowstone National Park in the western United States), or deep-sea hydrothermal vents. The numbers of species found in these places may be smaller than almost anywhere else, yet the species are quite distinctive. One such species is the bacterium *Thermus aquaticus*, found in the hot springs of Yellowstone. From this organism was isolated Taq polymerase, a heat-resistant enzyme crucial for a DNA-amplification technique widely used in research and medical diagnostics.

More generally, areas differ in the biodiversity of species found only there. Species whose geographical distribution is confined to a given area are called endemic species. On remote oceanic islands, almost all the native species are endemic. The Hawaiian Islands, for example, have about 1,000 plant species – a small number compared with those at the same latitude in continental Central America. Almost all the Hawaiian species, however, are found only there, whereas the species on continents may be much more wide-spread. Endemic species are much more vulnerable to human activity than are more widely distributed species, because it is easier to destroy all the habitat in a small geographic range than in a large one.

In addition to diversity among species, the concept of biodiversity also includes the genetic diversity within species. One example is our own species, for we differ in a wide variety of characteristics that are partly or wholly genetically determined: in height, weight, skin and eye colour, behavioural traits, and resistance to various diseases, for example. Likewise, genetic variety within a plant species may include the differences in individual plants that confer resistance to different diseases. For plants that are domesticated, such as rice, these differences may be of considerable economic importance, as they are the source of new disease-resistant domestic varieties.

The idea of biodiversity also encompasses the range of ecological communities that species form. A common approach to quantifying this type of diversity is to record the variety of ecological communities an area may contain. It is generally accepted that an area having, say, both forests and prairies is more diverse than one with forests alone, because each of these assemblages is expected to house different species. This conclusion, however, is indirect – i.e. it is likely to be based on differences in vegetation structure or appearance rather than directly on lists of species.

"Forest" and "prairie" are examples of names applied to ecological assemblages defined in a variety of ways, methods, and terms, and there is no clear consensus regarding what constitutes an assemblage. Technical terms that imply different degrees to which assemblages can be divided spatially include association, habitat, ecosystem, biome, life zone, ecoregion, landscape, and biotype. There is also no agreement on the boundaries of assemblages – such as where the forest biome ends and the prairie biome begins. Nonetheless, especially when these approaches are applied globally, as with the ecoregions used by the WWF, they provide a useful guide to biodiversity patterns.

The catalogue of Earth's biodiversity is very incomplete. About 1.5 million species have scientific names. Estimates of the total number of living species cluster around 10 million, which means that most species have not been discovered and described. (These estimates omit bacteria because of the practical problems in defining bacterial species.) Simply counting species must be, at best, an incomplete measure of biodiversity, since most species cannot be counted within a reasonable time. At the present rate of describing new species, it would take about 1,000 years to complete the catalogue of scientific names. Of the 1.5 million species now described, perhaps two-thirds are known from only one location and many from examining only one individual or a limited number of individuals, so knowledge of the genetic variation within species is even more constrained. From just a few well-studied species, it is clear that genetic variability can be substantial and that it differs in extent between species.

Species extinctions

In a given area, overall biodiversity can either increase or decrease, depending on the availability of resources and the interactions between species. In recent decades, the activities associated with the increasing human population have caused the biodiversity of many areas to decline.

Species extinction is the most obvious aspect of the loss of biodiversity. For example, species form the bulk of the examples in a comprehensive assessment of the state of the planet published in the early twenty-first century by the Millennium Ecosystem Assessment, an international effort coordinated by the United Nations Environment Programme. But biodiversity loss is more than just a matter of species extinction levels. Even a species that survives extinction can lose much of its genetic diversity as local, genetically distinct populations are lost from most of the species' original range. Furthermore, ecosystems may shrink dramatically in area and lose many of their functions, even if their constituent species manage to survive. Conservation is involved with studying all these kinds of losses, understanding the factors responsible for them, developing techniques to prevent losses, and, whenever possible, restoring biodiversity.

Rates of natural and present-day species extinction

According to the best estimates of the world's environmental experts, human activities have driven species to extinction at rates that are perhaps 1,000 times the natural, or background, rate. Future rates of extinction are likely be even higher. To show how these conclusions have been drawn, it is necessary to attempt to answer some difficult questions. How many species are there? How fast were species disappearing before human activity became pervasive? How fast are they becoming extinct at present? And what does the future hold for extinctions if current trends continue?

How many species are there?

Any absolute estimate of extinction rate, such as the number of extinctions per year, requires knowledge of how many species there are. As noted above, this number is not known with any great degree of certainty, and the problems of estimating it are formidable. Of the 1.5 million species described by taxonomists, only about 100,000 – comprising terrestrial vertebrates, some flowering plants, and attractive and collectible invertebrates such as butterflies and snails – are popular enough for taxonomists to know well. Birds are exceptionally well known: there are roughly 10,000 bird species, with only one or two new species being added each year.

Those who describe species cannot always be certain that the specimen in hand has not been given a name by someone else in a different country and sometimes even in a different century. Consequently, some taxonomic groups may have more names assigned to them than constituent species, which would result in erroneously high species estimates. Potentially much more serious as a source of error is the fact that some species groups have relatively few named members compared with the numbers that experts think exist in those groups. For example, taxonomists have only sparsely sampled some potentially rich communities, such as the bottom of the deep ocean and the canopies of rainforests.

One estimate of how many species might still be undescribed involves a comparison of fungi and flowering plants (angiosperms). In Britain, where both groups are well known, there are six times as many named species of fungi as of flowering plants. If this ratio applies worldwide, the world total of about 300,000 species of flowering plants, which are fairly well known globally, predicts a total of about 1.8 million species of fungi, which are not. Because only about 70,000 species of fungi currently have names, the prediction would suggest that these named species constitute only one twenty-fifth of the species that exist. On the other hand, samples from poorly known parts of the world suggest that the named species constitute a much larger fraction. Although this estimate of the number of species of fungi is almost certainly too high, it nonetheless suggests that there are large numbers of unknown species.

For insects, there are about 1 million described species, yet estimates of how many insect species exist range from 10 to 100 times this number. One such estimate was made by the entomologist Terry Erwin, by means of collecting a large sample of canopy-dwelling beetles from one species of tropical tree. The tree species had 163 beetle species specific to it. There are about 50,000 tree species in the world's tropical forests, so simple multiplication predicts roughly 8 million species of tropical forest-canopy beetles. Since 40 per cent of described insects are beetles, the total number of tropical forest-canopy insects could be 20 million. Adding half that number for insect species found on the ground beneath the trees gives an estimated grand total of 30 million species of tropical forest insects alone.

This estimate is critically dependent on the first number – the 163 species found only on the one species of tree. Furthermore, other calculations suggest that the estimate is too high. The number of insect species in tropical forests is more likely to be between 7 and 15 million – which still means that most of Earth's species would be tropical forest insects. Given the considerable uncertainties, many scientists consider that the total of all species is very roughly 10 million, a number that probably translates to "somewhere between 5 and 20 million with arguments for numbers both lower and higher".

An obvious concern follows regarding the usefulness of such calculations as a basis for assessing the loss of species. Any absolute estimate of species extinctions must be extrapolated from the 100,000 well-known species of living plants and animals, to the roughly 1.5 million described species, to the likely grand total of very roughly 10 million. Because of uncertainties about the total number of living species, published statements regarding the total number of species that become extinct per year or per day can vary a hundredfold.

Another approach to assessing species loss is to derive relative estimates – that is, estimates of the proportion of well-known species that become extinct in a given interval. Estimating such proportions, however, raises a critical concern of its own – namely, are these proportions actually typical of the great majority of species that are still undescribed? They are likely to be so if extinction rates in widely different species groups and regions turn out to be broadly similar.

There also is another way in which estimates of extinctions can be made relative. Extinctions have always been a part of

Earth's history. It is possible to make any estimates of current or future extinction relative to that history.

Calculating background extinction rates

To quantify the effect of modern human activity on the loss of species it is necessary to determine how fast species disappeared in the absence of that activity. Studies of marine fossils show that species last about 1–10 million years. We might assume that all these extinctions happened independently and gradually – i.e. in the "normal" way – rather than catastrophically (as they did at the end of the Cretaceous Period about 65 million years ago, when dinosaurs and many other land and marine animal species disappeared). On that basis, if one followed the fates of 1 million species, one would expect to observe about 0.1–1 extinction per year – in other words, one species going extinct every 1–10 years.

Human lifespans provide a useful analogy to the foregoing. If humans live for about 80 years on average, then 1 in 80 individuals should die each year under normal circumstances. If, however, many more than 1 in 80 were dying each year, then something would be abnormal. There might be an epidemic, for instance.

To make comparisons of present-day extinction rates conservative, we would assume that the normal rate is just one extinction per million species per year. This then is the benchmark – the background rate against which to compare modern rates.

This estimate of the background extinction rate derives from the abundant and widespread species that dominate the fossil record. By contrast, the species most likely to become extinct today are rare and local. Thus, the fossil data might underestimate background extinction rates. Importantly, however, these estimates can be supplemented from knowledge of speciation rates – the rates that new species come into being – of those species that often are rare and local. These rates cannot be much less than the extinction rates, or there would be no species left.

To explore the idea of speciation rates, one can refer again to the analogy of human lifespans and ask: How old are my living siblings? The answer might be anything from that of a newborn to that of a retiree living out his or her last days. The average age will be midway between them – that is, about half a lifetime. Ask the same question for a mouse, and the answer will be a few months; of long-living trees such as redwoods, perhaps a millennium or more. The age of one's siblings is a clue to how long one will live.

The equivalent of siblings on a species level are the species' closest living relatives in the evolutionary tree (models that reconstruct the evolutionary history of groups of organisms) – something that can be determined by differences in their DNA. The closest relative of human beings is the bonobo (*Pan paniscus*), whereas the closest relative of the bonobo is the chimpanzee (*P. troglodytes*). Taxonomists call such related species "sister taxa", following the analogy that they are splits from their "parent" species.

The greater the differences between the DNA of two living species, the more ancient the split from their common ancestor. Studies show that these accumulated differences result from changes whose rates are, in a certain fashion, fairly constant – hence the concept of the "molecular clock", which allows scientists to estimate the time of the split from knowledge of the species' DNA differences. For example, from a comparison of their DNA, the bonobo and the chimpanzee appear to have split 1 million years ago, and humans split from the line containing the bonobo and chimpanzee about 6 million years ago.

The advantage of using the molecular clock to determine speciation rates is that it works well for all species, whether common or rare. It works for birds and, as in the above example, for forest-living apes, for which very few fossils have been recovered. The conclusion that the bonobo and chimpanzee split a million years ago suggests that the lifespans of such species are (at least in the absence of modern human actions that threaten them) on million-year timescales. This is just one example, however. Is there evidence that speciation can be much more rapid?

Until recently, there seemed to be an obvious example of a high rate of speciation – a "baby boom" – of bird species. Its existence allowed for the possibility that the high rates of bird extinction that are observed today might be just a natural pruning of this evolutionary exuberance.

On either side of North America's Great Plains are 35 pairs of sister taxa including western and eastern bluebirds (*Sialia mexicana* and *S. sialis*), red-shafted and yellow-shafted flickers (both considered subspecies of *Colaptes auratus*), and ruby-throated and black-chinned hummingbirds (*Archilochus colubris* and *A. alexandri*). According to the rapid-speciation interpretation, a single mechanism seemed to have created them all. Each pair of sister taxa had one parent species ranging across the continent. Then a major advance in glaciation during the latter part of the Pleistocene Epoch, some tens of thousands of years ago, split each population of parent species into two groups. Each pair of isolated groups evolved to become two sister taxa, one in the west and the other in the east. Finally, the ice retreated, and, as the continent became warm enough, about 10,000 years ago, the sister taxa expanded their ranges and, in some cases, met once again.

The story, while compelling, is now known to be wrong. Molecular data show that, on average, the sister taxa split

2.45 million years ago. This means that the average species lifespan for these taxa is not only very much older than the rapid-speciation explanation for them requires but is also considerably older than the 1-million-year estimate for the extinction rate suggested above as a conservative benchmark.

Molecular-based studies find that many sister species were created a few million years ago, which suggests that species should last "a few million" years too. Indeed, they suggest that the background rate of one extinction among a million species per year may be too high. Nevertheless, this rate remains a convenient benchmark against which to compare modern extinctions.

Factors that cause extinction

A number of factors related to human activity are contributing to species extinction and the decline of overall biodiversity. The predominant cause is considered to be habitat loss. It is not hard to see why: clear a forest, destroy a coral reef, or dam a river, and the species found there are likely to be lost. These are instances of local extinctions, however, and their occurrence does not mean that the species involved will go extinct everywhere. Nonetheless, habitat destruction has caused numbers of extinctions in various parts of the world. Some scientists use the term "extirpation" for local extinctions, reserving "extinction" to mean global extinction. In this section the term "extinction" has the global meaning.

Extinction may also result from the interactions of introduced exotic species with native species, from overharvesting by humans, and from climatic changes occurring at the global scale.

Habitat loss

In some parts of the world, for example in areas of tropical rainforests, the role of habitat loss in extinction of species is fairly clear cut. In other places, however, human actions have extensively modified terrestrial habitats, yet these areas are not centres of extinction. To learn what makes centres of human-caused extinctions special, one must ask what their common features are.

Endemic species

Clearly, many of these places and their species are well known. The seventeenth-century Dutch artist Rembrandt van Rijn painted birds of paradise and marine cone shells gathered from the early European exploration of the Pacific, testifying to people's fascination for attractive and interesting specimens from "exotic" locales. Victorians filled their cabinets with such curiosities – birds, mammals, marine and terrestrial snail shells, and butterflies – and painted tropical flowers and grew them in their hothouses. Yet the natural history of North America and Europe are also very well known. In fact, what is special about the places with particularly high numbers of extinctions is that each area holds a high proportion of species restricted to it – that is, endemic species.

Remote islands have many terrestrial endemics – for instance, more than 90 per cent of the plants and land birds of the Hawaiian Islands live nowhere else. Some continental areas are rich in endemic species too. About 70 per cent of the flowering plants in South Africa's fynbos (shrubland) and nearly three quarters of Australia's mammals are endemic. In contrast, many areas have almost no endemic species: for example, only about 1 per cent of Europe's birds are found only there.

The simplest model of extinction would be to assume that within a species group – mammals, for instance – all species had roughly the same risk of extinction. Were this to be true, then the more species that live in a region, the more would be likely to go extinct. This is not the case, however. For example, the Hawaiian Islands and Britain in their pasts held very roughly the same number of breeding land birds. Yet the former have lost more than 100 species, while the latter has lost only a few – and those still survive on the European continent. The difference is that almost all the Hawaiian birds (for instance, honeycreepers such as the apapane and iiwi) were endemic, while only one of Britain's birds (the Scottish crossbill) is. Thus, the number of extinctions in an area depends only very weakly on its total number of species but strongly on its total number of endemics. Areas rich in endemic species are where extinctions will concentrate, unless they are so remote that human actions do not harm them.

Small range size

The size of a species' geographic range is by far the best explanation for differences in rates of extinction. Species with small ranges are much more vulnerable than those with large ranges, simply because it is much easier to destroy the former than the latter.

In terms of range size, the distribution of life on land has several remarkable features. First, many terrestrial species have very small range sizes relative to the average range size. In the Americas, for example, one in ten species of birds and over half the species of amphibians have geographic ranges smaller than the state of Connecticut (5,006 sq mi/12,966 sq km), and half the bird species have ranges smaller than the states of Washington (68,498 sq mi/169,639 sq km), Oregon (97,047 sq mi/251,351 sq km), and California (158,633 sq mi/410,858 sq km) combined. The average range size is very much larger, for some species have huge ranges. The American robin (*Turdus migratorius*), for example, breeds almost everywhere in the United States from Alaska to Florida to California, across all of continental Canada, and in much of Mexico.

Second, for many kinds of terrestrial organisms, species with small ranges typically have lower local population densities than do widespread species. For example, the American robin is generally a locally common bird across its entire range. But those species with ranges smaller than the size of Connecticut are generally very hard to find even in the midst of their ranges.

Third, terrestrial species with small ranges are geographically concentrated. North America, for instance, has almost no bird species with small ranges. Such species live almost exclusively in the tropics. The greatest concentrations of bird species are in the Amazon lowlands, with secondary centres in Central America and the forests along the coast of Brazil. But it is the Andes and the coastal forests that have the most bird species with small geographic ranges – in other words, these areas are the centres of endemism. It is in these areas that threatened species are concentrated.

Terrestrial hotspots

In the 1990s a team of researchers led by the British environmental scientist Norman Myers identified 25 terrestrial "hotspots" (also known as "biodiversity hotspots") of the world –areas on land where species with small geographic ranges coincide with high levels of modern human activity (see Figure 7.1). Originally these hotspots encompassed about 17 million square km (6.6 million square miles) of the roughly 130 million square km (50 million square miles) constituting Earth's ice-free land surface. Species ranges are so concentrated geographically in these regions that, out of a total of about 300,000 flowering-plant species described worldwide, more than 133,000 occur only there. The comparable numbers

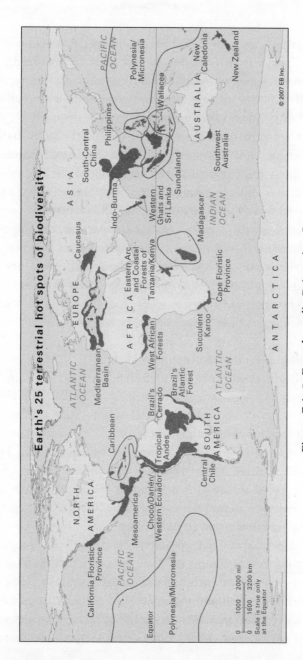

Figure 7.1 © Encyclopædia Britannica, Inc.

for birds are roughly 2,800 of 10,000 species worldwide (of which roughly two thirds are restricted to the land); for mammals, 1,300 of roughly 5,000 worldwide; for reptiles, roughly 3,000 of about 8,000 worldwide; and for amphibians, 2,600 of roughly 5,000 worldwide.

The hotspots have been sites of unusual levels of habitat destruction. Only about 12 per cent of the original habitat of these areas has survived to the beginning of the twenty-first century. Of this remaining habitat, only about two fifths is protected in any way. Sixteen of the 25 hotspots are forests, most of them tropical forests. For comparison, the relatively less-disturbed forests found in the Amazon, the Congo region, and New Guinea have retained about half their original habitats. As a consequence of these high levels of habitat loss, the 25 hotspots are locations where the majority of threatened and recently extinct species are to be found.

Predictions of extinctions based on habitat loss

Worldwide, about 6 per cent of the land surface is protected by some form of legislation, though the figure for the 25 hotspots is only 4.5 per cent of their original extent. (Such numbers are misleading, however, in that some areas are protected only on paper as their habitats continue to be destroyed.) These statistics lead naturally to the question of how many species will be saved if, say, 4.5 per cent of the land worldwide is protected. The answer turns out to be close to 50 per cent – a result that needs some explanation.

The land that countries protect, often as national or regional parks, frequently comprises "islands" of original habitat surrounded by a "sea" of cropland, grazing land, or cities and roads. On a real island the number of species that live there depends on its area, with a larger island housing more species than a smaller one. Many studies involving a wide range of animals and plants show that the relationship between area and species number is remarkably consistent. An island half the size of another will hold about 85 per cent of the number of species.

The next question is whether this relationship of species to area holds for "islands" of human-created habitat. If one were to conduct an experiment to test such an idea, one would take a continuous forest, cut it up into isolated patches, and then wait for species to become locally extinct in them. After sufficient time, one could count the numbers of species remaining and relate them to the area of the patches in which they survived. In the last decades of the twentieth century, scientists undertook exactly such an experiment, making use of government-approved forest clearing for cattle ranching, in the tropical forests around Manaus in Brazil. More generally, human actions have repeated this experiment across much of the planet in an informal way. Counts of species in areas of different sizes confirm the species-to-area relationship.

How can these results regarding local habitat "islands" be applied to global extinctions? The mostly deciduous forests of eastern North America provide a case history. The birds of the region have been well described, beginning with the explorations of the naturalist and artist John James Audubon in the first quarter of the nineteenth century, when the area was still mostly forested. Audubon shot and painted many species, including four that are now extinct – the Carolina parakeet (*Conuropsis carolinensis*), Bachman's warbler (*Vermivora bachmanii*), the passenger pigeon (*Ectopistes migratorius*), and (if not extinct, then very nearly so) the ivory-billed woodpecker (*Campephilus principalis*). A fifth species, the red-cockaded woodpecker (*Picoides borealis*), is threatened. Including the above, about 160 species of birds once lived in these eastern forests.

European settlement cleared these forests, and, at the low point in about 1870, only approximately half the forest remained. The species-to-area relationship predicts that about 15 per cent of the 160 bird species – that is, about 24 species – would become extinct. Why is this number much larger than the three to five extinctions or near-extinctions observed? The answer can be found by first supposing that all the eastern forests had been cleared, from Maine to Florida and westward to the prairies. Only those species that lived exclusively within these forests – that is, the endemics – would have gone extinct. Species with much larger ranges, such as the American robin mentioned above, would have survived elsewhere. How many species then were originally endemic to the forests of eastern North America? The answer is about 30, the rest having wide distributions across Canada and, for some, into Mexico. The species-to-area relationship predicts that 15 per cent of these 30 species, or 4.5 species, should go extinct, which is remarkably close to the observed number.

Eastern North America clearly is not a place where species with small ranges are concentrated, but the species-to-area predictions work in other places too. A variety of studies have examined birds and mammals on the islands of Southeast Asia, from Sumatra westward – one of the biodiversity hotspots (Sundaland). Both here and for birds in Brazil's Atlantic Forest, which is another hotspot, the extent of deforestation and the species-to-area relationship accurately predict the number of threatened species rather than extinct ones. Because the deforestation of these areas is relatively recent, many of their species have not yet become extinct.

The close match between the numbers of extinctions predicted by the species-to-area relationship and the numbers of species already extinct (as in eastern North America) or nearly extinct (as in more recently destroyed habitats) allows simple calculations that have worldwide import. As discussed above, the hotspots retain only 12 per cent of their original habitats, and only about 4.5 per cent of the original habitats are protected. As predicted by the species-to-area relationship, natural habitats that shrink to 4.5 per cent of their original extent will lose more than half of their species. Since these habitats once supported 30–50 per cent of terrestrial species, very roughly one quarter of all terrestrial species will be likely to become extinct worldwide.

In fact, species losses will probably be even greater because this calculation does not include nonendemic species – those that live both inside and outside the hotspots. For example, many of the species that live in the relatively less-disturbed tropical moist forests of the Amazon or the Congo region are the same as those that live in the adjacent hotspots. Human actions are clearing about 10 per cent of the original area of these forests every decade, with half of the area already gone. Species losses in these forests are still relatively few, but the rate will increase rapidly as the last remaining forests dwindle. If only the same percentage of these forests is protected as is the case for the hotspots, then they too will lose half their species.

In summary, many scientists believe that habitat destruction will put somewhere between a quarter and a half of all species worldwide on an inexorable path to extinction and will do so within the next few decades. If that proves true, extinction rates by the mid-twenty-first century will be several thousand times the benchmark rate of one extinction per million species per year.

Fire suppression as habitat loss

While most of the terrestrial hotspots are tropical moist forests, four areas – the California Floristic Province, the Cape Floristic Province in South Africa, the Mediterranean Basin, and Southwest Australia (see Figure 7.1) – are shrublands. They are also places where people live and grow crops; all four regions are noted for their wines, for example. Not only does this human activity convert land directly, but it also leads to the suppression of fire, especially near people's homes. This alteration of natural fire regimes by the reduction in fire frequencies leads to changes in vegetation, especially to the loss of the native fire-resistant species. Globally, huge areas of grasslands and shrublands would become heavily canopied forests were all fires suppressed. The effects of changes in fire frequencies on species losses have not yet been calculated.

Habitat loss in marine environments

The seas cover more than two thirds of Earth's surface, yet only 210,000 of the 1.5 million species that have been described are marine animal and plant species. Because the oceans are still poorly explored, the count of marine species may be even more of an underestimate than that of land species. For example, the Census of Marine Life, a decade-long international programme begun at the start of the twenty-first century, added 13,000 new marine species to the total count over the first four years of the effort. As on land, the peak of marine biodiversity lies in the tropics. Coral reefs represent just 0.2 per cent of the ocean surface, yet they account for almost 100,000 species. Between 4,000 and 5,000 species of fish – perhaps a third of the world's marine fish – live on coral reefs. The frequently cited metaphor that "coral reefs are the rainforests of the sea" underscores their importance for conservation.

When numbers of described marine species are mapped on a worldwide scale, it becomes clear that the global centre of marine biodiversity encompasses the waters of the Philippine and Indonesian islands. Numbers of species drop steeply to the east across the Pacific, and less steeply to the west across the Indian Ocean. In the Atlantic Ocean the highest levels of biodiversity are in the Caribbean. Fish, corals, molluscs, and lobsters all show similar patterns in the distributions of their species. Mirroring the patterns on land, the places with the most species are often not the places with the most endemic species. Major centres of endemism for fish, corals, molluscs, and lobsters include the Philippines, southern Japan, the Gulf of Guinea, the Sunda Islands in Indonesia, and the Mascarene Islands.

With the major exception of the Great Barrier Reef of Australia, most coral reefs are off the coasts of developing countries. Rapidly increasing human populations and poverty put increasing fishing pressure on nearshore reefs. In addition, in their efforts to sustain declining fish catches, people resort to extremely damaging fishing methods such as dynamite and poisons. Coral reefs are also threatened by coastal development, pollution, and global warming. Human activities threaten some three fifths of the world's reefs, with the highest damage being concentrated in areas with high rates of deforestation and high run-off from the land. As the tropical forests are destroyed, rains erode soils and wash the sediments down rivers into the sea, damaging the local coral reefs. Thus, the destruction of some of the most important terrestrial habitats – in the Caribbean and Southeast Asia – contributes to the destruction of some of the most important marine habitats offshore.

Whereas damage to coral reefs contributes to the loss of species, the greatest physical damage to ocean ecosystems involves the effects of bottom trawling, a commercial fishing method that involves dragging a cone-shaped bag of netting along the seabed. Damage from bottom trawling occurs over larger areas of the Earth than does tropical deforestation, and it involves even greater and more-frequent disturbances, albeit ones not easily seen. Bottom trawling disturbs about 15 million square km (6 million square miles) of the world's sea floor each year. This area of ocean is only about 4 per cent of the world total, but its small proportion belies its significance. About 90 per cent of the ocean consists of deep waters so poor in nutrients that they are the equivalent of the land's deserts. Almost all of the world's fisheries are concentrated in the 30 million square km (12 million square miles) of nutrient-rich waters on the continental shelf, plus a few upwellings. On average, the ocean floor of these productive waters is trawled roughly every two years.

The otter trawl is the most widely used bottom-fishing gear. As it is dragged forward, a pair of flat plates called otter boards – one on each side of the trawl net and weighing several tons – spreads horizontally to keep the mouth of the trawl open. A long rope with steel weights keeps the mouth open along its bottom edge. This heavy structure ploughs the ocean floor as it moves, creating furrows and crushing, burying, and exposing marine life. This activity destroys bottom-dwelling species including corals, brachiopods (lamp shells), molluscs, sponges, sea urchins, and various worms that live on rocks or pebbles on the sea bed. It is also damaging to other species, such as polychaete worms that burrow into the sea bed. Some species – deep-sea corals, for example – are extremely slow-growing, and are unable to recover before bottom trawls plough the area once again.

Habitat loss in freshwater environments

Freshwater ecosystems are divided into two major classes: flowing (such as rivers and streams) and static (such as lakes and ponds). Although the distribution of species in freshwater ecosystems is not as well known as for marine and terrestrial ecosystems, it is still clear that species are similarly concentrated. For fish, the major tropical rivers such as the Amazon River and its tributaries hold a large fraction of the world's freshwater fish species. Tropical lakes, particularly those in the Rift Valley of East Africa, also have large numbers of endemic species.

Riverine habitats have been extensively modified by damming and by channelization (the practice of straightening rivers by forcing them to flow along predetermined channels). A global survey published in the early twenty-first century revealed just how few of the world's large rivers are natural. In the contiguous United States almost every large river has been modified extensively. (Some, such as the Tennessee River, have been converted almost completely by dams to a series of artificial lakes.) Much the same is true in Europe. In terms of the area of their basins, more than half of the world's rivers are extensively modified. The water of some rivers barely reaches its final destination: this is the case for the Colorado River in the United States, which empties into the Gulf of California, and for the Amu Darya in Central Asia, which empties into the rapidly shrinking Aral Sea. Along their routes the water is used for agriculture or is lost as evaporation from dams. Large wild rivers are typical only of Arctic regions in Alaska, Canada, and Siberia – places so far away from urban centres that there has been no incentive to control their waters.

As a result of the massive changes to the world's rivers, large proportions of species living in rivers have become extinct or are expected to do so soon. For example, the Mississippi and St Lawrence river basins were home to 297 North American species of bivalve molluscs. Of these, 21 have become extinct in the past century and another 120 species are in danger of extinction. Of the approximately 950 species of freshwater fish of the United States, Canada, and Mexico, about 40 have become extinct in the past century. They have disappeared from a wide range of habitats: northern lakes, southern streams, wetlands, and particularly springs.

Pollution

Pollution is a special category of habitat loss: it involves chemical destruction rather than the more obvious physical destruction. Pollution occurs in all habitats – land, sea, and fresh water – and in the atmosphere. Atmospheric pollution is discussed in Chapter 6. Global warming (see Chapter 1) is one consequence of the increasing pollution of the atmosphere by emissions of carbon dioxide and other greenhouse gases.

Pollution is a global-scale problem, no less so for rivers and marine life. Wastes are often dumped into rivers, and they end up in estuaries and coastal habitats – regions that support the most diverse shallow-water ecosystems and the most productive fisheries. Rivers receive pollution directly from factories that dump a wide variety of wastes into them. They also receive run-off – rainwater that has passed over and through the soil while moving toward the rivers. In fact, water entering rivers after it has been used for irrigation has passed through the soil more than once – first as run-off, which is then returned to the land for irrigation, whereupon it soaks through the soil again.

Some polluted river water eventually reaches freshwater wetlands. In the case of the Florida Everglades, run-off from the agricultural areas upstream adds unwanted nutrients to an ecosystem that is naturally nutrient-poor. As it does so, the vegetation changes, and species not common in the Everglades begin to take over the habitats.

Other polluted waters reach estuaries on the way to the oceans. Estuaries are among the most polluted ecosystems on Earth. On entering the oceans, the polluted waters can in turn harm the ecosystems there. The Mississippi River, for example, drains a basin of more than 3 million square km (1.2 million square miles), delivering its water, sediments, and nutrients and other pollutants into the northern Gulf of Mexico. Fresh water is less dense than salt water and floats on top of it. This upper layer contains the nitrogen and phosphorus fertilizers that have run off croplands, and they fertilize the ocean's phytoplankton, causing excessive population growth in these species. As the masses of phytoplankton die, sink, and decompose, they deplete the water's oxygen. Ocean bottom dwellers such as shrimp, crabs, starfish, and marine worms suffocate and die, creating a "dead zone".

Such conditions also occur in Europe's Baltic, Adriatic, and Black seas. Within a few decades the Baltic has gone from being naturally nutrient-poor and diverse in species to being nutrient-rich and degraded in its ecosystems. In the Adriatic Sea, rising nutrient levels have generated a large increase in phytoplankton. Nutrients in the run-off flowing into the Black Sea seem to be contributing factors in the invasion and subsequent massive increase since the 1980s of the comb jelly *Mnemiopsis leidyi*. This has caused the decline of native species and fisheries.

Similar nutrient enrichment has led to increasing frequencies of toxic blooms of microscopic organisms such as *Pfiesteria piscicida* in the eastern United States, a dinoflagellate that kills fish and has been reported to cause skin rashes and other maladies in humans.

Rising levels of pollution may have also contributed to a wave of outbreaks of diseases affecting marine life. Caribbean coral reefs have been particularly affected, with successive waves of disease propagating throughout the region in recent decades. The result has been large declines in two species of major reef-building corals, *Acropora cervicornis* and

A. palmata, and the herbivorous sea urchin *Diadema antillarum*. Their combined loss has transformed Caribbean reefs from high-coral, low-algae ecosystems to high-algae, low-coral ones. The latter type of ecosystem supports far fewer species.

Introduced species

Many examples of species extinction implicate other species that have been introduced by humans. These may deplete or destroy populations of native species by predating on them, outcompeting them, or causing the spread of diseases to which they may be particularly susceptible.

Throughout history, humans have deliberately spread species as they moved into new areas. One example is the colonization of Pacific islands, such as the Hawaiian Islands, New Zealand, and Easter Island, by the Polynesians over the last 2,000 years. These settlers introduced pigs and rats to islands far too remote to have acquired native land mammals. The rats would have found the native birds, their eggs, and their young to be easy pickings, and the pigs would have destroyed the ground cover of the forests. With only Stone Age technology, the settlers may have exterminated as many as 2,000 bird species – some 17 per cent of the world total. Locally, they often exterminated all the bird species they encountered.

A more recent case is that of New Yorkers in the 1890s, who wanted all the birds in Shakespeare's works to inhabit the city's Central Park – resulting in the introduction of the starling (*Sturnus vulgaris*) to North America. Through the centuries hunters have demanded exotic birds and mammals to shoot, fishermen have wanted challenging fish, and gardeners have wanted beautiful flowers. The consequences in some cases have been devastating. Cacti and the shrub *Lantana camara*, for example, which were introduced as ornamental plants, have destroyed huge areas of grazing land worldwide.

Not all introductions are deliberate: several species of rats, for instance, were probably unwelcome hitchhikers on ocean voyages. Introduction of the tree snake that destroyed Guam's birds (see below) was also accidental.

Almost every type of living organism has been moved deliberately or accidentally. A study published in the 1990s surveyed species that travelled in the ballast tanks of ocean-going ships. Such ships take on seawater as ballast in one port and release it in other ports, sometimes half a world away. In one example, water taken aboard in Japan and released in Oregon contained 360 species, including microscopic single-celled animals and plants, jellyfish and corals, molluscs, various kinds of marine worms, and fish – an almost complete catalogue of major kinds of marine organisms.

Although not all species devastate the communities they enter, the example of what happened on Guam after the introduction of the tree snake *Boiga* illustrates just how damaging a species can be. Birds started disappearing from the central region of the island in the early 1960s. At the same time, the snakes started causing blackouts as they crawled up power poles and short-circuited the suspended lines, and many people who kept chickens noticed that snakes were killing the birds in their coops. Before *Boiga* made its appearance, ten species of native birds lived in the forests of Guam – the Micronesian starling (*Aplonis opaca*), the Mariana crow (*Corvus kubaryi*), the Micronesian kingfisher (*Halcyon cinnamomina*), the Guam flycatcher (*Myiagra freycineti*), the Guam rail (*Rallus owstoni*), the rufous fantail (*Rhipidura rufifrons*), the bridled white-eye (*Zosterops conspicillatus*), the white-throated ground dove (*Gallicolumba xanthonura*), the Mariana fruit dove (*Ptilinopus roseicapilla*), and the cardinal honeyeater (*Myzomela cardinalis*). An 11th species, the island swiftlet (*Aerodramus vanikorensis*), nested in caves and fed over the forest. In addition, the island supported a few introduced bird species, birds that nested in wetland habitats, and some seabirds.

The decline of the ten forest birds followed similar patterns – first on the central part of the island, where the main town and port are, and then on southern Guam by the late 1960s. By 1983 scientists could find the birds only in a small patch of forest in the north of the island, and by 1986 they were gone from there as well. The non-native birds suffered a similar pattern of decline, and many of the waterbirds and seabirds that nested on the island also suffered. The starling has survived on a small island, Cocos, off the south coast of Guam, and the swiftlet still nests on the high walls of a cave. But the snake has eliminated every species for which it could easily kill the eggs and young in the nest. Of the ten forest species, the rail and the flycatcher were endemic to the island, while the kingfisher is a distinctive subspecies. The kingfisher and the rail survive in captivity.

Islands such as Guam seem to be particularly vulnerable to introduced species. People have often brought domestic cats with them to islands, where some of the animals have escaped to form feral populations. Cats are efficient predators on vertebrates that have had no prior experience of them; they have caused the extinction of some 30 island bird species worldwide. Introduced herbivores, particularly goats, devastate native plant communities because the plants have had no chance to adapt to these new threats.

Although islands are particularly vulnerable, introduced species also wreak large-scale changes on continents. An exceptionally clear example is the loss of the American chestnut (*Castanea dentata*) in eastern North America after the accidental introduction of the fungus *Endothia parasitica*, which causes chestnut blight. The chestnut, a once-abundant tree, was removed with surgical precision over its entire range beginning in the early twentieth century. All that survive are small individuals that tend to become infected as they get older.

One mechanism by which introduced species can cause extinction is hybridization. In general, species are considered to be genetically isolated from one another – they cannot interbreed to produce fertile young. In practice, however, the introduction of a species into an area outside its range sometimes leads to interbreeding of species that would not normally meet. If the resident species is extremely rare, it may be genetically swamped by the more abundant alien. One example of a species threatened by hybridization is the white-headed duck (*Oxyura leucocephala*). The European population of this species lives only in Spain, where habitat destruction and hunting once reduced it to just 22 birds. With protection, it recovered to about 800 individuals, but it is now threatened by a related species, the ruddy duck (*O. jamaicensis*). This bird is native to North America, was introduced to Great Britain in 1949, and spread to the continent including Spain, where it hybridizes with the much rarer white-headed duck. Similar cases across the world involve other species of ducks, frogs, fish, cats, and wolves.

Overharvesting
Overharvesting, or overfishing in the case of fish and marine invertebrates, depletes some species to very low numbers and drives others to extinction. In practical terms, it reduces valuable living resources to such low levels that their exploitation is no longer sustainable. While the most familiar cases involve whales and fisheries, species of trees and other plants, especially those valued for their wood or for medicines, also can be exterminated in this way.

Whaling

Whaling offers an example of overharvesting that is interesting not only in itself but also for demonstrating how poorly biodiversity has been protected even when it is of economic value. The first whalers are likely to have taken their prey close to shore. Right whales were the "right" whales to take because they are large and slow-moving, feed near the surface and often inshore, float to the surface when harpooned, and were of considerable commercial value for their oil and baleen (whalebone). The southern right whale (*Eubalaena australis*), for example, is often seen in shallow, sheltered bays in South Africa and elsewhere. Such behaviour would make any large supply of raw materials a most tempting target. Whalers had nearly exterminated the North Atlantic species of the northern right whale (*Eubalaena glacialis*) and the bowhead whale (Greenland right whale; *Balaena mysticetus*) by 1800. They succeeded in exterminating the Atlantic population of the grey whale (*Eschrichtius robustus*). Whalers then moved on to species that were more difficult to kill, such as the humpback whale (*Megaptera novaeangliae*) and the sperm whale (*Physeter macrocephalus*).

The Napoleonic Wars gave whales a respite, but with the peace of 1815 came a surge of whalers into the Pacific Ocean, inspired by the stories of James Cook and other explorers. The first whalers arrived in the Hawaiian Islands in 1820, and by 1846 the fleet had grown to nearly 600 ships, the majority from New England. The catch on each whaling voyage averaged 100 whales, though a voyage could last as long as four years.

In the late 1800s, steamships replaced sailing ships, and gun-launched exploding harpoons replaced hand-thrown lances. The new technology allowed whalers to kill what until then had been the "wrong" whales – fast-swimming species such as the blue whale (*Balaenoptera musculus*) and fin whale (*B. physalus*). Whalers killed nearly 30,000 blue whales in 1931 alone. With World War II came another lull, but the catch of blue whales rose to 10,000 in 1947. The fin whale was next, with the annual catch peaking at 25,000 in the early 1960s. Then came the smaller sei whale (*B. borealis*, which no one had bothered to kill until the late 1950s) and finally the even smaller minke whale (*B. acutorostrata*), which whalers still hunt despite an international moratorium in effect since 1986 that seeks to curb commercial whaling.

The story of whaling is, in brief, the rapid depletion and sometimes extermination of one population after another, starting with the easiest species to kill and progressing to the most difficult. That whales are economically valuable raises the obvious question of why, for most of its history, the industry made no attempts to harvest whales sustainably.

Overfishing

Overfishing is the greatest threat to the biodiversity of the world's oceans, and contemporary information published for fisheries in the United States can serve as an example of the magnitude of the problem. Congress requires the National Marine Fisheries Service (NMFS) to report regularly on the status of all fisheries whose major stocks are within the country's exclusive economic zone, or EEZ. (Beyond its territorial waters, every coastal country may establish an EEZ extending 370 km [200 nautical miles] from shore. Within the EEZ the coastal state has the right to exploit and regulate fisheries and carry out various other activities to its benefit.) The areas involved are considerable, covering portions of the Atlantic, the Caribbean, the Gulf of Mexico, and the Pacific from off San Diego to the Bering Sea out to the west of the Hawaiian island chain, along with the islands constituting the western part of the former Trust Territory of the Pacific Islands. At the turn of the twenty-first century, the NMFS deemed some 100 fish stocks to be overfished and a few others close to being so, while some 130 stocks were not thought to be overfished. For about another 670 fish stocks, the data were insufficient to allow conclusions. Thus, a little under half of the stocks that could be assessed were considered to be overfished. For the major fisheries – those in the Atlantic, the Pacific, and the Gulf of Mexico – two thirds of the stocks were overfished.

As to the hundreds of stocks about which fisheries biologists know too little, most of them are not considered economically important enough to warrant more investigation. One species, the barn-door skate (*Raja laevis*), was an incidental catch of western North Atlantic fisheries in the second half of the twentieth century. As the name suggests, this is a large fish, too big to go unrecorded. Its numbers fell every year until, by the 1990s, none were being caught.

Logging and collecting
Similar cases of overharvested species are found in terrestrial ecosystems. For example, even when forests are not completely cleared, particularly valuable trees such as mahogany may be selectively logged from an area, eliminating both the tree species and all the animals that depend on it. Another example is the coast sandalwood (*Santalum ellipticum*), a tree endemic to the Hawaiian Islands that was almost completely eliminated from its habitats for its wood and fragrant oil.

Some species are overharvested not to be killed but to be kept alive and sold as pets or ornamental plants. Many species of parrots worldwide, for example, are in danger because of the pet trade, and the survival of cacti and orchid species is threatened by collectors. Aquatic species are also threatened in this way: seahorses, for example, are collected for aquariums, as well as for use traditional medicines; this adds to the decline in their populations caused by overfishing and habitat loss.

Secondary extinctions

Once one species goes extinct, it is likely that there will be other extinctions or even an avalanche of them. Some cases of these secondary extinctions are simple to understand – for example, for every bird or mammal that goes extinct, one or several species of parasite will also probably disappear. From well-studied species, it is known that bird and mammal species tend to have parasites on or inside them that can live on no other host.

Other extinctions cause changes that can be quite complicated. Species are bound together in ecological communities to form a "food web" of species interactions. Once a species is lost, those species that fed on it, were fed on by it, or were otherwise benefited or harmed by that species, will all be affected, for better or worse. These species, in turn, will affect yet other species. Ecological theory suggests that the patterns of secondary extinction are complex and thus may be difficult to demonstrate.

The most easily recognizable secondary extinctions should be seen in species that depend closely on each other. In the Hawaiian Islands, anecdotal evidence for secondary extinctions comes from a consideration of nectar-feeding birds. Before modern human activity on the islands, there were three nectar-feeding Hawaiian honeycreepers – the mamo (*Drepanis pacifica*), the black mamo (*Drepanis funerea*), and the iiwi (*Vestiaria coccinea*) – that had long downward-curving beaks, adapted to inserting into appropriately long and curved flowers. The first two birds are extinct, whereas the third is extinct on two islands, is very rare on a third, and has declined on others.

The extinctions of these honeycreepers may have followed the destruction of important nectar-producing plants by introduced goats and pigs. Many of the native lobelias on the Hawaiian Islands, such as those of the genera *Trematolobelia* and *Clermontia*, have clearly evolved to be pollinated by the three honeycreepers, and the plants were once important components of the forest's understory. About a quarter of these plant species are now extinct, at a rate that plainly exceeds extinction rates for the rest of the islands' flora. Their demise may perhaps be because they were so vulnerable to introduced mammalian herbivores. It is not certain, however, whether the plants disappeared first and then their bird pollinators, or vice versa.

Similarly, some surviving Hawaiian birds seem to be unusually specialized feeders and to be threatened as a consequence of the loss of their food sources. For example, another rare honeycreeper, the akiapolaau (*Hemignathus munroi*), is an insectivore that feeds on insects mainly on large koa (*Acacia koa*) trees. Today, however, few koa forests remain, because the trees have been overharvested for their attractive wood. Yet another Hawaiian honeycreeper, a seed-eating species called the palila (*Loxioides bailleui*), is endangered because it depends almost exclusively on the seeds of one tree, the mamane (*Sophora chrysophylla*), which is grazed by introduced goats and sheep.

Stories of secondary extinctions are nearly always unsatisfactory anecdotes because of the difficulty of teasing apart the various explanations for what happened in the past. There is abundant evidence from small-scale ecological experiments that a change in one species' numbers (including its complete elimination) will cause cascading effects in the abundance of other species. The stories are plausible enough that particular attention should be paid to the future consequences of contemporary extinctions.

Climate change

As has been discussed in earlier chapters, changes to the Earth's climate, in particular global warming, are now known to be taking place as a result of increases in greenhouse gases caused by human activity. Since the turn of the twenty-first century, the relationship between global warming and our planet's biodiversity has been clarified significantly. Although the precise effects of climate change on species extinction rates are still uncertain, they will almost certainly be large.

It has been estimated that one fifth to one third of all plant and animal species are likely to be at an increased risk of extinction if global average surface temperatures rise another 1.5–2.5°C (2.7–4.5°F) by the year 2100. This temperature range falls within the scope of the lower emissions scenarios projected by the Intergovernmental Panel on Climate Change (IPCC) (see Chapters 8 and 10). Species-loss estimates climb to as much as 40 per cent for a warming in excess of 4.5°C (8.1°F) – a level that could be reached in the IPCC's higher-emissions scenarios. A 40 per cent extinction rate would be likely to lead to major changes in the food webs within ecosystems and have a destructive impact on ecosystem function.

Surface warming in temperate regions is likely to lead to changes in various seasonal processes – for instance, earlier leaf production by trees, earlier greening of vegetation, altered timing of egg-laying and hatching, and shifts in the seasonal migration patterns of birds, fishes, and other migratory animals. In high-latitude ecosystems, changes in the seasonal patterns of sea ice threaten predators such as polar bears and walruses. (Both species rely on broken sea ice for their hunting activities.) Also in the high latitudes, a combination of warming waters, decreased sea ice, and changes in ocean salinity and circulation is expected to lead to reductions or redistributions in populations of algae and plankton. As a result, fish and other organisms that forage upon algae and plankton may be threatened. On land, rising temperatures and changes in precipitation patterns and drought frequencies will probably alter patterns of disturbance by fires and pests.

Other likely impacts on the environment include the destruction of many coastal wetlands, salt marshes, and mangrove swamps as a result of rising sea levels, and the consequent loss of certain rare and fragile habitats, which are often home to specialist species that are unable to thrive in other environments. For example, certain amphibians limited to isolated tropical cloud forests either have become extinct already or are under serious threat of extinction. Cloud forests – tropical forests that depend on persistent condensation of moisture in the air – are disappearing as optimal condensation levels move to higher elevations in response to warming temperatures in the lower atmosphere.

In many cases a combination of stresses caused by climate change as well as human activity represents a considerably greater threat than either climatic stresses or nonclimatic stresses alone. A particularly important example is coral reefs, which contain much of the ocean's biodiversity. Rising ocean temperatures increase the tendency for coral bleaching (a condition where zooxanthellae, or yellow-green algae, living in symbiosis with coral either lose their pigments or abandon the coral polyps altogether), and they also raise the likelihood of greater physical damage by progressively more destructive tropical cyclones. In many areas coral is also under stress from increased ocean acidification, marine pollution, run-off from agricultural fertilizer, and physical damage by boat anchors and dredging.

Another example of how climate and nonclimatic stresses combine is the threat to migratory animals. As these animals attempt to relocate to regions with more favourable climate conditions, they are likely to encounter impediments such as highways, walls, artificial waterways, and other man-made structures.

It is now clear that most species are shifting their geographic ranges toward cooler places and are starting important events such as breeding, migration, and flowering earlier in the year. (Such seasonal natural phenomena are called phenological events.) One study, published in the early twenty-first century, found that more than 80 per cent of nearly 1,500 species of animals and plants from a wide variety of habitats worldwide were changing in the ways expected from global warming. Another study, published at about the same time, found consistent northward movement of the northern boundaries of animal and plant species in Europe and North America. Such geographic shifts of species alter important ecological interactions with their prey, predators, competitors, and diseases. Some species may benefit, while others can be harmed – as, for example, when migratory insect-eating birds arrive too late to exploit the emergence of moth larvae that they typically feed to their young.

Even if scientists know that species are likely to move in the direction of cooler habitats – toward the poles or up mountain-sides – they will find the exact changes difficult to predict because so many different factors are involved. For example, one contemporary study examined the changes in the ranges and feeding ecology of butterflies in Britain in the last decades of the twentieth century. Two species were found to have increased in range, as expected, but by unexpected amounts. One species lived in isolated habitat patches and could not fly the distance to some of the patches that were suitable for populating. Rising temperatures increased the number and density of suitable patches, allowing the butterfly to reach the distant patches by making use of intervening newly available ones as "stepping-stones". The second butterfly species was able to exploit a previously unused food plant that grew in shady places that formerly had been too cool. With that change in diet, the butterfly was able to greatly expand its range. In both cases, the species benefited from the warming climate, at least at the northern edges of their ranges. It takes little imagination to foresee the consequences if, for example, the mosquito *Aedes aegypti*, which carries dengue hemorrhagic fever, were to expand its range to an unexpected degree across the southern United States, which is its present northern limit.

Can anything useful be inferred about the likely consequences of climate change in terms of species extinction? Assuming that the change has simple effects, scientists can predict where a species should be in the future if it is to live in the same range of climatic conditions that it does now. The real concern is that this range of suitable conditions, or the species' "climate envelope", may shrink to nothing as conditions change – that is, there may be no suitable conditions for it in the future.

Other things being equal, one would expect that species with small geographic ranges will be more likely to be affected than species with large ranges. For a species with a large range, global change may cause the species to disappear from the south and appear further north, but the change in the size of the range may be quite small. The species may be present in many of the same areas both before and after changes in global temperature.

The picture is less hopeful for species that have very small ranges. Many of these, for example, live in mountainous areas. To date, human activity has had relatively small effects on such species because of the impracticalities of cultivating land in mountains, specially on steep slopes. But it is these species –

living on cool mountain tops that are now becoming too warm for them – that have nowhere else to go. Such arguments lead scientists to believe that extinctions caused by rising temperatures will be additional to those caused by habitat loss. Rough calculations suggest that rising temperatures may threaten about a quarter of the species in terrestrial hotspots.

Conservation

Species extinction is the most obvious aspect of the loss of biodiversity and forms the bulk of the examples in most assessments of biodiversity; however, the subject of conservation is broader than this. A species that survives extinction can lose much of its genetic diversity as local, genetically distinct populations are lost from most of the species' original range. Furthermore, ecosystems may shrink dramatically in area and lose many of their functions, even if their constituent species manage to survive. Conservation is involved with studying all these kinds of losses, understanding the factors responsible for them, developing techniques to prevent losses, and, whenever possible, restoring biodiversity. Several conservation techniques are described below.

Preventing the loss of biodiversity

A thorough knowledge of the factors that cause extinction and the vulnerability of different species to them is an essential part of conserving species. Conservation is largely about removing or reducing those factors, and doing so for the most vulnerable species and in the places where species are most vulnerable. Much of the task of conservation professionals is to protect habitats large enough to house viable populations of species, first deciding where the priorities should be and sometimes restoring habitats that already have been destroyed. Local conservation groups often spend time removing introduced species, which can mean physically weeding invasive plants or trapping invasive animals. These activities must be accompanied by efforts to prevent the introduction of new threats. Others work to reduce harvesting directly or to reduce the incidental catch of nontarget species. Sometimes, when working with very rare species, scientists may not know the exact causes of their decline, which can lead to intense arguments about exactly how to proceed.

Habitat protection

Because the loss of habitat is the primary reason that species are lost, both locally and globally, protecting more habitat emerges as the most important priority for conservation. This simple idea raises difficult questions, however, over which habitats should be prioritized.

If reserves were judiciously placed over the identified hotspots of biodiversity, a large fraction of species might be saved. Presently, the allocation of reserves around the world is poor. Reserves larger than 100,000 square km (40,000 square miles) are generally in high mountains, tundras, and the driest deserts – areas that are not particularly species-rich. On the other hand, hotspots such as Madagascar and the Philippines protect less than 2 per cent of their land.

The same issues hold on smaller scales, as illustrated by a study reported in the late 1990s. The Agulhas Plain at the southern tip of Africa is one of the world's "hottest" spots for concentrations of vulnerable plant species. An area only some 1,500 square km (600 square miles) in size was found to house 1,751 plant species, 99 of them endemic. While most of the state forests and private nature reserves in the area are coastal, most of the hotspot's endemic plants live inland. Given that new reserves must be created if these plants are to survive, where should they be situated to encompass the maximum number of species at minimum cost?

Fortunately, the data available to make these decisions included a knowledge of the distribution of plant species over the Agulhas Plain in fairly good detail – the kind of information not likely to be available in most hotspots. This allowed the plain's plant-species composition to be divided into a grid of cells, each 3×3 km. Computer algorithms (systematic problem-solving methods) were then used to select sets of cells from the grid according to their complementary species composition – that is, the aim was to encompass as many species or as many endemics as possible in as small an area (as few grid cells) as possible.

Habitat connections

Habitats that are not completely destroyed may be fragmented to the degree that individual fragments are too small to hold viable populations of many species – they may suffer from inbreeding or other increased demographic risks. Yet in total the fragments may actually be of sufficient area to support these species. An obvious conservation intervention is to find ways of connecting fragments by "corridors". These corridors can be created from currently unprotected land between existing re-serves or by restoring land between existing habitat fragments. Corridors are particularly important in the protection of species whose habitats are threatened by global warming and who must therefore migrate to survive.

Many contemporary efforts to create corridors are small-scale; they can be as simple as hedgerows that connect woodlots, a strategy that is likely to work for some small species. Other efforts are far more ambitious. One of the earliest large efforts involved a plan to connect various parks and other protected areas in Florida by corridors of land that would have to be purchased or otherwise protected from development. From this initial effort, a conservation group, the Wildlands Project, has developed an extensive set of plans for many areas in North America. The intention is to set priorities for the acquisition of land in order to create a network of corridors that would eventually link together large parts of the continent.

A regional example of corridor creation is in Costa Rica. Situated in the Caribbean lowlands, La Selva Biological Station is one of the major centres for research on tropical forests. Occupying an area of about 16 square km (6 square miles), the station is bordered on the south by Braulio Carrillo National Park, a much larger area of forest covering 460 square km (180 square miles). The national park extends to La Selva through a forest corridor that descends in elevation from nearly 3,000 metres (nearly 10,000 feet) at Barva Volcano down to 35 metres (115 feet) above sea level at La Selva. The corridor had been threatened by agricultural development until conservation groups, realizing that La Selva would become an isolated forest "island", purchased the corridor to protect it.

Habitat management

Once protected, areas must often be managed in order to maintain the threatened species within them. Management may involve the removal of introduced species, as mentioned earlier. It can also involve restoring natural ecological processes to the area. Original fire and flooding regimes are examples of such processes, and they are often controversial because human actions can alter them significantly.

Fire control

Despite the often valid reasons for suppressing wildfires, the practice can change vegetation dramatically and sometimes harm species in the process. Human activities have changed fire regimes across large areas of the planet, including some biodiversity hotspots. Getting the fire regimes right can be essential for conserving species.

An example of a species for which the control of fire regimes has proved both possible and essential is Kirtland's warbler (*Dendroica kirtlandii*). This endangered species nests only in the Upper Peninsula of Michigan – an exceptional case of a bird species with a tiny geographic range well outside the tropics. The bird places its nest in grasses and shrubs below living branches of jack pines (*Pinus banksiana*) that are between 5 and 20 years old. The region's natural wildfires originally maintained a sufficient area of young jack pines. But modern practices suppressed fires, and the habitat declined. Active management of fires to ensure that there are always jack pines of the right age has steadily increased the population of warblers since the early 1990s, when it had included only about 200 singing males. By the early years of the twenty-first century, there were about 1,300 singing males.

Flood control

Humans also control water levels, and the resulting changes can have important consequences for endangered species. An example of a species so affected is the Cape Sable seaside sparrow (*Ammodramus maritimus mirabilis*) found in the Florida Everglades. The Everglades once stretched from Lake Okeechobee in the north to Florida Bay in the south. Water

flowed slowly over a wide area, and its levels varied seasonally: summer rains caused the levels to rise in late summer and
early autumn; then the dry season dropped the levels to their
lowest in late May. Under this natural regime, some prairies
were continuously flooded for many years, drying out in only
the driest years, whereas others were flooded for only a few
months each year. It is in these drier prairies that the sparrows
nest from about the middle of March until the water floods
their nests in the summer.

Water-management actions have diverted the flow of water
to the west of its natural path, making the western part of the
Everglades unnaturally wet in some years during the bird's
nesting season. Unnatural flooding during the four years that
followed 1992 reduced the bird population in the west to less
than 10 per cent of its 1992 level. In the east the prairies have
become unnaturally dry and so have become subject to an
increased number of fires, which has jeopardized the sparrow
population there. It is clear that rescuing the Cape Sable
seaside sparrow requires the restoration of the natural water
levels and flows within the Everglades and the consequent
return to natural fire regimes.

Habitat restoration

Once a habitat has been destroyed, the only remaining
conservation tool is to restore it. The problems involved
may be formidable, and they must include actions for dealing
with what caused the destruction. Restorations are massive
ecological experiments; as such, they are likely to meet with
different degrees of success in different places. Restoration of
the Everglades, for example, requires restoring the natural
patterns of water flow to thousands of square kilometres of
southern Florida.

A case history of habitat restoration comes from the Mid-western United States. In Illinois, natural ecosystems cover less than 0.1 per cent of the state, so restoration is almost the only conservation tool available. North Branch is a 20-km (12-mile) strip of land running northward from Chicago along the north branch of the Chicago River. Early in the twentieth century, it was protected from building but later abandoned. Beginning in the 1970s, a group of volunteers first cleared out introduced European buckthorn (*Rhamnus cathartica*), removed abandoned cars, and planted seeds that they had gathered from combing the tiny surviving remnants of original prairies in places such as along railways and in cemeteries. The initial effort to replant the prairie species failed, however: animals ate the small growing plants. The solution was to restore the natural fire regime, although controlled burns – setting fires safely near homes – posed a difficult technical challenge. But once it was accomplished, the results were immediate and dramatic. The original prairie plants flourished and the weeds retreated, although with an important exception. Under the native trees grew non-native thistles and dandelions.

The original habitats, locally called barrens, constituted a visually striking and ecologically special habitat. Restoring them was a particular challenge, and the main conservation problem was finding the right mix of species. One recommendation was to use remnant barrens as models, but the North Branch volunteers rejected them as being too degraded. In the early twentieth century, naturalists had speculated that barrens were special because they lacked some characteristic prairie species and, at the same time, had their own distinctive species. The volunteers examined habitat descriptions in old treatises of local plants, looking for such species, many of which certainly would now be scarce. The key discovery was a list of barrens' plants published by a country doctor in 1846. The scientific names had changed in 150 years, but, by tracking them through the local literature, the volunteers found many of their putative barrens-specific plants on the doctor's list. Using this information, the group succeeded in restoring barrens sites that by 1991 held 136 native plants, including whole patches consisting of species that, until restoration, had been locally rare.

This example illustrates two rules of ecological restoration. The first is that one must be able to save all the component species of the original ecological community. The second is that one needs to know which species belong and which do not. Without the right species mix, restorers must constantly weed and reseed.

Putting a price on conservation

Estimating the economic value of biodiversity is controversial. The value of whales, for example, once would have been just the price that traders would have paid for their parts, mostly oil and baleen. Today whales have an obvious economic value in terms of what tourists spend to watch them from boats – vessels that often now sail from the same ports that once supported whaling. Less obvious as an economic factor are the deeply held beliefs that hunting whales is cruel or that whales have a right to exist. Many people throughout the world object to whaling even though they may rarely or never have seen whales. However, economists incorporate these ideas into their calculations of value by asking how much the public would pay to protect whales.

A relevant real-world example is found in the aftermath of the oil spill from the tanker *Exxon Valdez*, which ran aground in Prince William Sound, Alaska, in 1989. When the spill polluted the spectacular scenery of Alaska's coast, many Americans felt that they had suffered a personal loss. Exxon was forced to pay for cleaning up the mess and was fined not only for harming people's livelihoods but also for incurring that loss.

The propriety of decisions that put monetary values on moral issues is contentious. However, in 1997 a team of ecologists and economists led by Robert Costanza attempted to do this on a grand scale when they reported their calculations of how much the diverse services provided by Earth's ecosystems are worth. Their approach was to group these services into 17 categories and to classify the planet's surface into 16 ecosystem types. Many combinations of services and ecosystem types were not estimated. Mountains, the Arctic, and deserts certainly have cultural values and devoted eco-tourists, but these were absent, as were urban areas with city parks. Nonetheless, the total value of $33 trillion per year at which the team arrived is roughly twice the value of the global economy.

A challenge for today's society is to comprehend the economic values of Earth's ecosystems, for often they are taken for granted. One effort to do this is found in the steps taken by New York City to improve its water supply by protecting the watersheds that supply that water. The cost of this protection was small compared with the alternative of greatly expanding the city's system of water-treatment plants. Another example, aimed at slowing the rate of deforestation, can be seen in the Costa Rican government's willingness to pay for protecting forest watersheds. A global example is the effort to reduce carbon emissions worldwide by rewarding those who protect existing forests or who restore forests on deforested land.

Governments and international agencies often encourage the destruction of biodiversity by providing financial subsidies in the interests of economic growth. For example, many countries massively support their national fishing fleets by subsidizing the price of fish. The former Soviet Union was such an example, but upon its break-up the Russian government withdrew the subsidies, with the result that much of the fishing fleet has rusted in harbour. A study conducted in the late twentieth century by the Food and Agriculture Organization of the United Nations estimated that the total annual value of the world fish catch was about $70 billion. Yet the cost of catching these fish was $92 billion, reflecting a loss of $22 billion. The cost included routine maintenance, fuel, insurance, supplies, and labour. The magnitude of the loss was certainly an underestimate, for any private company would also include in its total cost the annual costs of paying off the loans for the fishing boats and other facilities. Those loans, the study concluded, would account for another $32 billion, making the annual deficit $54 billion per year.

Agricultural subsidies worldwide are even larger – a global review of all subsidies estimates that they amount to trillions of dollars per year. The review suggests that a powerful strategy for conservation would be to examine whether these subsidies are sensible, given concerns about the environment and the need to be fiscally responsible.

PART 3

IDEAS, ARGUMENTS, AND PROGRESS

8

ENVIRONMENTALISM: PAST, PRESENT, AND FUTURE

The changes that are occurring to Earth's climate, and in its land cover, atmosphere, oceans and rivers as a result of human activity, have now reached the top of the political agenda. While predictions will always be speculative, scientists are now fairly certain of the kinds of broad changes that will occur in the natural world as a result of expected temperature rises, and of the implications of this for agriculture, population displacement, and human health. The question of what should be done, however, is more difficult to agree on – since it requires agreement on the relative responsibility, and accountability, of different governments, and indeed of different parties within countries. Consensus must also be reached on the kinds of species and natural systems to preserve – which raises fundamental philosophical questions.

Environmental ethics and ideas

Ethics

The problem of the nature of intrinsic value is one of many difficult ethical questions provoked by environmental issues. Many philosophers in the past have agreed that human experiences have intrinsic value, and utilitarians (who hold that an action is right if it tends to promote happiness and wrong if it tends to produce the reverse) have always accepted that the pleasures and pains of animals, at least, are of some intrinsic significance. But this does not explain why it is so bad if, for example, orang-utans become extinct or a rainforest is cut down. Are these things to be regretted only because of the experiences that would be lost to humans or other sentient beings? From the late twentieth century, some philosophers defended the view that trees, rivers, species (considered apart from the individual animals of which they consist), and perhaps even ecological systems as a whole have a value independent of any instrumental value they may have. There is, however, no agreement on what the basis for this value should be.

Concern for the environment also raises the question of obligations to future generations. How much do human beings living now owe to those not yet born? For those who hold a social-contract ethic (whereby rights and duties are defined by a hypothetical agreement) or for the ethical egoist (for whom moral decisions are determined only by self interest), the answer would seem to be, nothing. Although humans existing in the present can benefit those existing in the future, the latter are unable to reciprocate. Most other ethical theories, however, do give some weight to the interests of future generations. From a utilitarian point of view, the fact that members of future generations do not yet exist is no reason for giving less consideration to their interests than to the interests of present generations – provided that one can be certain that future generations will exist and will have interests that will be affected by what one does. In the case of, say, the storage of radioactive wastes or the emission of gases that contribute to climate change, it seems clear that what present generations do will indeed affect the interests of generations to come.

A related issue, and one that applies in particular to climate change, is the question of shared use of a common resource. This is the problem of the "tragedy of the commons", the subject of an essay by Garrett Hardin, written in 1968. It describes what happens when a number of individuals have unlimited access to a resource – for example, a common on which to graze their animals. In practice, everyone takes as much as they can, and the result is overexploitation and degradation of the resource so that it is no use to anyone. This predicament has obvious relevance to today's world in respect of resources such as clean air and water and the preservation of biodiversity – but it also relates to the emission of greenhouse gases. Climate change is often conceived of as an issue of global equity. How much of a scarce resource (the capacity of the atmosphere safely to absorb waste gases produced by human activity) may each country use? Are industrialized countries justified in using far more of this resource, on a per capita basis, than developing countries, considering that the human costs of climate change will fall more heavily on developing countries because they cannot afford the measures needed to mitigate them?

Most philosophers (and many ordinary people) agree that these are important moral issues. These questions become even more complex when one considers that the size of future generations can be affected by government population policies and by other less-formal attitudes toward population growth and family size. The notion of overpopulation is in part a philosophical issue. What is optimum population? Is it the population size at which the average level of welfare will be as high as possible? Or is it the size at which the total amount of welfare – the average multiplied by the number of people – is as great as possible? These views carry the implication that we are obliged to bring more people into the world as long as they will be happy (the "total" view), or happier than average (the "average" view). Philosophers have given much thought to finding alternatives, but those suggested had their own difficulties, and the question has remained one of the most challenging conundrums in applied ethics. Nevertheless, the consideration of these and other dilemmas has influenced each stage in the development of environmentalism.

Key ideas

Apocalyptic environmentalism

The vision of the environmental movement of the 1960s and early 1970s was generally pessimistic, reflecting a pervasive sense of "civilization malaise" and a conviction that Earth's long-term prospects were bleak. Works such as Rachel Carson's *Silent Spring* (1962), Garrett Hardin's *The Tragedy of the Commons* (1968), Paul Ehrlich's *The Population Bomb* (1968), Donella H. Meadows' *The Limits to Growth* (1972), and Edward Goldsmith's *Blueprint for Survival* (1972) suggested that the planetary ecosystem was reaching the limits of what it could sustain. This so-called apocalyptic, or survivalist, literature encouraged reluctant calls from some environmentalists for increasing the powers of centralized governments over human activities that are deemed environmentally harmful – a viewpoint expressed most vividly in Robert Heilbroner's *An Inquiry into the Human Prospect* (1974), which argued that human survival ultimately required the sacrifice of human freedom. Counter-arguments, such as those presented in Julian Simon's and Herman Kahn's *The Resourceful Earth* (1984), emphasized humanity's ability to find or to invent substitutes for resources that were scarce and in danger of being exhausted. (See Chapter 10 for more about key environmental thinkers.)

The Gaia hypothesis

In 1979 James Lovelock put forward his "Gaia hypothesis", a concept that has had a profound influence on environmental thought. In essence, the theory holds that the Earth is a living, self-regulating entity capable of re-establishing an ecological equilibrium, even without the existence of human life. The physical conditions of the Earth's surface, oceans, and atmosphere, it suggests, have been made fit and comfortable for life and have been maintained in this state by the biota themselves. Evidence includes the relatively constant temperature of the Earth's surface, which has been maintained for the past 3.5 billion years despite a 25 per cent increase in energy coming from the sun during that period. The remarkable constancy of the Earth's oceanic and atmospheric chemistry for the past 500 million years also is invoked to support the theory. Integral to the hypothesis is the crucial involvement of the biota in the cycling of various elements vital to life. The role that living things play in both the carbon and sulphur cycles is a good example of the importance of biological activity and the complex interrelationship of organic and inorganic elements in the biosphere.

In his initial proposition, Lovelock suggested that the biospheric transformations of the atmosphere support the biosphere in an adaptive way through a sort of "genetic group selection". This idea generated extensive criticism and spawned a steady stream of new research that has enriched the debate and advanced both ecology and environmental science.

The Greek word Gaia, or Gaea, meaning "Mother Earth", is Lovelock's name for Earth, which is envisioned as a "superorganism" engaged in planetary biogeophysiology. The goal of this superorganism is to produce a homeostatic, or balanced, Earth system through "a feedback of cybernetic systems". The scientific process of research and debate will eventually resolve the issue of the reality of the "Gaian homeostatic superorganism", and Lovelock has since revised his hypothesis to exclude goal-driven genetic group selection. Nevertheless, it is now an operative norm in contemporary science that the biosphere and the atmosphere interact in such a way that an understanding of one requires an understanding of the other. Furthermore, the reality of two-way interactions between climate and life is well recognized.

Timeline of major events

1962 Publication of Rachel Carson's *Silent Spring*

1972 United Nations Environment Programme founded

1979 Publication of James Lovelock's Gaia hypothesis

1988 UN Intergovernmental Panel on Climate Change (IPCC) established

1990 IPCC publishes first report

1992 United Nations Framework Convention on Climate Change adopted in Rio de Janeiro, Brazil

1995 IPCC publishes second report

1997 Kyoto Protocol adopted

2001 IPCC publishes third report

2007 IPCC awarded Nobel Peace Prize (with Al Gore) IPCC publishes fourth report UN Climate Change Conference in Bali, Indonesia, debates post-Kyoto agreement All major nations except US ratify Kyoto Protocol

History of environmentalism

Concern for the impact on human life of problems such as air and water pollution dates back to at least Roman times. Pollution was associated with the spread of epidemic disease in Europe between the late fourteenth and the mid-sixteenth century, and soil conservation was practised in China, India, and Peru as early as 2,000 years ago. In general, however, such concerns did not give rise to public activism.

The contemporary environmental movement arose primarily from concerns in the late nineteenth century about the protection of the countryside in Europe and the wilderness in the United States, and the health consequences of pollution during the Industrial Revolution. In opposition to the dominant political philosophy of the time, liberalism – which held that all social problems, including environmental ones, could and should be solved through the free market – most early environmentalists believed that government rather than the market should be charged with protecting the environment and ensuring the conservation of resources. An early philosophy of resource conservation was developed by Gifford Pinchot (1865–1946), the first chief of the US Forest Service, for whom conservation represented the wise and efficient use of resources.

Also in the United States, at about the same time, a more strongly biocentric approach arose in the preservationist philosophy of the naturalist John Muir (1838–1914) and of Aldo Leopold (1887–1948), a professor of wildlife management who was pivotal in the designation of Gila National Forest in New Mexico in 1924 as America's first national wilderness area. Leopold introduced the concept of a "land ethic", arguing that humans should transform themselves from conquerors of nature into citizens of it. His essays, compiled posthumously in *A Sand County Almanac* (1949), had a significant influence on later biocentric environmentalists, and fed into the concept of "deep ecology" (or "ecosophy") conceived by the Norwegian environmental philosopher Arne Naess during the 1970s, which asserted that every living creature in nature is equally important to Earth's precisely balanced system.

Environmental organizations established from the late nineteenth to the mid-twentieth century were primarily middle-class lobbying groups concerned with nature conservation, wildlife protection, and the pollution that arose from industrial development and urbanization. There were also scientific organizations concerned with natural history and with biological aspects of conservation efforts.

Beginning in the 1960s, the various philosophical strands of environmentalism were given political expression through the establishment of "green" political movements in the form of activist nongovernmental organizations and environmentalist political parties. Despite the diversity of the environmental movement, four pillars provided a unifying theme to the broad goals of political ecology: namely, protection of the environment, grassroots democracy, social justice, and nonviolence. However, for a small number of environmental groups and individual activists who engaged in ecoterrorism, violence was viewed as a justified response to what they considered the violent treatment of nature by some interests, particularly the logging and mining industries. The political goals of the contemporary green movement in the industrialized West focused on changing government policy and promoting environmental social values. In the less-industrialized or developing world, environmentalism has been more closely involved in "emancipatory" politics and grassroots activism on issues such as poverty, democratization, and political and human rights, including the rights of women and indigenous peoples. Examples of such movements include the Chipko movement in India, which linked forest protection with the rights of women, and the Assembly of the Poor in Thailand, a coalition of movements fighting for the right to participate in environmental and development policies.

The early strategies of the contemporary environmental movement were self-consciously activist and unconventional, involving direct-protest actions designed to obstruct and to draw attention to environmentally harmful policies and projects. Other strategies included public-education and media campaigns, community-directed activities, and conventional lobbying of policy makers and political representatives. The movement also attempted to set public examples in order to increase awareness of and sensitivity to environmental issues. Such projects included recycling, green consumerism ("buying green"), and the establishment of alternative communities, including self-sufficient farms, workers' cooperatives, and cooperative-housing projects.

The electoral strategies of the environmental movement included the nomination of environmental candidates and the registration of green political parties. These parties were conceived of as a new kind of political organization that would bring the influence of the grassroots environmental movement to bear directly on the machinery of government, make the environment a central concern of public policy, and render the institutions of the state more democratic, transparent, and accountable. The world's first green parties – the Values Party, a nationally based party in New Zealand, and the United Tasmania Group, organized in the Australian state of Tasmania – were founded in the early 1970s. The first explicitly green member of a national legislature was elected in Switzerland in 1979; later, in 1981, four greens won legislative seats in

Belgium. Green parties also have been formed in the former Soviet bloc, where they were instrumental in the collapse of some communist regimes, and in some developing countries in Asia, South America, and Africa, though they have achieved little electoral success there.

The most successful environmental party has been the German Green Party (die Grünen), founded in 1980. Although it failed to win representation in federal elections that year, it entered the Bundestag (parliament) in both 1983 and 1987, winning 5.6 per cent and 8.4 per cent of the national vote, respectively. The party did not win representation in 1990, but in 1998 it formed a governing coalition with the Social Democratic Party, and the party's leader, Joschka Fischer, was appointed as the country's foreign minister. Throughout the last two decades of the twentieth century, green parties won national representation in a number of countries and even claimed the office of mayor in European capital cities, including Dublin and Rome, in the mid-1990s.

By this time green parties had become broad political vehicles, though they continued to focus on the environment. In developing party policy, they attempted to apply the values of environmental philosophy to all issues facing their countries, including foreign policy, defence, and social and economic policies.

Despite the success of some environmental parties, environmentalists remain divided over the ultimate value of electoral politics. For some, participation in elections is essential because it increases the public's awareness of environmental issues and encourages traditional political parties to address them. Others, however, have argued that the compromises necessary for electoral success invariably undermine the ethos of grassroots democracy and direct action. This tension has perhaps been most pronounced in the German Green Party. The party's

Realos (realists) accepted the need for coalitions and compromise with other political parties, including traditional parties with views sometimes contrary to that of the Green Party. By contrast, the Fundis (fundamentalists) maintained that direct action should remain the major form of political action and that no pacts or alliances should be formed with other parties. Likewise, in Britain, where the Green Party has achieved success in some local elections but failed to win representation at the national level (though it did win 15 per cent of the vote in the 1989 European Parliament elections), this tension was evidenced in disputes between so-called "electoralists" and "radicals".

The implementation of internal party democracy has also caused fissures within environmental parties. In particular, earlier strategies, such as continuous policy involvement by party members, grassroots control over all party institutions and decisions, and the legislative rotation of elected members to prevent the creation of career politicians, were sometimes perceived as unhelpful and disruptive when green parties won representation to local, national, or regional assemblies.

By the late 1980s environmentalism had become a global as well as a national political force. Some environmental non-governmental organizations (e.g. Greenpeace, Friends of Earth, and the World Wide Fund for Nature [WWF]) established a significant international presence, with offices throughout the world and centralized international headquarters to coordinate lobbying campaigns and to serve as campaign centres and information clearing houses for their national affiliate organizations. Transnational coalition building was, and remains, another important strategy for environmental organizations and for grassroots movements in developing countries, primarily because it facilitates the exchange of information and expertise, but also because it strengthens lobbying and direct-action campaigns at the international level.

Through its international activism, the environmental movement has influenced the agenda of international politics. Although a small number of bilateral and multilateral international environmental agreements were in force before the 1960s, since the 1972 United Nations Conference on the Human Environment in Stockholm the variety of multilateral environmental agreements has increased to cover most aspects of environmental protection as well as many practices with environmental consequences, such as the trade in endangered species, the management of hazardous waste (especially nuclear waste), and armed conflict. The changing nature of public debate on the environment was reflected also in the organization of the 1992 United Nations Conference on Environment and Development (Earth Summit) in Rio de Janeiro, Brazil, which was attended by some 180 countries and various business groups, nongovernmental organizations, and the media.

In the twenty-first century, the environmental movement has combined the traditional concerns of conservation, preservation, and pollution remediation with more contemporary concerns with the environmental consequences of economic practices as diverse as tourism, trade, financial investment, and the conduct of war. Environmentalists are likely to intensify the trends of the late twentieth century, during which some environmental groups increasingly worked in coalition not just with other emancipatory organizations, such as human rights and indigenous-peoples groups, but also with corporations and other businesses.

Key organizations and events

United Nations Environment Programme

The United Nations Environment Programme (UNEP) is the environmental voice of the United Nations, and is concerned with environmental issues at the global and regional level. As one of the three implementing agencies of the Global Environment Facility (GEF), it helps developing countries and those with transitional economies to implement environmentally sound policies, assists them in meeting the cost of doing so, and encourages sustainable development. UNEP was established after the 1972 UN Conference on the Human Environment, in Stockholm, Sweden, and is based in Nairobi, Kenya. Its mandate is to coordinate the development of environmental policy consensus by keeping the global environment under review and bringing emerging issues to the attention of governments and the international community.

The activities of UNEP cover a wide range of issues relating to the atmosphere and to marine and terrestrial ecosystems. The organization has played a major role in developing international environmental conventions, promoting environmental science and information, and working with governments and regional institutions on the development and implementation of policy. It works with environmental nongovernmental organizations and helps to develop international environmental law.

Intergovernmental Panel on Climate Change

Established by a United Nations panel in 1988, the Intergovernmental Panel on Climate Change (IPCC) is headquartered with the World Meteorological Organization in Geneva, Switzerland. The IPCC assesses peer-reviewed literature and

industry practices to determine the impact of, and possible
responses to, climate change. While it produces no research of
its own, its members – divided into three working groups and a
task force – assemble reports from hundreds of scientists and
policymakers from around the globe. These are analysed and
distributed as special papers or as more comprehensive assessment reports.

The first IPCC report, published in 1990, stated that a
considerable quantity of data showed that human activity
affected the variability of the climate system; nevertheless,
the authors of the report could not reach a consensus on
the causes and effects of global warming and climate change at
that time. The 1995 IPCC report stated that the balance of
evidence suggested "a discernible human influence on the
climate". The 2001 IPCC report confirmed earlier findings
and presented stronger evidence that most of the warming over
the previous 50 years was attributable to human activities. The
2001 report also noted that observed changes in regional
climates were beginning to affect many physical and biological
systems and that there were indications that social and economic systems were also being affected.

The IPCC's Fourth Assessment, issued in 2007, reaffirmed
the main conclusions of earlier reports, but the authors also
stated – in what was regarded as a conservative judgement –
that they were at least 90 per cent certain that most of the
warming observed over the previous half-century had been
caused by the release of greenhouse gases through a multitude
of human activities. Both the 2001 and 2007 reports stated
that during the twentieth century there had been an increase in
the global average surface temperature of 0.6°C (1.1°F), within a margin of error of 0.2°C (0.4°F). Whereas the 2001 report
forecasted an additional rise in average temperature by 1.4–
5.8°C (2.5–10.4°F) by 2100, the 2007 report refined this

forecast to an increase of 1.8–4.0°C (3.2–7.2°F) by the end of the twenty-first century. These forecasts were based on examinations of a range of scenarios that characterized future trends in greenhouse gas emissions.

Each IPCC report has helped to build a scientific consensus that elevated concentrations of greenhouse gases in the atmosphere are the major drivers of rising near-surface air temperatures and their associated ongoing climatic changes. In this respect, the current episode of climatic change, which began about the middle of the twentieth century, is seen to be fundamentally different from earlier periods, in that critical adjustments have been caused by activities resulting from human behaviour rather than non-anthropogenic factors. The IPCC's 2007 assessment projected that future climatic changes could be expected to include continued warming, modifications to precipitation patterns and amounts, elevated sea levels, and "changes in the frequency and intensity of some extreme events". Such changes would have significant effects on many societies and on ecological systems around the world.

In October the IPCC and former US vice president Al Gore were jointly awarded the Nobel Prize for Peace for their "efforts to build up and disseminate greater knowledge about man-made climate change, and to lay the foundations for the measures that are needed to counteract such change".

Earth Summit, Rio de Janeiro, 1992

On a global scale, climate-change policy is guided by two major treaties: the United Nations Framework Convention on Climate Change (UNFCCC) of 1992 and the associated 1997 Kyoto Protocol to the UNFCCC (named after the city in Japan where it was concluded).

The UNFCCC was negotiated between 1991 and 1992. It was adopted at the United Nations Conference on Environment and Development in Rio de Janeiro, Brazil, in June 1992 and became legally binding in March 1994. In Article 2 the UNFCCC sets the long-term objective of "stabilization of greenhouse gas concentrations in the atmosphere at a level that would prevent dangerous anthropogenic interference with the climate system". Article 3 establishes that the world's countries have "common but differentiated responsibilities," meaning that all countries share an obligation to act – although industrialized countries have a particular responsibility to take the lead in reducing emissions because of their relative contribution to the problem in the past. To this end, the UNFCCC Annex I lists 35 specific industrialized countries and countries with economies in transition plus the European Community (EC), and Article 4 states that these countries should work to reduce their anthropogenic emissions to 1990 levels. However, no deadline is set for this target; moreover, the UNFCCC does not assign any specific reduction commitments to non-Annex I countries (that is, developing countries).

Kyoto Protocol, 1997

The follow-up agreement to the UNFCCC, the Kyoto Protocol, was negotiated between 1995 and 1997 and was adopted in December 1997. The Kyoto Protocol regulates six greenhouse gases released through human activities: carbon dioxide, methane, nitrous oxide, perfluorocarbons, hydrofluorocarbons, and sulphur hexafluoride.

Under the Kyoto Protocol, Annex I countries are required to reduce their aggregate emissions of greenhouse gases to 5.2 per cent below their 1990 levels by no later than 2012. Toward this goal, the protocol sets individual reduction targets for each

Annex I country. These targets require the reduction of greenhouse gases in most countries, but they also allow increased emissions from others. For example, the protocol requires the then 15 member states of the EU and 11 other European countries to reduce their emissions to 8 per cent below their 1990 emission levels, whereas Iceland, a country that produces relatively small amounts of greenhouse gases, may increase its emissions as much as 10 per cent above its 1990 level. The protocol requires three countries – New Zealand, Ukraine, and Russia – to freeze their emissions at 1990 levels.

The Kyoto Protocol outlines five requisites by which Annex I parties can choose to meet their 2012 emission targets. First, it requires the development of national policies and measures that lower domestic greenhouse gas emissions. Second, countries may calculate the benefits from domestic carbon sinks that soak up more carbon than they emit. Third, countries can participate in schemes that trade emissions with other Annex I countries. Fourth, signatory countries may create joint implementation programmes with other Annex I parties and receive credit for such projects that lower emissions. Fifth, countries may receive credit for lowering the emissions in non-Annex I countries through a "clean development" mechanism, such as investing in the building of a new wind-power project.

In order to go into effect, the Kyoto Protocol had to be ratified by at least 55 countries, including enough Annex I countries to account for at least 55 per cent of that group's total greenhouse gas emissions. More than 55 countries quickly ratified the protocol, including all the Annex I countries except for Russia, the United States, and Australia. It was not until Russia, under heavy pressure from the EU, ratified the protocol that it became legally binding in February 2005.

The most-developed regional climate-change policy to date has been formulated by the EU, in part to meet its commitments under the Kyoto Protocol. By 2005 the 15 EU countries that have a collective commitment under the protocol reduced their greenhouse gas emissions to 2 per cent below their 1990 levels, though it is not certain that they will meet their 8 per cent reduction target by 2012. In 2007 the EU set a collective goal for all 27 member states to reduce their greenhouse gas emissions by 20 per cent below 1990 levels by the year 2020. As part of its effort to achieve this goal, the EU in 2005 established the world's first multilateral trading scheme for carbon dioxide emissions, covering more than 11,500 large installations across its member states.

In the United States, by contrast, President George W. Bush and a majority of senators rejected the Kyoto Protocol, citing the lack of compulsory emission reductions for developing countries as a particular grievance. At the same time, US federal policy does not set any mandatory restrictions on greenhouse gas emissions, and US emissions increased over 16 per cent between 1990 and 2005. Partly to make up for a lack of direction at the federal level, many individual US states have formulated their own action plans to address global warming and climate change and have taken a host of legal and political initiatives to curb emissions. These initiatives include capping emissions from power plants, establishing renewable portfolio standards requiring electricity providers to obtain a minimum percentage of their power from renewable sources, developing vehicle emissions and fuel standards, and adopting "green building" standards.

Climate Change Conference, Bali, 2007

Countries differ in opinion on how to proceed with international policy after the commitment period of the Kyoto Protocol ends in 2012. The EU supports a continuation of a legally based collective approach in the form of another treaty, but other countries, including the United States, support more voluntary measures, as in the Asia-Pacific Partnership on Clean Development and Climate that was announced in 2005. Long-term goals formulated in Europe and the United States seek to reduce greenhouse gas emissions by up to 80 per cent by the middle of the twenty-first century. Related to these efforts, the EU set a goal of limiting temperature rises to a maximum of 2°C (3.6°F) above pre-industrial levels. (Many climate scientists and other experts agree that significant economic and ecological damage will result should the global average of near-surface air temperatures rise more than 2°C [3.6°F] above pre-industrial temperatures in the next century.) Despite differences in approach, countries have begun negotiations on a new treaty, based on an agreement made at the United Nations Climate Change Conference in 2007 in Bali, Indonesia, that will replace the Kyoto Protocol after it expires.

Future climate-change policy and practice

As public policies relative to global warming and climate change continue to develop globally, regionally, nationally, and locally, they fall into two major types. The first type, mitigation policy, focuses on different ways to reduce emissions of greenhouse gases. As most emissions come from the burning of fossil fuels for energy and transportation, much of the mitigation policy focuses on switching to less-carbon-intensive energy sources (such as wind, solar, and hydropower), improving energy efficiency for vehicles, and supporting the development of new technology. In contrast, the second type, adaptation policy, seeks to improve the ability of various societies to face the challenges of a changing climate. For example, some adaptation policies are devised to encourage groups to change agricultural practices in response to seasonal changes, while other policies are designed to prepare cities located in coastal areas for elevated sea levels.

In either case, long-term reductions in greenhouse gas discharges will require the participation of both industrial countries and major developing countries. In particular, the release of greenhouse gases from Chinese and Indian sources is rising quickly, in line with the rapid industrialization of those countries. In 2006 China overtook the United States as the world's leading emitter of greenhouse gases in absolute terms (though not in per capita terms), largely because of China's increased use of coal and other fossil fuels. Indeed, all the world's countries are faced with the challenge of finding ways to reduce their greenhouse gas emissions while promoting environmentally and socially desirable economic development (known as "sustainable development" or "smart growth"). Whereas some opponents of those calling for corrective action continue to argue that the costs of short-term mitigation measures will be too high, a growing number of economists and policymakers argue that it will be less costly, and possibly more profitable, for societies to take early preventive action than to address severe climatic changes in the future. Many of the most harmful effects of a warming climate are likely to take place in developing countries. Combating such effects in developing countries will be especially difficult, as many of these countries are already struggling and possess a limited capacity to meet challenges from a changing climate.

It is expected that each country will be affected differently by the expanding effort to reduce global greenhouse gas emissions. Countries that are relatively large emitters will face greater reduction demands than will smaller emitters. Similarly, countries experiencing rapid economic growth are expected to face growing demands to control their greenhouse-gas emissions as they consume increasing amounts of energy. Differences will also occur across industrial sectors and even between individual companies. For example, producers of oil, coal, and natural gas – which in some cases represent significant portions of national export revenues – may see reduced demand or falling prices for their goods as their clients decrease their use of fossil fuels. In contrast, many producers of new, more climate-friendly technologies and products (such as generators of renewable energy) are likely to see increases in demand.

In addition, a growing number of the world's cities are initiating a multitude of local and subregional efforts to reduce their emissions of greenhouse gases. Many of these municipalities are taking action as members of the International Council for Local Environmental Initiatives and its Cities for Climate Protection programme, which outlines principles and steps for taking local-level action. In 2005 the US Conference of Mayors adopted the Climate Protection Agreement, in which cities committed to reduce emissions to 7 per cent below 1990 levels by 2012. Many private firms are developing corporate policies to reduce greenhouse gas emissions. One notable example of an effort led by the private sector is the creation of the Chicago Climate Exchange as a means for reducing emissions through a trading process.

To address global warming and climate change, societies must find ways to fundamentally change their patterns of energy use in favour of less-carbon-intensive energy generation, transportation, and forest and land-use management. A growing number of countries have taken on this challenge, and there are many things individuals can do too. For instance, consumers now have more options to purchase electricity generated from renewable sources. Additional measures that would reduce personal emissions of greenhouse gases and also conserve energy include the operation of more energy-efficient vehicles, the use of public transportation when available, and the transition to more energy-efficient household products. Individuals might also improve their household insulation, learn to heat and cool their residences more effectively, and purchase and recycle more environmentally sustainable products.

9

KEY ENVIRONMENTAL THINKERS

Writings on humanity's modern relationship with the Earth and its natural resources started to emerge around 200 years ago, with the change in that relationship that began with the Industrial Revolution. As long ago as the eighteenth century it was observed that resources were finite and that our exploitation of them would necessarily be limited. Another key theme of early thinkers, galvanized by the rapid progress of industrialization and the associated destruction of wilderness, was humanity's spiritual relationship with the land. From the early twentieth century, scientific advances made possible an increasingly in-depth study of environmental and climatic phenomena, and this, stimulated by the accelerating global effects of human economic activity, has generated a canon of scientific environmental work that is growing every year. The following are just a few of the major thinkers and scientists who have studied and influenced environmental thought in our time.

Thomas Malthus (1766–1834)

The British economist and demographer Thomas Malthus is best known for his theory that population growth will always tend to outstrip food supply, and that betterment of humankind is impossible without stern limits on reproduction. This thinking is commonly referred to as Malthusianism.

In 1798 Malthus published anonymously the first edition of *An Essay on the Principle of Population as It Affects the Future Improvement of Society, with Remarks on the Speculations of Mr. Godwin, M. Condorcet, and Other Writers*. The work received wide notice. Briefly, crudely, yet strikingly, Malthus argued that human hopes for social happiness must be vain, for population will always tend to outrun the growth of production. The increase of population will take place, if unchecked, in a geometric progression, while the means of subsistence will increase in only an arithmetic progression. Population will always expand to the limit of subsistence and will be held there by famine, war, and ill health. "Vice" (which included, for Malthus, contraception), "misery", and "self-restraint" alone could check this excessive growth.

John Muir (1838–1914)

The Scottish-born naturalist John Muir was an early advocate of US forest conservation and largely responsible for the establishment of the Sequoia and Yosemite national parks in California. In his early career he worked on mechanical inventions, but in 1867, after an industrial accident nearly cost him an eye, he devoted himself to nature. He walked from the Middle West to the Gulf of Mexico, keeping a journal, *A Thousand-Mile Walk to the Gulf* (published posthumously in 1916) as he went. In 1868 he went to the Yosemite Valley,

California, from which he took many trips into Nevada, Utah, Oregon, Washington, and Alaska, inspired by his interest in glaciers and forests. He was the first to attribute the spectacular Yosemite formations to glacial erosion, a theory now generally accepted.

As early as 1876, Muir urged the federal government to adopt a forest conservation policy. The Sequoia and Yosemite national parks were established in 1890. Early in 1897, President Grover Cleveland designated 13 national forests to be preserved from commercial exploitation; business interests induced Congress to postpone the effect of that measure, but two eloquent magazine articles by Muir, in June and August 1897, swung public and Congressional opinion in favour of national forest reservations. In 1908 the government established the Muir Woods National Monument in Marin County, California.

Aldo Leopold (1887–1948)

The American Aldo Leopold was a professor of wildlife management and a prolific writer for scientific journals and conservation magazines. He played a key role in the designation of America's first national protected area, Gila National Forest in New Mexico. In his fifties Leopold became increasingly concerned with disseminating the conservation message to the general public, and concentrated his efforts on writing essays to inspire people about the value of the natural world and the imperative to preserve it. His *A Sand County Almanac*, published in 1949, a year after his death, is a classic text in conservation ethics. Leopold conceived the idea of the "land ethic", which defined a new relationship between people and the land. "A thing is right," he said, "when it tends to preserve the integrity, stability and beauty of the biotic community. It is wrong when it does otherwise."

Rachel Carson (1907–1964)

Rachel Carson was an American biologist well known for her writings on environmental pollution and the natural history of the sea. She took her bachelor's degree in biology, and subsequently taught for five years at the University of Maryland. From 1936 Carson worked as an aquatic biologist with the US Bureau of Fisheries, where she remained until 1952. An article in *The Atlantic Monthly* in 1937 served as the basis for her first book, *Under the Sea-Wind*, published in 1941. It was widely praised, as were all her books, for its remarkable combination of scientific accuracy and thoroughness with an elegant and lyrical prose style. *The Sea Around Us* (1951) became a national bestseller, won a National Book Award, and was eventually translated into 30 languages. Her third book, *The Edge of the Sea*, was published in 1955.

Carson's prophetic *Silent Spring* (1962) was first serialized in *The New Yorker* and then became a bestseller, creating worldwide awareness of the dangers of environmental pollution. She stood behind her warnings of the consequences of indiscriminate pesticide use, despite the threat of lawsuits from the chemical industry and accusations that she engaged in "emotionalism" and "gross distortion". Some critics even claimed that she was a communist. Carson died before she could see any substantive results from her work on this issue, but she left behind some of the most influential environmental writing ever published.

Hubert Lamb (1913–1997)

The British climatologist Hubert Lamb founded the Climate Research Unit at the University of East Anglia in 1972, where his pioneering research on climate change has been crucial in bringing the issue into the political arena. He is noted for taking a stance against the prevailing view in the 1960s that, notwithstanding seasonal and some interannual variation, climate is generally constant. Lamb is credited for his role in setting the research agenda that has brought climate change to the prominence it has today.

James Lovelock (1919–)

The British environmentalist James Lovelock is famed for his "Gaia hypothesis" (see Chapter 8), which asserted that the Earth is a living, self-regulating "superorganism" that sustains the conditions necessary for life through a series of feedback systems involving the biosphere, atmosphere, oceans, and soil. The theory was published in 1979 in *Gaia: A New Look at Life on Earth*. It generated considerable controversy, and Lovelock later modified some of the more extreme claims of his original proposition, but the hypothesis has nonetheless been highly influential with regard to contemporary scientific views on the interrelationship of biophysical systems.

Edward O. Wilson (1929–)

The world's leading authority on ants, the American Edward O. Wilson is also the foremost proponent of sociobiology – the study of the genetic basis of the social behaviour of all animals, including humans. His book *The Diversity of Life* (1992) recounts the origin of biodiversity, its maintenance within biological communities, and the threats posed by current human activities.

Crispin Tickell (1930–)

In the late 1970s, the British diplomat and environmentalist Crispin Tickell published *Climatic Change and World Affairs*. Tickell sounded a warning: "A shift of 2°C in mean temperatures leads either to ice ages or to melting of the polar ice caps, either of which would destroy much of present civilization."

At that time the global warming concerns arising from the burning of fossil fuels were still a decade away, but the report provided the impetus to research the possible links between the burning of fossil fuels and global warming.

Wallace Broecker (1931–)

The American Dr Wallace S. Broecker is a key contemporary thinker in the interpretation of the Earth as a biological, chemical, and physical system. In his early career he studied ocean circulation patterns using data on the radiocarbon content of seawater, and helped to date the end of the last ice age by radiocarbon dating of marine shells. In the 1970s Broecker was one of the leaders of the Geochemical Ocean Sections programme, which studied the world's oceans using radiocarbon dating. In the 1980s, he used radioactive fallout from nuclear tests carried out 20 years previously to examine the rate of uptake of carbon dioxide by the ocean. He also devised his theory of global ocean circulation, often termed Broecker's Conveyor Belt.

Much of Broecker's work is concerned with the carbon cycle and the ocean's influence on atmospheric carbon dioxide levels. Among his discoveries is the fact that past significant climate shifts may have happened much more rapidly than had previously been thought possible – an example being the Younger Dryas event that occurred around the end of the last ice age. He has received numerous awards, among the more recent the 1996 Blue Planet Prize from the Asahi Glass Foundation (which supports scientific and technological research to solve environmental problems), and the 2008 Benjamin Franklin Medal in Earth and Environmental Science.

Paul Ehrlich (1932–)

Paul Ehrlich is an American biologist and educator who in 1990 shared Sweden's Crafoord Prize (established in 1980 and awarded by the Royal Swedish Academy of Sciences, to support those areas of science not covered by the Nobel Prizes) with the biologist E.O. Wilson. Ehrlich received early inspiration to study ecology when in his high school years he read William Vogt's *Road to Survival* (1948), an early study of the problem of rapid population growth and food production. Though much of his research was done in the field of entomology, Ehrlich's overriding concern became unchecked population growth. He was concerned that humanity should treat Earth as a "spaceship" with limited resources and a heavily burdened life-support system; otherwise, he feared, "mankind will breed itself into oblivion." He has published a distillation of his many articles and lectures on the subject in *The Population Bomb* (1968) and written hundreds of papers and articles on the subject.

Jared Diamond (1937–)

Jared Mason Diamond is professor of Geography and Physiology at the University of California, Los Angeles. He is the author of, amongst others, *Guns, Germs, and Steel* which won the Pulitzer Prize in 1998 and *Collapse: How Societies Choose to Fail or Succeed* (2005). In *Collapse*, Diamond explores the decline of a variety of societies, including the Maya and Ancestral Pueblo (Anasazi) civilizations as well as Norse and Inuit settlements in Greenland, where the former disappeared and the latter flourished as temperatures decreased. *Collapse*, like his earlier work, centres on an analysis of environmental factors and rejects previous explanations that humans alone were responsible for a society's fate. Of the five factors Diamond identifies as instrumental to the collapse of a society, the squandering of natural resources is the most common. By contrast, societies that successfully managed their resources are shown to have prospered. For Diamond, the behaviour and fates of previous civilizations can teach us much about modern society.

Sir Nicholas Shackleton (1937–2006)

A British geologist who was a pioneer in the study of paleoclimatology and in the understanding of the mechanisms behind global warming, Shackleton was an expert in paleoceanography – the analysis of the composition of tiny marine fossils in ocean sediments as a way of learning about the climate conditions that prevailed when they were alive. He also ascertained how changes in the amount of carbon dioxide in the atmosphere affect the warming and cooling of the Earth. In 1976 he and two colleagues, James Hays of Columbia University, New York City, and John Imbrie of Brown University, Providence, Rhode Island, demonstrated that over the past million years regular variations in the Earth's orbit around the sun caused changes in the Earth's climate, including the periodic occurrence of ice ages.

Wangari Maathai (1940–)

Wangari Muta Maathai is a Kenyan politician and environmental activist who was awarded the 2004 Nobel Prize for

Peace. She is the first black African woman to win a Nobel Prize.

While working with the National Council of Women of Kenya, she developed the idea that village women could improve the environment by planting trees to provide a fuel source and to slow the processes of deforestation and desertification. The Green Belt Movement, an organization she founded in 1977, had, by the early twenty-first century, planted some 30 million trees and inspired similar grassroots movements in other African countries. Maathai was elected to Kenya's National Assembly with 98 per cent of the vote in 2002, and in 2003 was appointed assistant minister of environment, natural resources, and wildlife. The Nobel Prize committee commended her "holistic approach to sustainable development that embraces democracy, human rights and women's rights in particular". She is the author of *The Green Belt Movement: Sharing the Approach and the Experience* (1988).

Rajendra K. Pachauri (1940–)

Rajendra Kumar Pachauri is an Indian economist and environmental scientist who became chair of the Intergovernmental Panel on Climate Change (IPCC) in 2002. He is also head of the Energy and Resources Institute in New Delhi, an institution that researches and promotes sustainable development.

In 2007, Pachauri accepted the Nobel Peace Prize on behalf of the IPCC. In his acceptance speech, he stated that climate change "raised the threat of dramatic population migration, conflict, and war over water and other resources as well as a realignment of power among nations."

Amory Lovins (1947–)

The American physicist Amory Lovins and his wife, Hunter Lovins, founded the Rocky Mountain Institute in 1982 as a research centre for the study and promotion of the "whole Earth" approach to architecture favoured by James Lovelock and the landscape architect Ian McHarg (see Chapter 10, page 424). Years before "green" building standards were published in the United States, the institute, which was housed in an energy-efficient and aesthetically appealing building, formulated the fundamental principle of authentic green architecture: to use the largest possible proportion of regional resources and materials. In contrast to the conventional, inefficient practice of drawing materials and energy from distant, centralized sources, the Lovins team chose to follow the "soft energy path" for architecture – i.e. drawing from the now-familiar list of alternative energy sources such as wind, solar, water, and geothermal energy.

The Asahi Glass Foundation awarded the two 2007 Blue Planet Prizes to Lovins and his compatriot Joseph L. Sax. Lovins was rewarded for his contribution to the protection of the environment through the improved energy efficiency advocated by his "soft energy path" and for his invention of an ultralight and fuel-efficient vehicle called the Hypercar.

Al Gore (1948–)

The former Vice President of the United States, Albert Arnold Gore Jr, favours strong measures to protect the environment. His ideas on this issue were first set out in his book *Earth in the Balance: Ecology and the Human Spirit* (1992). Following his narrow defeat in the 2000 US presidential election, Gore devoted much of his time to environmental issues. He discussed global warming in the 2006 documentary *An Inconvenient Truth* and in its companion book. The film won an Academy Award for best documentary. In 2007 Gore was awarded, jointly with the IPCC, the Nobel Peace Prize for his efforts to raise awareness about global warming.

10

RESPONSES TO CHANGE

The computer-based climate models used by the Intergovernmental Panel on Climate Change (IPCC) to predict climate change have, over several decades of development, "consistently provided a robust and unambiguous picture of significant climate warming in response to increasing greenhouse gases." The greatest uncertainties in the climate models, however, lie in predictions of human behaviour. These variables include economic and population growth, both of which will affect greenhouse gas emissions and their cumulative concentrations. For this reason the IPCC has used several different emissions scenarios, based on different assumptions concerning global and regional development. For the six major scenarios, the range of the projected rise in global average annual temperature is between 1.8°C and 4°C (3.2–7.2°F) over the next century if no measures are taken to reduce greenhouse gas emissions. All of the models predict that the currently observed changes to the physical world brought on by global warming will continue and will accelerate over the coming decades. Surface warming of 2°C (3.6°F) is considered by many scientists to be the threshold above which pervasive and extreme climatic effects will occur.

Although some uncertainties linger at the regional and local levels, the general trends are now fairly well understood. The largest unknown is how people and governments will respond to the situation. Humans and other living beings are not inexperienced in adapting to change, and human adaptation can be achieved through a variety of means, such as technology, resource management, modification of behaviour, or social policy. The immediate consequences of climate change may be addressed in this way, and some adaptation is already taking place on an ad hoc basis. According to the IPCC's Fourth Assessment, however, "more extensive adaptation than is currently occurring is required to reduce vulnerability to climate change. There are barriers, limits and costs, which are not fully understood." Climate change is projected to put severe stress on the capability of supplying such necessities as water, food, and health care.

In order to stabilize the climate change that is being driven by global warming, mitigation efforts seek to reduce the concentration of greenhouse gases in the atmosphere. In the words of the IPCC, "Mitigation efforts over the next two to three decades will have a large impact on opportunities to achieve lower stabilization levels." Mitigation can be approached through alternative sources of energy with reduced or zero emissions, including renewable sources of energy; technologies that improve energy efficiency; making voluntary changes in lifestyle and energy use behaviour; and carbon capture and storage. One form of mitigation is an emphasis on sustainable development, including the use of green architecture to design buildings that make efficient use of energy and water. A further important strategy to promote energy conservation is to put a price on carbon. By assigning costs to carbon dioxide emissions and placing a value on the reduction of these emissions, a carbon market can operate in which carbon "credits" are bought and sold to provide economic incentives to meet emission regulations.

The IPCC has attempted to assess the potential costs of mitigation. Although the question is complex, there is some agreement that it would be in the order of 1 per cent of global gross domestic product (GDP). Some studies have also tried to assess the economic cost to society from the impacts of climate change on the assumption that no mitigation attempts are made. Although there is less certainty about these costs, there is agreement that they would very probably outweigh the cost of mitigation (for example, 1–5 per cent of GDP globally, with the cost rising as high as 25 per cent of GDP for less-developed countries).

Although governments and other legislative bodies have a major role to play in climate change mitigation, much can also be done at an individual level. There is growing public awareness of the environmental crisis and the urgency with which steps must be taken to avert it, and people are increasingly willing to make changes in their lifestyles. These changes may relate to housing and to personal energy use, and might involve, for example, investment in alternative energy sources. This chapter discusses ways in which energy consumption may be reduced through a different approach to building, and explores the various alternatives to fossil fuels. It describes carbon trading and carbon capture – solutions that are relatively new and whose potential may yet be realized. It ends with a global perspective on the speed with which industrial development can take hold, and a discussion of the urgency with which it is necessary to act to avoid the consequences.

Green architecture

Green design was a pervasive topic in boardrooms and living rooms in 2007, particularly as the costs of maintaining the status quo became apparent. The building of shelter (in all its forms) consumes more than half of the world's resources – translating into 16 per cent of the Earth's freshwater resources, 30–40 per cent of all energy supplies, and 50 per cent by weight of all the raw materials withdrawn from the Earth's surface. Architecture is also responsible for 40–50 per cent of waste deposits in landfills and 20–30 per cent of greenhouse gas emissions. Unfortunately, too many architects working since the post-World-War-II building boom have been content to erect emblematic civic and corporate icons that celebrate profligate consumption and globalization. More recently, however, designers and users have begun to evaluate a building on its environmental integrity, as embodied in the way it is designed and operated. The green movement is changing the message of architecture from "egocentric" to "ecocentric".

Development of eco-awareness in building design

Environmental advocacy in the US gained its first serious momentum as part of the youth movement of the 1960s. In rebellion against the perceived evils of high-rise congestion and suburban sprawl, some of the earliest and most dedicated eco-activists moved to rural communes, where they lived in tent-like structures and geodesic domes. In a sense, this initial wave of green architecture was based on admirable characteristics of Native Americans' lifestyle and its minimal impact on the land. At the same time, by isolating themselves from the greater community, these youthful environmentalists were ignoring one of ecology's most important principles: that interdependent elements work in harmony for the benefit of the whole.

Influential pioneers who supported a more integrative mission during the 1960s and early 1970s included the architectural critic and social philosopher Lewis Mumford, the landscape architect Ian McHarg, and the scientist James Lovelock. They led the way in defining green design, and they contributed significantly to the popularization of environmental principles. For example, in 1973 Mumford proposed a straightforward environmental philosophy: "The solution of the energy crisis would seem simple: transform solar energy via plants and produce enough food power and manpower in forms that would eliminate the wastes and perversions of power demanded by our high-energy technology. In short, plant, eat, and work!"

McHarg, who founded the department of landscape architecture at the University of Pennsylvania, laid the ground rules for green architecture in his seminal book *Design with Nature* (1969). Envisioning the role of human beings as stewards of the environment, he advocated an organizational strategy, called "cluster development", that would concentrate living centres and leave as much natural environment as possible to flourish on its own terms. This "whole Earth" concept also became the basis of Lovelock's Gaia hypothesis (see Chapter 8).

The development of green awareness was accelerated by the destructive politics and economics of the 1970s in the US, where the lack of business regulation meant that there were no limits to the consumption of fossil fuels. Meanwhile, the 1973 OPEC (Organization of the Petroleum Exporting Countries) oil crisis brought the cost of energy into sharp focus and reminded the global community that it depended on a very small number of petroleum-producing countries for its supplies. This crisis, in turn, brought into relief the need for diversified sources of energy and spurred corporate and government investment in solar, wind, water, and geothermal sources of power (see "Alternative energy sources", page 431).

For American architects and builders, a significant milestone was reached in 1994 with the formulation of Leadership in Energy and Environmental Design (LEED) standards, established and administered by the US Green Building Council. These standards provided measurable criteria for the design and construction of environmentally responsible buildings. There are four basic areas within which they must qualify, as follows.

Buildings must conserve energy, for example, by being oriented to take full advantage of seasonal changes in the sun's position. They must use diversified and regionally appropriate sources of energy, which may, depending on the location, include solar, wind, geothermal, biomass, water, natural gas, and even, if necessary, petroleum and nuclear power.

Buildings must be constructed with recycled, renewable, or low-embodied-energy materials (those involving minimal energy in their extraction and production) that are locally sourced and free from harmful chemicals. Supplies must be evaluated on the basis of their entire production cycle, including nonpolluting raw ingredients, durability of the product, and the potential for recycling. Materials should be evaluated in terms of their distance from origin, taking into account the energy consumed in transport.

Buildings must conserve and monitor water usage and supplies. Grey water (i.e. previously used, for example for laundry) should be cleansed and recycled, and building-by-building catchments for rainwater should be installed.

Finally, whenever possible, existing buildings should be reused and the surrounding environment preserved. Designs should make use of "Earth shelters" (constructions built into the ground), roof gardens, and extensive planting throughout and around buildings.

The 1980s and early 1990s brought a new surge of interest in the environmental movement and the rise to prominence of a group of socially responsive and philosophically oriented green architects. The American architect Malcolm Wells opposed the legacy of architectural ostentation and aggressive assaults on the land in favour of the gentle impact of underground and earth-sheltered buildings – exemplified by his own house in Brewster, Massachusetts, built in 1980. The low impact, in both energy use and visual effect, of a structure that is surrounded by earth creates an almost invisible architecture and represents a green ideal. As Wells explained, this kind of underground building "offers huge fuel savings and a silent, green alternative to the asphalt society." Other key figures in this movement are the physicist Amory Lovins and his wife, Hunter Lovins, discussed in Chapter 9.

In 1975 the American architect Pliny Fisk III launched the Center for Maximum Potential Building Systems (Max Pot) in Austin, Texas. In the late 1980s the centre joined with others to support an experimental agricultural community called Blueprint Farm, in Laredo, Texas. Its broader mission – with applications to any geographic location – was to study the correlations between living conditions, botanical life, the growing of food, and the economic-ecological imperatives of construction. The facility was built as an integrative prototype, recognizing that nature thrives on diversity. Fisk concluded that single-enterprise and monoculture territories were environmentally dysfunctional – because, for example, all of the crop's predators converge, natural defences are overwhelmed, and chemical spraying to eliminate insects and weeds becomes mandatory. In every respect Blueprint Farm stood for diversified and unpredictable community development. The crops were varied, and the buildings were constructed from steel gathered from abandoned oil rigs and combined with such enhancements as earth berms (raised terraces or barriers), sod roofs, and straw bales. Photovoltaic panels, evaporative cooling, and wind power were incorporated in this utopian demonstration of farming and green community standards.

The American architect William McDonough rose to green-design fame in 1985 with his Environmental Defense Fund Building in New York City. That structure was one of the first civic icons for energy conservation, due to the architect's close scrutiny of all of its interior products, construction technology, and air-handling systems. Since then McDonough's firm has established valuable planning strategies and built numerous green buildings – most significantly, the Herman Miller factory and offices (Holland, Michigan, 1995), the corporate offices of Gap, Inc. (San Bruno, California, 1997), and Oberlin College's Adam Joseph Lewis Center for Environmental Studies (Oberlin, Ohio, 2001).

McDonough's main contribution to the evolution of sustainable design is his commitment to what he has called "ecologically intelligent design" – a process that involves the cooperation of the architect, corporate leaders, and scientists. This design principle takes into account the "biography" of every aspect of manufacture, use, and disposal: the choice of raw ingredients, transport of materials to the factory, fabrication process, durability of goods produced, usability of products, and recycling potential. McDonough's latest version of the principle – referred to as "cradle-to-cradle" design – is modelled after nature's own waste-free economy and makes a strong case for the goal of reprocessing, in which every element that is used in or results from the manufacturing process has its own built-in recycling value.

Principles of green

The advances in research and in building techniques achieved by the above-mentioned leaders in green design have been compiled into a number of databases containing various methods of environmental construction and sustainable materials. Some of these methods and materials have been in use for thousands of years, yet remain the basis for contemporary advances in environmental technology. The following list includes the essential green-design principles for residences in the new millennium.

- *Energy sources.* Whenever feasible, build homes and communities that supply their own power. Such buildings may operate entirely independent of the regional power grid, or they may be able to feed excess energy back on to the grid. Wind and solar power (see "Alternative energy sources", below) are the usual alternatives.
- *Energy conservation.* Weatherize buildings for maximum protection against the loss of warm or cool air. It is now possible to buy extremely dependable moisture-resistant insulating materials that do not cause indoor humidity problems. Laminated glass has also been radically improved in recent years; some windows provide the same insulation value as traditional stone, masonry, and wood construction.
- *Reuse of materials.* Do the research to find recycled building materials. Although such products were scarce in the early 1990s, today numerous companies specialize in salvaging refuse from demolition sites.
- *Safety of materials.* Thoroughly research the chemical composition and off-gassing (emission of volatile/noxious compounds) characteristics of all products to be used in building.

- *Siting.* Consider using underground or earth-sheltered architecture, which can be ideal for domestic living. Starting at a depth of about 1.5 m (5 ft) below the surface, the temperature is a constant 52 degrees – which makes Earth itself a dependable source of climate control.

Challenges to architectural change

If architecture is to become truly green, then a revolution of form and content – including radical changes in the entire look of architecture – will have to occur. The environmental movement in the twenty-first century will meet resistance, because its proponents appear to be asking society to scale back the benefits of industrialization. The ultimate success of green architecture is likely to require that advocates achieve a broad-based philosophical consensus and provide the same kind of persuasive catalyst for change that the Industrial Revolution offered in the nineteenth century. This means shaping a truly global (as well as optimistic and persuasive) philosophy of the environment. The architecture profession will have to abandon the past century's specialization and reliance on technology. Integrated thinking can produce a productive checklist of grass-roots-originated, community-oriented, and globally unifying objectives.

Designers in the twenty-first century can make better use of ideas from larger fields of environmental science and technology. Already there exists a rich reservoir of inspiration from science and nature – cybernetics, virtual reality, biochemistry, hydrology, geology, and cosmology, to mention a few. Furthermore, as the Industrial Revolution was a generator of change in many fields in the nineteenth century, so can the information revolution, with its model of integrated systems, serve as a conceptual framework for a new approach to architecture and design in the broader environment.

In architecture, context has meaning well beyond the siting of individual structures. Once community governments have used their legislative power to insist on state-of-the-art green standards, they should do everything possible to encourage appropriate artistic responses to regional attributes such as surrounding topography, indigenous vegetation, and cultural history. For instance, communities might encourage innovative fusions of architecture with landscape – where trees and plants become as much a part of architectural design as construction materials – so that buildings and their adjacent landscapes essentially merge.

Continuing advances in environmental technology have significantly strengthened the goals of sustainable architecture and city planning over the last decade, but there is still a tendency for many people to feel that the environmental crisis is far beyond their comprehension and control. At the same time, if the message of the gurus of green technology encourages the public to transfer all responsibility to engineering and science, then the social and psychological commitment needed for change is threatened as well. Technological solutions must be viewed as only one tool in the green crusade.

Increasing numbers of people are seeking new symbiotic relationships between their homes and the broader ecology. This growing motivation is one of the most promising signs of hope in the development of a consensus philosophy of the environment. If successful, it will confirm anthropologist Margaret Mead's optimistic observation: "Never doubt that a small group of thoughtful, committed citizens can change the world. Indeed, it is the only thing that ever has."

Alternative energy sources

The awareness that the burning of fossil fuels is adversely affecting the environment is compounded by growing concern over the prospect of rapidly dwindling reserves of these fuels. These worries have prompted efforts to develop viable alternative energy sources. Sources of renewable energy such as solar, wind, water, and geothermal power are becoming increasingly attractive because of their inexhaustible supply and nonpolluting nature. However, significant limitations on the harnessing of these energy sources still prevent their use for large-scale power production. Another form of renewable energy, biofuel, was strongly promoted in the last decade or so as an alternative to fossil fuels, but has recently come under criticism because of the hidden environmental costs of its production. Furthermore, nuclear power has received renewed attention as an alternative to fuels that generate greenhouse gases in their consumption.

Solar power

The sun is an extremely powerful energy source, and sunlight is by far the largest source of energy received by the Earth, but its intensity at the Earth's surface is actually quite low. Nonetheless, the total amount of solar energy received by Earth is vastly in excess of the world's current and expected energy requirements. If suitably harnessed, sunlight has the potential to satisfy all future energy needs.

The sunlight that reaches the Earth's surface consists of nearly 50 per cent visible light, 45 per cent infrared radiation, and smaller amounts of ultraviolet and other forms of electromagnetic radiation. This radiation can be converted into either thermal energy (heat) or electrical energy, though the former is easier to accomplish. Two main types of devices are used to capture solar energy and convert it to thermal energy: flat-plate collectors and concentrating collectors. Because the intensity of solar radiation at the Earth's surface is so low, both types of collectors must be large in area. Even in sunny parts of the world's temperate regions, for instance, a collector must have a surface area of about 40 square metres (430 square feet) to gather enough energy to serve the energy needs of one person.

The most widely used flat-plate collectors consist of a blackened metal plate, covered with one or two sheets of glass, that is heated by the sunlight falling on it. This heat is then transferred to air or water, called carrier fluids, that flow past the back of the plate. The heat may be used directly, or it may be transferred to another medium for storage. Flat-plate collectors are commonly used for hot-water heating and house heating. The storage of heat for use at night or on cloudy days is usually accomplished by using insulated tanks to store the water heated during sunny periods. Such a system can supply a home with hot water drawn from the storage tank, or, with the warmed water flowing through tubes in floors and ceilings, it can provide space heating. Flat-plate collectors typically heat carrier fluids to temperatures ranging between 66–93°C (150–200°F). The efficiency of such collectors (i.e. the proportion of the energy received that they convert into usable energy) ranges from 20 per cent to 80 per cent, depending on the design of the collector.

When higher temperatures are needed, a concentrating, or focusing, collector is used. These devices concentrate sunlight received from a wide area onto a small blackened receiver, thereby considerably increasing the light's intensity in order to produce high temperatures. The arrays of carefully aligned mirrors or lenses used in these so-called solar furnaces can focus enough sunlight to heat a target to temperatures of 2,000°C (3,600°F) or more. This heat can be used to study the properties of materials at high temperatures, or it can be used to operate a boiler, which in turn generates steam for a steam-turbine–electric-generator power plant. The solar furnace has become an important tool in high-temperature research. For producing steam, the moveable mirrors are so arranged as to concentrate large amounts of solar radiation upon blackened pipes through which water is circulated and thereby heated.

Solar radiation may be converted directly into electricity by solar cells (photovoltaic cells). In such cells, a small electric voltage is generated when light strikes the junction between a metal and a semiconductor (such as silicon) or the junction between two different semiconductors. The power generated by a single photovoltaic cell is typically only about 2 watts. By connecting large numbers of individual cells together, however, as in solar-panel arrays, hundreds or even thousands of kilowatts of electric power can be generated in a solar electric plant. The energy efficiency of most present-day photovoltaic cells is only about 15–20 per cent, and, since the intensity of solar radiation is low to begin with, huge and costly assemblies of such cells are required to produce even moderate amounts of power. Consequently, photovoltaic cells that operate on sunlight or artificial light have so far found major use only in low-power applications – for example, as power sources for calculators and watches,. Larger units have been used to provide power for water pumps and communications systems in remote areas, and for weather and communications satellites.

Solar energy is also used on a small scale for purposes other than those described above. In some countries, for instance, specially designed solar ovens are employed for cooking, and solar energy is used to produce salt from seawater by evaporation.

The potential for solar energy is enormous, since about 200,000 times the world's total daily electric-generating capacity is received by the Earth every day in the form of solar energy. Unfortunately, though solar energy itself is free, the high cost of its collection, conversion, and storage still limits its exploitation.

Wind power

Although wind is intermittent and diffuse, it contains tremendous amounts of energy. Sophisticated wind turbines have been developed to convert this energy to electric power. For example, more than 15,000 wind turbines are now in operation in Hawaii and California at specially selected sites. Their combined power rating of 1,500 megawatts is roughly equal to that of a conventional steam-turbine power installation.

Modern wind turbines extract energy from the wind, mostly for electricity generation, by rotation of a propeller-like set of blades that drive a generator through appropriate shafts and gears. The older term "windmill" is often still used to describe this type of device, although electric power generation rather than milling has become the primary application. Interest in wind turbines for electricity generation was rekindled by the oil crisis of the mid-1970s, caused by the abrupt curtailment of oil shipments from the Middle East to many of the highly industrialized nations of the world. However, high initial costs, intermittent operation, and maintenance costs have prevented wind turbines from becoming a significant factor in commercial power production.

Not all the kinetic energy of the wind can be extracted by wind turbines, because there must be a finite velocity as the air leaves the blading. It can be shown that the maximum efficiency (energy extracted divided by energy available in the captured wind area) obtainable is about 59 per cent, although actual wind turbines extract only a portion of this amount. Currently, the maximum efficiency obtainable with a propeller-type windmill is roughly 47 per cent; this occurs when the propeller-tip speed is between five and six times the wind velocity. For a given rotor speed, it drops rapidly as the wind velocity decreases. The theoretical maximum energy obtainable from a rotor with a diameter of 30 metres in a wind with a speed of 14 metres per second is about 690 kilowatts. When these limitations are coupled to the need for suitable sites with steady winds, it becomes apparent that wind turbines alone will not play a major role in meeting the power demands of an industrialized nation.

Hydroelectric power

Converting the energy in moving water to electricity has been a long-standing technology, yet hydroelectric power plants are estimated to provide only about 2 per cent of the world's energy requirements. The technology involved is simple enough: hydraulic turbines change the energy of fast-flowing or falling water into mechanical energy that drives power generators, which produce electricity. Hydroelectric power plants, however, generally require the building of costly dams. Another factor that limits any significant increase in hydroelectric power production is the scarcity of suitable sites for additional installations except in certain regions of the world.

In some coastal areas, for example in the Rance River estuary in Brittany, France, hydraulic turbine-generator units have been used to harness the great amount of energy in ocean tides. At most such sites, however, the capital costs of constructing dam-like structures with which to trap and store water are prohibitive.

Nonetheless, hydroelectric power is a preferred energy source in areas with heavy rainfall and with hilly or mountainous regions that are in reasonably close proximity to the main load centres. Some large hydroelectric sites that are remote from load centres may be sufficiently attractive to justify the long high-voltage transmission lines. Small local hydro sites may also be economical, particularly if they combine storage of water during light loads with electricity production during peaks.

Geothermal power

Geothermal power is obtained by using heat from the Earth's interior. Most geothermal resources are in regions of active volcanism. Hot springs, geysers, pools of boiling mud, and fumaroles (vents of volcanic gases and heated groundwater) are the most easily exploited sources of such energy. The ancient Romans used hot springs to heat baths and homes, and similar uses are still found in some geothermal regions of the world, such as Iceland, Turkey, and Japan. The greatest potential for geothermal energy, however, lies in the generation of electricity. Geothermal energy was first used to produce electric power at Larderello, Italy, in 1904. By the late twentieth century, geothermal power plants were in operation in Italy, New Zealand, Japan, Iceland, Mexico, the United States, and elsewhere, and many others were under construction in other countries. The principal US plant, located at The Geysers, north of San Francisco, can generate up to 1,900 megawatts, although production may be restricted to prolong the life of the steam field.

The most useful geothermal resources are hot water and steam trapped in subsurface formations or reservoirs, with temperatures of 80–350°C (176–662° F). Water and steam hotter than 180°C (356°F) are the most easily exploited for electric-power generation and are utilized by most existing geothermal power plants. In these plants the hot water is converted to steam, which is then used to drive a turbine whose mechanical energy is then converted to electricity by a generator. Hot, dry subsurface rocks may also become more widely used as a source of geothermal energy once the technical problems of circulating water through them for heating and conversion to steam are completely resolved.

Biofuels

Biofuels are fuels that are derived from biomass – that is, plant material or animal waste. Since such materials can be replenished readily, biofuels are attractive as a renewable source of energy. Some long-exploited biofuels, such as wood, can be used directly as a raw material that is burned to produce heat. The heat, in turn, can be used to run generators in a power plant to produce electricity. A number of existing power facilities burn grass, wood, or other kinds of biomass.

Liquid biofuels are of particular interest because of the vast infrastructure already in place to use them, especially for transportation. The liquid biofuel in greatest production is ethanol (an alcohol, C_2H_5OH), which is made by fermenting starch or sugar. In the United States – the leading producer – ethanol biofuel is made primarily from corn grain, and is typically blended with petrol to produce a fuel that is 10 per cent ethanol. In Brazil, which had been the leading producer until 2006, ethanol biofuel is made primarily from sugar cane, and is commonly used as 100-per-cent ethanol fuel or in petrol blends containing 85 per cent ethanol. The second-most-common liquid biofuel is biodiesel, which is made primarily from oily plants (such as soya or oil palm) and to a lesser extent from other sources (such as cooking waste from restaurants). Biodiesel, which has found greatest acceptance in Europe, is used in diesel engines, usually blended with petroleum diesel in various percentages.

Other biofuels include methane gas (CH_4), which can be derived from the decomposition of biomass in the absence of oxygen, and methanol (CH_3OH), butanol ($C_4H_{10}O$), and dimethyl ether (CH_3OCH_3), which are in development. Much focus is on the development of methods to produce ethanol from biomass that has a high content of cellulose. This cellulosic ethanol could be produced from abundant low-value material, including wood chips, grasses, crop residues, and municipal waste. The mix of commercially used biofuels will undoubtedly shift as these fuels are developed, but the range of possibilities presently known could furnish power for transportation, heating, cooling, and electricity.

Biofuels supply environmental benefits but can also have serious drawbacks. As a renewable energy source, plant-based biofuels in principle make little net contribution to greenhouse gas emissions because the carbon dioxide (CO_2) that enters the air during their combustion will have been removed from the air earlier when they grew. Such a material is said to be carbon neutral. In practice, however, the industrial production of agricultural biofuels can result in additional emissions of greenhouse gases that can offset the benefits of using a renewable fuel. These emissions include CO_2 from the burning of fossil fuels to produce the biofuel: in the use of farming equipment, fertilizer manufacture, transportation of the biofuel, and distillation. With these factors taken into account, ethanol made from corn represents a relatively small energy gain; the energy gain from sugar cane is greater and that from cellulosic ethanol may be greater still.

Land use is also a major factor in evaluating the benefits of biofuels. Corn and soya beans are important foods, and their use in producing fuel can therefore affect the economics of food price and availability. In 2007 about one fifth of US corn output was to be used for biofuel, and one study showed that even if all US corn land was used to produce ethanol, it could replace only 12 per cent of petrol consumption. Crops grown for biofuel can also compete for the world's natural habitats. For example, emphasis on ethanol derived from corn is shifting grasslands and brushlands to corn monocultures, and emphasis on biodiesel is bringing down ancient tropical forests to make way for palm plantations. Loss of natural habitat can change hydrological systems, increase erosion, and generally reduce biodiversity and wildlife areas. The clearing of land can also result in the sudden release of a large amount of CO_2 as the plant matter it contained decays or is burned.

In September 2007 the use of biofuels as a substitute for fossil fuels in order to combat global warming came under criticism from a team of researchers led by Nobel Prize winner Paul J. Crutzen. The group calculated that biofuel production and consumption could actually result in the release of more greenhouse gases than they saved. For example, the fertilizer used to grow biofuel crops leads to the release of nitrous oxide, a gas whose greenhouse effect is 300 times greater than that of CO_2. The group also determined that biodiesel made from rapeseed oil (canola) released the equivalent of up to 1.7 times more greenhouse gases than conventional diesel, and that fuels derived from sugar cane and corn released the equivalent of 0.5–0.9 and 0.9–1.5 times more greenhouse gases than petrol, respectively.

On September 11, 2007, a report from the Organisation for Economic Co-operation and Development warned that subsidizing biofuels would lead to surging food prices and, potentially, to the destruction of natural habitats.

A boom in the production of biofuel was under way in 2007, especially in the United States, where in January of that year approximately 75 refineries for producing the ethanol from corn were being built or expanded. This construction, not including additional facilities in preparation, was expected to double existing capacity, and the demand for corn pushed its price so high that US farmers planted more land with the crop than they had done in a generation. Biofuel was perceived as a beneficial alternative to petroleum and other fossil fuels.

Some of the disadvantages of biofuel production described above apply mainly to low-diversity biofuel sources – corn, soya beans, sugar cane, and oil palms – which are traditional agricultural crops. A recently proposed alternative would be to use high-diversity mixtures of species: the North American tall-grass prairie being a specific example. Converting de-graded agricultural land presently out of production to such high-diversity biofuels could increase wildlife area, reduce erosion, cleanse waterborne pollutants, store CO_2 from the air as carbon compounds in the soil, and ultimately restore fertility to degraded lands. Such biofuels could be burned directly to generate electricity or converted to liquid fuels as technologies develop.

The proper way to grow biofuels to serve all needs simul-taneously will continue to be a matter of much experimenta-tion and debate, but the fast growth in biofuel production is likely to continue. In the European Union, for example, 5.75 per cent of transport fuels are to be biofuels by 2010, with 10 per cent of its vehicles to run exclusively on biofuels by 2020. In December 2007, the US President George W. Bush signed into law the Energy Independence and Security Act, which mandated the use of 136 billion litres (36 billion gallons) of biofuels annually by 2020 – more than a sixfold increase over 2006 production levels. The legislation required, with certain stipulations, that 79 billion litres (21 billion gallons) of the amount be biofuels other than corn-derived ethanol. In addition, the law continued government subsidies and tax incentives for biofuel production. Some observers hoped that the law would encourage the commercialization of technology for producing cellulosic ethanol, for which there were a number of pilot plants in the United States.

The distinctive promise of biofuels not shared by other forms of renewable energy, such as solar power, is that in combination with an emerging technology called carbon capture and storage (see below), biofuels are capable of perpetually removing CO_2 from the atmosphere. Under this vision, biofuels would remove CO_2 from the air as they grew, energy facilities would capture that CO_2 when the biofuels were later burned for power, and then the captured CO_2 would be stored in long-term repositories. With proper planning and the necessary technological advances, therefore, biofuels may have the potential to help create the conditions necessary for a sustainable world.

Nuclear power

Nuclear power is produced by reactions involving the cores, or nuclei, of atoms – in contrast to ordinary chemical reactions, which involve only the atoms' orbital electrons. One method of releasing nuclear energy is by controlled nuclear fission, and nuclear fission reactors now operate in many parts of the world. Another method, controlled nuclear fusion, has not yet been perfected. Nuclear reactions generate a huge amount of energy, but opposition to the industry arises from concerns over its safety, due to the risk of environmental contamination from radioactive waste. The accidental leakage of waste from nuclear power plants poses a particular risk to the health of humans and the wider environment. The most well-known such accident occurred at the Chernobyl plant in Russia in 1986.

Many countries now possess nuclear power plants, although the degree to which these contribute to national power production varies considerably. France, for example, generates 80 per cent of its power from nuclear, while in the US and Britain, nuclear supplies 20 per cent. Some countries are now reconsidering the advantages of nuclear as an alternative to fossil fuels: in March 2008 Britain announced a nuclear expansion programme. Germany, in contrast, intends to phase out nuclear power by 2021.

In Sweden, the public voted in 1980 to phase out nuclear energy, and 25 years later, in 2005, the Barsebäck 2 nuclear reactor was finally shut down. Vattenfall, the Swedish energy company that ran Barsebäck, announced plans to replace it with the largest wind farm in northern Europe. Despite this, an opinion poll showed that approximately 80 per cent of Swedes favoured continuing the use of nuclear power, and only 10 per cent supported phasing it out. People feared that abandoning nuclear power would make it necessary to import fossil-fuel-generated power from elsewhere in Europe.

Carbon capture and storage

Carbon capture and storage (CCS) is a technological solution to the increased quantity of CO_2 being released into the atmosphere. It involves capturing the gas from large-scale sources such as power plants and sequestering (storing) it in long-term repositories. One method of capture is to absorb the

CO_2 into a suitable solvent. Storage repositories may include geological formations beneath the land, or sediments of the deep ocean; partially depleted oil and gas reservoirs are being considered as potential sites. Conceivably, the CO_2 could also be stored as solids such as carbonates. The technology has been developed to some extent, but CCS is still in a relatively embryonic stage and does not yet offer a solution to carbon emissions on a large scale.

Emissions trading and carbon credits

Emissions trading is a means to provide an economic incentive to industry to reduce its emissions of pollutants such as CO_2. A limit, or cap, is set on the amount of the pollutant that may be discharged into the atmosphere by a business, and businesses are issued with permits, or "credits", that represent the amount they are permitted to emit. If a company needs to exceed this quantity of emissions, it may buy credits from another business whose emissions are less than its allowance.

The European Union plan for the trading of carbon emissions came into force on January 1, 2005. Approximately 12,000 industrial installations, which accounted for about half of all EU CO_2 output, had their emissions capped. Any factory that emitted more than a specified amount of CO_2 would be penalized unless it covered the excess by purchasing "carbon credits" from other factories that emitted less than their allowance. The European Commission had endorsed the national allocation plans for all but four countries (Italy, Poland, the Czech Republic, and Greece), and the plan for Spain was conditional pending minor changes. The first stage, to run from 2005 to 2007, covered the cement, glass, paper-and-pulp, electric-power, and iron-and-steel industries. The second stage, to run from 2008 to 2012, was expected to impose tighter restrictions and would perhaps be extended to cover additional producers, such as the chemical and aluminium industries and, possibly, the aviation industry. The opening price was about 8 euros per tonne of CO_2, and the trade in unused emission allowances was expected eventually to be worth billions of euros.

On December 20, 2006, the European Commission proposed bringing airline emissions from flights between EU airports into the EU carbon-emissions trading scheme in 2011, and in 2012 to include all airline flights into or out of EU airports. In 2007 EU transport ministers approved the proposal, and the EU Environment Committee later recommended that the plan begin in 2010 for all flights to or from EU airports. On September 28 in Montreal, however, a majority of delegates to the 36 meeting of the International Civil Aviation Organization (ICAO) rejected the EU proposals. ICAO Assembly President Jeffrey Shane said that members did not object to the concept of using emissions-trading schemes to combat climate change, but objected to the unilateral imposition of such schemes, which he said had to be agreed to by airlines.

Global economic change

All countries emit greenhouse gases, but highly industrialized countries and more populous countries emit significantly greater quantities than others. Countries in North America and Europe that were the first to undergo the process of industrialization have been responsible for most greenhouse gas emissions, in absolute cumulative terms, since the mid-eighteenth century. The United States, which comprises approximately 5 per cent of the global population, emitted almost 21 per cent of global greenhouse gases in 2000. The same year, the then 25 member states of the European Union, possessing a combined population of 450 million people – approximately 7.5 per cent of global population – emitted 14 per cent of all anthropogenic greenhouse gases. This figure was roughly the same as the proportion released by the 1.2 billion people of China. In 2000 the average American emitted 24.5 tons of greenhouse gases, the average person living in the EU released 10.5 tons, and the average person living in China discharged 3.9 tons.

Until 2007, the United States was the world's largest emitter of greenhouse gases. In June of that year, however, figures were released showing that, in absolute terms (though not per capita) it had been overtaken by China. According to the Netherlands Environmental Assessment Agency, China produced 6,200 million tonnes of carbon dioxide in 2006, compared with 5,800 million tonnes from the US. Britain produced about 600 million tonnes.

The figures are a stark illustration of the speed with which economic development can take place. China is not the only developing country to be emerging among the major economies of the world – India too is growing rapidly, with its own increasing greenhouse gas emissions. Much current international debate is concerned with the relative responsibilities of different countries with regard to the curbing of emissions: the United States, in particular, has justified its abstention from the Kyoto agreement in terms of the disparity of emissions controls for developed and developing countries. It should be borne in mind, however, that the older industrialized countries have reaped the benefits of the cheap labour and production industries in China. In this sense, economic development may be seen not in terms of discrete, national boundaries, but as a global phenomenon in which we all play a part. The story of China's economic growth, described below, is taken by many to represent, on an accelerated timescale, the economic and ecological changes in which every industrialized country has been involved since the Industrial Revolution.

The China of today is a far cry from the country that in the 1950s the Swedish Nobel-Prize-winning economist Gunnar Myrdal predicted would remain mired in poverty. As the world's fourth-largest economy and third-largest trading country, China in 2007 accounted for approximately 5 per cent of world gross domestic product (GDP) and had recently graduated in status to a middle-income country. Beijing was also emerging as a key global aid donor. China now supplies more than a third of the world's steel, half of its cement, and about a third of its aluminium.

China's achievements in poverty reduction from the post-Mao-Zedong era, in terms of both scope and speed, have been impressive: about 400 million people have been lifted from poverty. The standard of living for many Chinese is improving, and this has led to a widespread optimism that the government's goal of achieving an overall well-off, or *Xiaokang*, society, is possible in the near future.

The figures illustrating China's remarkable economic achievements, however, conceal huge and outstanding challenges that, if neglected, could jeopardize those gains. Many local and foreign-development analysts agree that what they consider to be China's unsustainable and reckless approach to growth is putting the country (and, some suggest, the world) on the brink of environmental catastrophe. To date, China has already been coping with limited natural resources that are fast disappearing. In addition, not everyone in China has shared the benefits of growth – about 135 million people, or one tenth of the population, still lived below the international absolute poverty line of $1 per day. There today exists significant inequality between urban and rural populations, as well as between the poor and the rich. The increasing number of protests in the country has been attributed to both environmental causes and experiences of injustice.

In 2007 China consumed more coal than the US, Europe, and Japan combined. Beijing was also the biggest emitter of sulphur dioxide, which contributes to acid rain. Chinese scientists have blamed the increase in emissions on rapid economic growth and the fact that China relies on coal for 70 per cent of its energy needs. The changing lifestyle of the increasing number of middle-class families contributes to the problem: in 2007, in Beijing alone, 1,000 new cars were added to the roads every day. More than 300,000 premature deaths annually are attributed to airborne pollution. Seven of the ten most polluted cities in the world are located in China.

China's environmental problems go beyond atmospheric emissions. The UN 2006 Human Development Report cited China's worsening water pollution and its failure to restrict heavy polluters. More than 300 million people, it said, lacked access to clean drinking water. About 60 per cent of the water in China's seven major river systems was classified as being unsuitable for human contact. China has about 7 per cent of the world's water resources and roughly 20 per cent of its population.

China is beginning to realize, however, that its growth path is not cost-free. The State Environmental Protection Administration and the World Bank have estimated that air and water pollution are costing China an increasingly high percentage of its GDP. Although the Chinese government is carrying the responsibility for fixing the overwhelming environmental consequences of China's breakneck growth, it may be argued that those who have benefited greatly from China's cheap labour and polluting industries – the transnational companies and industrialized countries – should share responsibility for mitigating the resulting environmental damage.

In 2004 the Chinese government began setting targets for reducing energy use and cutting emissions. At first, the idea of adopting a slower growth model and the predictions about the looming environmental disaster were not received with enthusiasm. By 2007, however, targets had been established for shifting to renewable energy, for employing energy conservation, and for embracing emission-control schemes. China aims to produce 16 per cent of its energy needs from renewable sources by 2020.

At the 2008 G8 summit in Japan member countries, including the US, signed up to a statement which vowed:

> We are committed to avoiding the most serious consequences of climate change and determined to achieve the stabilization of atmospheric concentrations of global greenhouse gases . . . Achieving this objective will only be possible through common determination of all major economies, over an appropriate time frame, to slow, stop and reverse global growth of emissions and move towards a low-carbon society.

INDEX

aerosols   23–7, 30–1, 118, 331
agriculture
  albedo   27, 114–15
  climate change   10, 208–9,
    210
  global warming, impact
    of   208–9
  land use   27, 133–4, 143,
    188–9, 202–3
air masses   78, 79–80, 91, 96
albedo   145, 277, 330
  agriculture   27, 114–15
  and climate   114–15
  planetary   14, 118
  snow/ice   25, 36–7, 128
Amazon   187, 191–2, 201–2,
  203, 205, 350, 351
ammonia   24, 65, 99, 105,
  106
AMO (Atlantic Multidecadal
  Oscillation)   30, 132
Antarctica   6, 96–7, 149, 248,
  249, 280, 282

anthropogenic activities
  see humans
anticyclones   246, 290,
  295–8
  Siberian   95, 309
  subtropical   87, 88, 91
  see also cyclones; wind
Arctic   6, 8, 38–9, 96–7, 192,
  272, 275, 277, 280–2
architecture, green   423–30
Atlantic Multidecadal
  Oscillation (AMO) 30,
  132
  atmosphere   98–102
  biosphere, influence
    on   108–11
  cyclones   258–9
  early   99, 220
  evolution and
    structure   100–2, 286–9
  and global warming   102–4
  humidity   59–63
  ocean interaction   249–55

planetary boundary
  layer   109–11
pressure   56–9, 168–9,
  289–90, 302–3
radiation   46, 61–2
solar radiation   43–5
tropospheric cycle, effect
  of   258–9
see also aerosols; greenhouse
gases; pollution, air

bioclimatology   123–4
biodiversity
  definition   336–40
  economic value of   382–4
  effect of climate change
    on   371–5
  hotspots   350–2
  species extinction   340–1
  see also conservation; habitat;
  species
biofuels   437–41
biogenic
  aerosols   25, 29
  gases   99–100, 102–114
  ice nuclei   66, 117–20, 210
biological pump   18, 37
biosphere
  carbon cycle   104–5
  cloud condensation
    nuclei   117–18
  control on surface friction and
    winds   115–17
  Earth's energy budget
    102–4
  temperature, controls
    on   111–14
  nitrogen cycle   105–6
  planetary boundary layer,
    influence on   108–11

sulphur cycle   106–7
birds   282, 346–7, 348,
  349–50, 353–4, 362,
  363–4, 369–70, 379–80
Bjerknes, J.   169, 170, 269
breezes   299–302
British Isles   50, 72, 116, 349
Broecker, Wallace   414–15
budgets
  Earth's energy   14, 48, 79,
    102–4
  radiation   45–6
  surface energy   46–7
buildings, eco-friendly   423–30

carbon
  in atmosphere   19, 34, 192,
    287
  cycle   28, 37–9, 104–5,
    189–92
  capture and storage   442–3
  dioxide (CO_2) 9, 13, 18–20,
    34, 37–9, 99, 104–5, 145,
    192, 206
  deforestation   188, 204–5
  dissolved organic (DOC)
    231–2
  emissions trading   443
  and forests   191–2
  monoxide (CO)   22, 99
  in oceans   9, 18, 37–8, 230,
    224, 278–9, 329–30
  sequestration   28
  soil   188, 205–6
Carson, Rachel   412
CFCs (chlorofluorocarbons)
    331–2
China   18, 26, 406, 445–8
cities   24, 121–3, 209–11, 286,
    327–9

Clausius-Clapeyron
    equation 34–5
climate
  classification of 74–84
  definition 40
  distribution of major
    types 84–98
  of early Earth 152–5
  and life 98–102
  urban 121–3
Climate Change Conference,
    Bali 404–5
climate change
  abrupt 155–8
  agriculture, emergence
    of 143
  civilization, since emergence
    of 133–9, 140
  effects of 4–11
  and extinction 371–5
  future policy 405–8
  through geologic time
    146–52
  within human lifespan
    125–32
  responses to 420–2
  sensitivity 34
  soil 205–6
climate types
  continental sub-arctic 95
  humid continental 94–5
  humid subtropical 91–2
  macroclimates 76, 97
  Marine West Coast 93–4
  Mediterranean 92–3, 130
  mesoclimates 76–7, 97
  mid-latitude steppe and
    desert 89–90
  mountain 97–8
  snow and ice 96–7

  trade wind littoral 86–7
  tropical monsoon 86–7
  tropical and sub-tropical
    desert 88–9
  tropical and sub-tropical
    steppe 89
  tropical wet-dry 87
  tundra 81, 96, 145, 192
  urban 121–3
  wet-equatorial 86
clouds 46, 54, 61–2, 88,
    294–5
  and aerosols 24, 26
  condensation nuclei 24,
    117–18, 123
  feedbacks 35–6
  planetary boundary
    layer 110
  precipitation, origins of
    63–9
  radiation, absorption of 43,
    44
condensation 59–60, 61, 63,
    65
conservation 375–7
continentality 49–50, 53
convection 47, 288–9
coral reefs 356–8, 361–2, 372
Coriolis force 241, 242, 243,
    245, 246, 260, 261, 289,
    321
currents, ocean
  Antarctic Circumpolar 248,
    249, 276
  causes of 241–3
  distribution of 239–41
  Ekman layer 243–4
  equatorial 244–5
  geostrophic 243
  Gulf Stream 259–62

gyres   246–9, 260, 275–6
Kuroshio   55, 249, 263, 264
temperature, effect on   53–5
wind-driven   244–9
cyclones   169–70, 290, 297,
    325, 333
  atmosphere, influence
    on   258–9
  extratropical   291–5
  oceans, effect on   257–8
  subtropical   70, 71, 254–8
  synoptic forecasting   165–6
  see also anticyclones
Dansgaard-Oeschger
    cycle   141, 156–7
deforestation   28, 114, 188,
    189, 199–205, 368–9, 383
deserts   88–90, 109, 189, 208
desertification   114, 116,
    118–19, 187, 205, 206–8
dew point   61, 112, 113
Diamond, Jared   415–16

Earth
  albedo   see main entry
    albedo
  energy budgets   14, 79,
    102–4
  orbit   29, 33, 41, 127, 145,
    147–8
  surface roughness   115–17
Earth Summit   401–2
Ehrlick, Paul   415
El-Niño–Southern Oscillation
    see ENSO
emissions trading   443–4
energy sources, alternative
  geothermal   436–7
  hydroelectric   435–6
  nuclear   441–2

  solar   431–4
  wind 434–5
ENSO (El-Niño–Southern
    Oscillation) 29, 31, 55,
    129–31, 138, 182, 246,
    252, 318
  and climate change   264–72,
    334
  and Peru 266–8
environmentalism
  ethics   388–90
  Gaia hypothesis   391–2
  history of   393–8
  key ideas   390–1
  key organizations   399–405
  key thinkers   409–19
  movement   395–8
Europe   27, 30, 53, 55, 124,
    142
  monsoonal tendencies
    315–16
evaporation   17, 27–8, 46–7,
    49, 108, 111
  humidity   62–3
  hydrologic cycle   214–17
evapotranspiration   28, 36–7,
    189
extinction   8, 12, 148, 205,
    335, 340–1
  background rates of   344–7
  causes of   347–62
  climate change, effect
    on   371–5
  hybridization   365
  logging, effect of   368
  overfishing   367–8
  overharvesting   365
  secondary   369–70
  species loss   343–4
  whaling   366–7

feedbacks
  carbon cycle   37–9
  clouds   35–6
  ice albedo   36–7
  water vapour   34–5
fishing   367–8, 384
fog   61, 93, 296, 300, 301
forests
  and albedo   27, 114
  Amazon   187, 191–2,
    201–2, 203, 205
  biogenic aerosols   25
  as carbon reservoirs
    191–2
  climate change, effects of   8,
    10, 372
  logging, effects of   368–9
  planetary boundary
    layer   109
  species loss   352, 353–5
  see also deforestation
fossil fuels   13, 15, 18, 20, 24,
  103, 104
fossils   139, 344
fungi   107, 342

Gaia hypothesis   391–2, 413,
  424
gases
  biogenic   99–100, 102–114
  carbon dioxide   9, 13,
    18–20, 34, 37–9, 99,
    104–5, 192
  chlorofluorocarbons   331–2
  fluorinated 23
  methane   20–2, 38, 99, 101,
    107, 145, 206
  nitrogen   103, 105–6
  nitrous oxide   23, 99, 105,
    206

  ozone see main entry ozone
  water vapour see main entry
    water vapour
glacial-interglacial
    periods   140–6, 150, 151,
    156
global warming
  aerosols   23–7
  atmosphere   102–3
  biodiversity, effect on   371–5
  causes of   11–13
  consequences of   205, 333–4
  deforestation   28, 114, 188,
    189, 208
  greenhouse effect   13–15,
    329–30
  and greenhouse gases   4, 13,
    14, 17–23, 104, 133, 145,
    153, 333, 406–8
  land use change   27–9
  melting icecaps   279–80
  natural influences on   29–33
  and oceans   232–3
  radiative forcing   16–17, 34
  stratospheric ozone
    depletion   29
  trends   420–2
Gore, Al   418–19
green
  architecture   421–30
  political parties   395–7
  principles   428–9
greenhouse effect   13–15,
  329–30
greenhouse gases   4, 13, 14,
  103, 104, 133, 145, 153,
  406–8
  atmosphere, evolution
    100–2
  biofuels   333, 438–9

Kyoto Protocol 402–4

Gulf Stream 53–4, 236, 246–7, 259–62

gyres 246–9, 260, 275, 276

habitat

connections 377–8

fire suppression 355–6

forests 368

freshwater 358–60

loss of 9, 336, 347–8, 352–62

management 378–82

marine 356–8

pollution 360–2

protection of 376–7

restoration of 380–2

hailstones 69, 179–80

Hawaiian Islands 348–9, 362, 368, 369–70

heat, sensible and latent 108, 111, 114, 119

heatwaves 11, 286

humans, effect on

agriculture 208–9

climate 133–9, 140

deforestation 199–205

desertification 206–8

fire suppression 355–6, 379

greenhouse gas emissions 4, 12, 13, 18–19, 20, 21, 22–7, 192, 333

hydrosphere, effect on 277–9

oceans 232–3, 357–8

ozone layer, damage to 331–2

pollution 211, 360–2

urbanization 209–10

humidity

atmosphere 59–63

in cities 121

and climate 60–2

climate types, distribution of 86, 88

definition 59–60

and evaporation 62–3

relative 35, 60, 65, 83

and transpiration 113–14

hydrogen 100, 101

hydrologic cycle 63, 120, 193, 210, 214–17, 279

hydrosphere 213–17, 277–9

ice

albedo 36–7

caps, melting 279–80

climate type 96–7

drift 275–6

nuclei, biogenic 66, 117–20, 210

sea 272–7, 282–4

icebergs 282–4

infra-red 14, 20, 103, 112

insects 342–3

insolation 14, 29, 33, 139, 143

Intergovernmental Panel on Climate Change see IPCC

intertropical convergence zone (ITCZ) 86, 87, 235, 303

IPCC (Intergovernmental Panel on Climate Change) 3–4, 16, 104, 188, 279, 283, 333, 399–401, 420–2

Fourth Assessment Report 3–4, 8, 12, 279, 283, 333, 400, 421

irradiance, solar 31–3

isotopes, cosmogenic 32

ITCZ (intertropical convergence
    zone)   86, 87, 235, 303

jet streams   33, 54, 55, 91,
    323–4

Köppen, Wladamir   80–3, 168
Köppen's climatic type
    distribution   84–98
Kyoto Protocol   402–4

Lamb, Hubert   412–13
land
    carbon cycle   189–92
    soil   192–9
    pollution   211
    use and climate change
        27–9, 199–211
latitudes   45, 49
    and ocean
        temperature   235–7
    precipitation
        distribution   69–71
    see also climate types;
        temperature
Leopold, Aldo   393, 411
lightning   69, 120
Little Ice Age   133–4, 265
logging see deforestation
Lovelock, James   368–9,
    391–2, 413, 424
Lovins, Amory   418, 426

Maathai, Wangari   417
Madden–Julian Oscillation
    (MJO)   317–18
Malthus, Thomas 410
marine life   9, 18, 37, 190–1,
    212, 228
    acid rain, effect on   330–1

condensation nuclei   118
    decline in species   335
    habitat loss   356–8
    overfishing   367–8
    pollution   361–2
    whaling   366–7
McDonough, William 427
McHarg, Ian   424
mesosphere   320, 324–6
methane   20–2, 38, 99, 101,
    107, 145, 206
Millenium Ecosystem
    Assessment   335, 340
MJO (Madden–Julian
    Oscillation)   317–18
monsoons   91, 92 , 304–5
    climate types   86–7
    in Europe and North
        America   315–6
    Indian Ocean   71, 86, 137,
        138–9, 250, 305–9
    Malaysian–Australian   310–13
    variability of   317–20
    West African   313–15
mountains
    breezes   301–2
    climate type   97–8
    and species extinction
        374–5
Muir, John 393, 410–11
Mumford, Lewis   424

NAO (North Atlantic
    Oscillation)   130, 131
nitrogen   103, 105–6
nitrogen cycle   105–6, 193
nitrous oxide   23, 99, 105,
    206
North Atlantic Oscillation
    (NAO)   130, 131

oceans 9
  atmospheric
    interactions 249–50
  and atmospheric
    pressure 57
  and carbon dioxide 18, 19,
    37–8, 212–13
  chemical composition
    of 225–33
  circulation 53–5, 238–9,
    244–9
  Clausius-Clapeyron
    equation 34–5
  condensation nuclei 118
  coral reefs 356–8, 361–2,
    372
  currents 53–5, 239–49
  cyclones, effect of 257–8
  distribution 217–19
  evolution 219–25
  habitat loss 356–8
  hydrologic cycle 212–17
  ice 272–7, 282–4
  ice caps, melting of 289–80
  oscillations 29–30
  overfishing 367–8
  and PBL 109
  physical properties of 233–8
  pollution 232–3, 361
  precipitation, influence
    on 71
  salinity 223, 226–7, 233–5
  sea aerosol 25
  sea level 6, 8, 279–80
  temperature 6, 37, 49–50,
    53–5, 118, 132, 235–8,
    250–5, 372
  tropospheric cycle 257–8
  waves 269–71
  wind 25, 54, 239, 244–9

orbit, Earth's 29, 33, 41, 127,
    145, 147–8
oxygen 99, 101–2, 103, 104,
    107, 153–4, 222, 287–8
  cycle 105
  in sea water 229–30
  ozone 22, 32, 44, 99, 107,
    287–8
  layer 326, 331–2
  stratospheric 29, 107

Pachauri, Rajendra K. 417–18
Pacific Decadal Oscillation
    (PDO) 131–2
palaeoclimatology 124–6
PBL (planetary boundary
    layer) 107, 108–11, 117
PDO (Pacific Decadal
    Oscillation) 131–2
Peru 53, 55, 266–8
photosynthesis 18, 38, 99,
    107, 190
  and atmospheric
    chemistry 153–4
planetary
  albedo 14
  boundary layer (PBL) 107,
    108–11, 117
Pliny III, Fisk 426
pollution, air 22, 111, 285,
    296,
  aerosols 24–6
  "Asian brown cloud"
    309–10
  climatic effects of 329–30
  land 211
  oceans 232–3
  smog 24, 286
  urban 121–2, 286, 327–30
prairies 381–2

precipitation 59–60, 81, 83, 119
  acid 106–7, 330–1
  amounts 72–3
  and climate distribution 86–98
  clouds 65–9
  definition 63
deserts 208
  effects of 73–4
  and humid climate 91–2
  hydrologic cycle 215–17
  measurement of 64
  planetary boundary layer 110
  recycled 120
  tropical cyclones 258–9
  in urban areas 121
  world distribution of 69–71
  see also climate types
pressure, atmospheric 56–9, 108–9, 168–9, 289–90, 302–3

radiation
  atmosphere 46, 61–2
  and absorption 103–4
  aerosols, effect of 23–7, 29
  balance 52
  budgets 45–6
  and evaporation 63
  infra-red 14, 103, 112
  and invisible vapour 61–2
  solar 13–14, 31–3, 36, 41–8, 123, 128, 213, 235
  terrestrial 13–14, 16
  thermal 45–6, 60, 121
radiative forcing 4, 16–17, 19–20, 22
  aerosols 23, 24, 26–7

Earth's orbit 33
  feedback mechanisms 34
  ozone 29
  solar 32
  volcanoes 30
rainfall see precipitation
rivers 224–53, 358–61

Sahel, Africa 10, 114, 116, 127, 209, 250–1
salinity
  ocean 223, 226–7, 233–5
  sea ice 273–4
sea
  aerosol 25
  breezes 299–301
  ice 272–7, 282–4
  level 6, 8, 12, 279–80
  see also oceans
seasons 31, 33, 43
  atmospheric pressure, effect on 57–9
  change in 8, 371–2
  climate classification 81, 84
  distribution of climate types 84–98
  monsoons 304–5
  in Holocene Epoch 139
  solar radiation 123
  temperature variations 50
  and variation in climate 127–9
seawater
  chemical composition 225–33
  physical properties 233–8
Shackleton, Sir Nicholas 416
sinks 18, 19, 21, 37, 145, 153, 188, 192
smog 24, 110, 286

snow   6, 12, 25, 43, 94, 95–7, 128, 277, 283
    see also ice soil   188, 192–3
    climate change   205–6
    climate influences on   195
    erosion   73, 188, 196–9, 205
structure   193–5
solar
    forcing   32
    output   31–3
    power   431–4
    radiation   13–14, 31–3, 36, 41–8 123, 128, 213, 235
    see also   albedo
solubility pump 18, 38
Southern Oscillation   268–72
species   339–40
    biodiversity   336–40
    effect of climate change on   8, 201, 281–2, 371–5
    extinction   341, 344–62
    hybridization   365
    introduced 362–5
    number of   339–44
    pet trade   368
stratosphere   30, 31, 32, 44, 324–6
stratospheric ozone   29, 107
sulphur
    aerosols   3, 24–5, 30, 31, 118
    cycle   106–7
    dioxide   24, 65, 99, 106, 117

temperature   40, 83
    in Archean Eon   101
    adiabatic changes   52, 301–2
    air   17, 23, 23, 34, 35, 41, 53

biosphere controls on   111–14
change   46–7, 279
continentality   49–50, 53
in desert climate   88–90
diurnal   50
and extinction   371
height variations   52–3
in humid climate   91–2
land surface   4, 6, 12–13, 28, 34, 36, 61–2
ocean currents, effect on   53–5
ocean surface   6, 37, 49–50, 53–5, 118, 132, 235–8, 250–5, 372
short-term changes in   55–6
solar radiation   111
urban   121–2
    see also climate types; environmentalism, key groups
thermohaline circulation   37–8, 263–4
thunderstorms   67–9, 110–11, 86, 90, 93, 94, 114, 165–6, 172, 189, 299
Tickell, Crispin   414
tornadoes   92, 94, 179–80, 299
transpiration   27–8, 47, 108, 111-14, 120
trees   25, 42, 135–6, 137, 142, 143, 364, 368
    see also deforestation; forests; vegetation
tropopause   320, 321
troposphere   31, 249–50, 257–8
tundra   81, 96, 145, 192
turbulence see convection

United Nations
  Environment Programme 3,
    207, 399
  Framework Convention on
    Climate Change 401–2
urbanization 24, 121–3,
  209–11, 286, 327–9
USA
  Atlantic Multidecadal
    Oscillation 132
  animal/bird species 346,
    349–50, 353–4
  architecture, green 423–7
  deforestation 200
  deglaciation period 143
  drought 135
  and El-Niño 129
  environmentalists 393–4
  Florida Everglades 379–80
  greenhouse gases 444–5
  habitat movement 379–80
  habitat restoration 381–2
  Kyoto Protocol 404, 445
  Los Angeles pollution 296,
    328–9
  Midwest 111–12, 113–14
  monsoonal tendencies
    315–16
  North Atlantic
    Oscillation 130
  overfishing 367–8
  precipitation 74
  temperature 111–12, 113–14
  volcanic eruptions 30
  weather forecasting 162,
    163, 165–7, 172, 175–6,
    177–8, 181

vegetation 350
  acid rain damage 331
  biodiversity 338
  and climate 80
  introduced species 362–3,
    364–5
  leaf fall 128
  leafing out 18, 111–12,
    113–14
  photosynthesis 18, 38, 99,
    107, 190–1
  surface roughness, Earth
    115–17
  transpiration 27–8, 47, 108,
    111–14, 120, 145
  trees 25, 42, 135–6, 137,
    142, 143, 364, 368
  vegetation-based climate
classification 80–3
  volcanoes 16, 19, 21–2,
    287, 327
  aerosols 25, 30–1, 130–1

water vapour 17, 46–7
  atmosphere 108
  biosphere 115–16
  feedback 34–5
  humidity 59–63
  precipitation 63–5
  temperature 112–13
waves 269–71, 321
weather forecasting
  applications of 162–4
  history of 159–61, 164–5
  long-range 180–4
  modern forecasting 170–7
  numerical 174–7
  short-range 177–80
  station networks 166–7
  synoptic reports 165–6, 171
Wells, Malcolm 426
whaling 366–7, 382

Wilson, Edward O. 413
wind 29, 31, 40, 41, 46, 49, 83
  anticyclone *see main entry*
  anticyclones
  atmospheric pressure 56–9,
    289–90
  biosphere 115–17
  cyclones *see main entry*
  cyclones
  and El-Niño–Southern
    Oscillation 129–30,
    268–70
  jet streams 33, 54, 55,
    323–4
  local systems 299

oceans 25, 54, 239, 244–9
power 434–5
precipitation 70–1
scale classes 298–9
soil erosion 197
stratospheric/
  mesospheric 324–6
trade 70, 303
upper level 320–6
in urban areas 122
in weather forecasting
  168–70
see also climate types;
  monsoons

ENCYCLOPÆDIA
Britannica®

Since its birth in the Scottish Enlightenment Britannica's commitment to educated, reasoned, current, humane, and popular scholarship has never wavered. In 2008, Britannica celebrated its 240th anniversary.

Throughout its history, owners and users of *Encyclopædia Britannica* have drawn upon it for knowledge, understanding, answers, and inspiration. In the Internet age, Britannica, the first online encyclopedia, continues to deliver that fundamental requirement of reference, information, and educational publishing – confidence in the material we read in it.

Readers of Britannica Guides are invited to take a FREE trial of Britannica's huge online database. Visit

http://www.britannicaguides.com

to find out more about this title and others in the series.